Supporting the Military-Affiliated Learner

Supporting the Military-Affiliated Learner

Communication Approaches to Military Pedagogy and Education

Edited by
Victoria McDermott
Leandra Hinojosa Hernández
Amy R. May

LEXINGTON BOOKS
Lanham • Boulder • New York • London

Published by Lexington Books
An imprint of The Rowman & Littlefield Publishing Group, Inc.
4501 Forbes Boulevard, Suite 200, Lanham, Maryland 20706
www.rowman.com

6 Tinworth Street, London SE11 5AL, United Kingdom

British Library Cataloguing in Publication Information Available

Library of Congress Cataloging-in-Publication Data

Names: McDermott, Victoria, Hernández, Leandra Hinojosa, editor. | May, Amy R., editor. | 1995– editor.
Title: Supporting the military-affiliated learner : communication approaches to military pedagogy and education / edited by Leandra Hinojosa Hernández, Amy R. May, Tori McDermott.
Other titles: Communication approaches to military pedagogy and education
Description: Lanham : Lexington Books, [2021] | Includes bibliographical references and index. | Summary: "Supporting the Military-Affiliated Learner challenges the academic community to 1) reevaluate how they support military-affiliated learners (MALs) and address how the military-civilian-academic divide causes disparities and barriers to MAL academic achievement and retention and 2) implement programs and develop strategies to facilitate equitable academic integration from application to graduation."—Provided by publisher.
Identifiers: LCCN 2020040465 (print) | LCCN 2020040466 (ebook) | ISBN 9781793618085 (cloth) | ISBN 9781793618092 (epub)
Subjects: LCSH: Soldiers—Education, Non-military—United States. | Veterans—Education—United States. | Communicative competence.
Classification: LCC U716 .S85 2021 (print) | LCC U716 (ebook) | DDC 371.828/35500973—dc23
LC record available at https://lccn.loc.gov/2020040465
LC ebook record available at https://lccn.loc.gov/2020040466

To all military-affiliated learners. You serve with honor and distinction, and as you transition to higher education, it's our responsibility to serve you with the same level of commitment.

Contents

Acknowledgments

First, we would like to thank our contributors for their hard work and dedication in helping us publish this edited volume. Through several rounds of workshopping and revision, you worked tirelessly to produce scholarship that has the potential to transform the experiences of MALs in higher education. (#citationactivism #thisisgettingexcessive #starbucks #Chadsaid #PedrosPeppers #Chad&PedroEditors4&5 #askmySGTMAJ #mixologist #PutItInaTable #FootnoteToEndnote #TORI! #ApplicationtoGraduation)

Thank you to the research participants who allowed us to learn from their lived experiences, from the struggles to the triumphs, and everything in between. Now, we challenge the readers to better support our MALs from application to graduation. This is our chance to better serve those who have served us.

Victoria: I would like to thank my husband, Chad, for his never-ending support of my research goals and career aspirations. Without you, this adventure in Alaska would not have been possible. I look forward to many, many more! Thank you to my family for the emotional support from afar. I am especially grateful to Amy and Lea for their continued support and mentorship. Thank you for believing in my abilities as an academic. Thank you to the other educators and scholars who have supported me throughout my educational career. Thank you to all my friends who kept me sane during this editing process, specifically my substitute husband Nike! Without all of your emotional support this would not have been possible. Thank you to all the military-affiliated learners I have met and have yet to meet! I hope to continue supporting you throughout all your educational endeavors.

Leandra: First and foremost, I would like to thank all of the military-affiliated learners I have been lucky enough to work with over the last several years at institutions such as Trident University International, National

University, Texas A&M University, the University of Houston, and now Utah Valley University. From 2008 to the present, I have had the privilege of working with, mentoring, and learning from MALs of all life backgrounds. Such interactions have challenged me, helped me grow, and filled me with pride. This book is for you all and for countless MALs we will work with and advocate for in the future. Second, I would like to thank Sarah De Los Santos Upton, my sister-scholar and academic life partner. Thank you for being my comrade and for your endless support of all of our writing projects. Your friendship is invaluable! Third, I would like to thank Tori, Amy, and Jennifer N. Belding for their military spouse sisterhood throughout this project and several others. Jenn, I had no idea our chance meeting on Facebook would have led to the creation of our first book *and* that it would introduce me to Tori and Amy! Thank you for being a writing sounding board and an amazing collaborator. Tori and Amy, thank you for seeing this project to fruition with me. I value your support and our friendship tremendously, and I am very excited for our future collaborations. I'm so proud of both of you, and I learn from you daily. Our conversations during 2020 kept me going during the pandemic and provided such an important source of social support and laughs. Fourth, I would like to thank my parents, Froilán and Ernestine, and my sister, Alexandria, for their love and continued support of all of my military and academic endeavors over the past several years. I always say it takes a village to publish a book, and I am so thankful you are part of my village. I couldn't do it without you all. Finally, Pedro, I had no idea when we met in college that you would have a career in the Navy that would take us around the world, position me as a MAL during my graduate work (an identity I never envisioned would become such a fundamental part of my life), and eventually lead us to where we are now. Thank you for your endless support of my academic goals and for always ensuring that I was equipped to write and do research with good music, countless books and cheeseburgers, and climbing trips to clear my mind. This book is for you all.

Amy: Alex, my DH, if it weren't for you, I wouldn't be here. "I make more money with a dependent" might have launched our journey, but here we are twenty years later forever changed by your service to the military. I am so proud of everything you've accomplished, and it has been my privilege to serve with you. Presley, thank you for always being there to offer a smile, word of encouragement or needed hug to keep me going. This book was driven, in part, by my desire to change the landscape of higher education for military dependents. I never want you to feel like being a child of the military hinders you in any way. Our daily Starbucks run kept me going. Mom, Dad, and Baboo, thank you for always saying "you can do anything" and making me believe. I love you! To my dear, sweet Vladimir, your countless hours of service to getting this book done are so noted. From your hours of attentive listening to frequent mental health breaks, I owe you. A whole bag of Beggin' Strips are coming your way, brah. To my mentor, Mike Allen, you took a chance on a military spouse who only had a few years to get it done. I was clueless, but you guided me every step of the way. You are the mentor I strive to be for my MALs. To all my fellow MALs, I have learned with you and from you for the past 10 years. I struggled in many ways to achieve my academic goals, but dedicated educators, support staff, and mentors guided my way. This book is driven, in part, by my desire to make it better for those who walk with me and for those who come behind me. Tori, I continue to be inspired by your enthusiasm for learning and passion for research. And your organizational skills made this book happen! We have a lot more to accomplish together, so "add it to the list!" Finally, Lea, you are an incredibly talented scholar, and I learned so much from you working on this book. Your critical love approach has challenged me to do more, be more, and to grow as an editor, educator, and human.

Introduction

The Current Status of Military Education in Higher Education

Victoria McDermott, Leandra Hinojosa Hernández, and Amy R. May

Military-affiliated learners[1] (MALs) are an underserved population of students in today's higher education classroom. In 2018, there were 2.1 million service members in the U.S. military with nearly 2.8 million dependents, including spouses and children (Defense Manpower Data Center, 2018) pursuing higher education amid the struggles of military life. While many faculty, staff, and administrators may have an idea of military-affiliated lifestyles and experiences more broadly, they may not fully appreciate everything that living a military lifestyle may entail (e.g., frequent relocations, strict moral code, extended periods of training in the field, and long-duration family separation periods). For MALs, their life is heavily influenced by the realities of military service, which also impacts their experiences in the college classroom. As colleges and universities actively recruit MALs and vie for their tuition dollars, developing an understanding of military culture and challenging associated stereotypes becomes critically important to supporting this population of learners and helping them achieve their educational goals. As such, this edited volume seeks (1) to explicate barriers that MALs face when trying to transition to, navigate, and succeed in higher education; (2) to explore strategies, programs, and interventions that have been successful in supporting MALs' self-efficacy and retention; (3) to provide a foundational understanding of military culture for communication instructors and educators to understand and encourage MALs' academic achievements; and (4) to develop best practice recommendations for resource allocation, instructional design, and intercultural communication strategies to better support MALs throughout campus communities. Through the analysis of MALs and the barriers they may face to academic success, we provide strategies, programs, and policy initiatives for supporting MALs from application to graduation. Ultimately, through this volume, we seek to cultivate a holistic, culturally inclusive, and responsive communication environment for MALs in higher education to encourage meaningful discourse among MALs, peers, educators, administrators, and policymakers. We hope this volume provides readers with tangible best practices for promoting and supporting all MALs in continuing and completing their higher education endeavors.

THE CURRENT CULTURAL LANDSCAPE

The landscape of higher education is changing rapidly. After decades of crippling budget cuts, a global pandemic, and an increase in calls for accountability by state legislators and accrediting bodies (Mitchell, Leachman, & Masterson, 2017), colleges and universities are being forced to reevaluate how they educate and serve "the most diverse generation in higher education history" (Levine & Dean, 2013, para. 31). One population of immediate concern is the MAL. Over 75 billion dollars[2] have been invested in this population of learners through the Post-9/11 GI Bill (CBO, 2019), making them increasingly attractive for colleges and universities struggling to fund programs and counteract dwindling enrollment. However, little is known about the MAL population (American Council on Education, 2015) and their unique needs in and out of the classroom; as such, faculty, staff, and administrators may be underprepared to serve this population and ensure their academic success.

MALs encounter and respond to many of the same issues and concerns as their civilian counterparts with added challenges from military-specific experiences that impact their transition to college and their preparedness to learn (Drummet, Coleman, & Cable, 2003). For service members, the military is unlike most jobs and requires a full-life commitment with extended trainings, long work hours, frequent relocations, and associated mental health and relational struggles that stem from deployments, combat, and extended family separation (Green, Nurius, & Lester, 2013; Hernández, 2019; Hernández & Belding, 2019; Padden, Connors, & Agazio, 2011). Unlike most careers that subscribe to the 9 a.m. to 5 p.m. structure, the military is a greedy career path and institution (Segal, 1986) that leaves little room for participating in cultures other than the military (Hernández & Belding, 2019; May & McDermott, 2019). This "greedy institution" significantly impacts not only the potential educational experiences of service members, reservists, and veterans, but presents potential barriers to their spouses and children as well

(Hernández, 2019; Sollitto & Cole, 2019). Spouses may have to juggle a single-parent lifestyle, frequent changes in jobs and institutions due to duty station reassignments, higher rates of unemployment and underemployment when compared to their civilian counterparts, and long periods without their military spouse. Children may have to frequently change schools and manage periods of separation from one or both of their parents. Moreover, due to the all-encompassing nature of the military, spouses and children are affected by military cultural norms, such as abiding by a strict code of conduct less an active duty or veteran spouse get in trouble, living on bases and posts isolated from civilian communities, and putting the needs of the military above all else.

For MALs, the transition out of military culture and into the classroom, if only for a few hours, can be challenging and uncertain. From their uniforms and the way they present themselves to their lived experiences, there are multifaceted differences that impact classroom dynamics for MALs. For example, academia provides less structure and requires more freewill and independence (May & McDermott, 2019), requiring MALs to shed strict compliance to hierarchy/chain of command, uniformity, and defined social norms to meet expectations and perform well (in this edited volume see May & McDermott, chapter 2). Furthermore, MALs may face stigma from non-military peers and/or from faculty, staff, and administrators who lack an in-depth understanding of the realities of military service (Griffin & Gilbert, 2012; Griffin & Gilbert, 2015). Additionally, MALs may be older, working full-time jobs, and parenting, in addition to representing many other historically underserved populations, such as first-generation college students, English as a second language (ESL) learners, and low-income, minoritized populations (American Council on Education, 2015). Furthermore, active-duty members may face barriers with dynamic work schedules and a lack of consistent access to communication technology (e.g., the internet, cell phones). Military spouses, although oftentimes more educated than their civilian counterparts, have a history of being unemployed and underemployed (Hernández, 2019; Lim & Schulker, 2015). Struggling to find their own identity outside of military life (Sollitto & Cole, 2019; in this edited volume, see Rossetto & Owlett, chapter 5) and with frequent changes of location, military spouses can feel isolated from traditional student peers in regards to their outside responsibilities, priorities, and commitments. While the layers of richness afforded by the experiences of MALs create vital opportunities for colleges and universities to promote student diversity of backgrounds, thoughts, and skills, without an in-depth and robust understanding of perceived institutional barriers and evidence-based andragogy, higher education is doomed to fail these students.

THE IMPORTANCE OF STUDYING AND SUPPORTING MALS

For educators and administrators, there is limited literature and resources available to learn more about MALs and the unique military circumstances that shape self-efficacy and educational success. Particularly in communication contexts, there is little research pertaining to communication between instructors and MALs *and* the experience of MALs in the communication classroom. There is even less information about how military students impact the classroom environment and how other students in the classroom sense-make their military-affiliated peers. Much of the research that does exist focuses overwhelmingly on veteran students. There are, however, other military-affiliated members affected by the military lifestyle—spouses, children, as well as active-duty and reservist members returning to higher education during military service—who also need support and understanding in the classroom. Furthermore, much of the current research on MALs stems solely from the perspective of the "deficit model," mainly highlighting the skills they may lack, the impact the military has had on their mental health, and their potential inability to connect with college peers (Barry, Whiteman, MacDermid Wadsworth, & Hitt, 2012; Hart & Thompson, 2016; Livingston, Havice, Cawthon, & Fleming, 2011; Rumann & Hamrick, 2009; Weiss, 2015). Overall, this research provides little information on the benefits and expertise MALs can bring to the classroom environment, that is, the assets they bring to college communities. These assets include extensive and rich work experiences, exposure to a variety of leadership styles and skills, and expertise in a wide range of occupational skills, ranging from healthcare to telecommunications, logistics, and acquisitions.

Moreover, there are limited resources for MALs pursuing a degree in higher education. While the military provides transitioning assistance back into civilian life ("The Army's Official Transition Website", n.d.), their help ends once the service member officially leaves active duty, leaving veterans to fend for themselves. Additionally, this assistance is only for active-duty members who are retiring, ultimately neglecting active-duty members who seek to take classes while in service, reservists, spouses, and children. Additionally, the available literature and resources for MALs in higher education provide mainly anecdotal support for military students to utilize in the classroom.

Collectively, there is a dearth of resources to prepare those affiliated with the military and members of the academy to effectively collaborate to support MALs' success in degree completion. For MALs, the transition into higher education can be difficult, fraught with roadblocks, frustration, and a lack of support, which can ultimately lead some to quit higher education (Alschuler & Yarab, 2016; Artino, 2009; Olsen, Badger, & McCuddy, 2014). Due to the dearth of research and resources for using an asset-based approach to understanding MALs, our volume seeks to provide faculty, staff, and administrators with strategies to effectively and holistically connect MALs' previous lived experiences, skills, and perspectives to course work, classroom environments, and institutions as a whole.

MILITARY LEXICON

Additionally, in this edited volume, we seek to dispel common stereotypes, misconceptions, and exclusive language that surround the MAL identity (in this edited volume, see Chiara, Chiara, & Ashlock, chapter 3; see Howe, chapter 4; and see Sciullo, chapter 9). Throughout this volume, we attempt to illuminate the contours of MAL identities in the higher education classroom. For example, we use the term Military-Affiliated Learners to encompass more than just active-duty service members and veterans. Spouses, children, and other family members of active-duty troops and veterans have both their lives and identities greatly impacted by the military lifestyle. As such, it is important to acknowledge the barriers to success in higher education that those affiliated with a military lifestyle might face.

Moreover, we challenge the use of terms such as post-traumatic stress disorder (PTSD) (Smith & Whooley, 2015). While both very real and extremely impactful to the lives of MALs, the language of "disorder" frames the military experience and subsequent mental health issues in a negative manner that may perpetuate stereotypes and may fuel bias against veterans and their lived experience. Instead, we encourage the use of Post-Traumatic Stress, removing the "disorder" and associated stigma and potential barriers to help-seeking behavior (Egan Chamberlin, 2010, 2; Vieira, 2020).[3]

Additionally, we seek to increase awareness of and challenge stereotypes of MALs in higher education classrooms that may inhibit their willingness to engage and succeed. We question the assumptions that all MALs identify as conservative, limiting how and what they are able to communicate in the graduate classroom with peers (in this edited volume, see Howe, chapter 4). We challenge the norm in policy debate that all debate ends in nuclear war, dismissing the lived experiences of those who have been impacted by the realities of war (in this edited volume, see Sciullo, chapter 9). We seek to utilize more inclusive pedagogical frameworks that acknowledge the diverse lived identities and experiences of MALs that are informed by, but not limited to, their MAL background (in this edited volume, see Hernández, De Los Santos Upton, & Aguirre, chapter 7). Overarchingly, we urge readers to reflect on and challenge their understanding of the MAL population, stereotypes, language usage, and misinterpretations. Through this volume, we seek to redefine the MAL identity, in ways that encourage, support, and enhance the diverse knowledge and lived experiences they possess.

BOOK ORGANIZATION

This book contains chapters from a variety of scholars from a range of disciplines with unique and varied experiences serving the MAL population. Below are descriptions of the chapters.

Exploring MAL Identities

Chapter 1: Setting the Stage: MALs in Higher Education

Although assumed to be a small and recent addition to the university environment, MALs have been impacting the formation of higher education since the implementation of the first GI bill in 1944. McDermott, Stewart, Walker, May, and Hernández explore the history of MALs entering and reforming the higher education classroom. Met originally with skepticism and backlash, MALs revolutionized the higher education system, both increasing its accessibility among the American population and redefining pedagogical approaches to teaching.

Chapter 2: Understanding Dual Identity Formation: Identifying Commonalities and Resolving "Differentness" for Military-Affiliated Learners

May and McDermott offer an in-depth look at the impact the military has on MALs' identity in higher education. Through qualitative survey responses, May and McDermott identify themes that impact MALs' identity formation in higher education, such as the willingness to disclose their affiliation and benefits MALs bring to the classroom. May and McDermott explore the nuances of the many possible MAL identities from activity duty and reservists to veterans and dependents, all within the higher education classroom. With this newfound understanding of MALs' expectations and perceived identity characteristics, May and McDermott provide recommendations for campus administrators and faculty to increase MALs' feelings of connectedness with campus communities to increase MALs' retention and success in higher education.

Chapter 3: What Is Impacting the Military-Affiliated Learner Experience in Higher Education?

Although this volume uses MALs to refer to Military-Affiliated Learners, the use of this term often does not take into account the nuances associated with MAL identities. Thus, to understand the veteran experience in higher education, Chiara, Chiara, and Ashlock explore the nuances that impact the formation of the veteran identity from stereotypical misconceptions of veterans to the impact of gender on integration into civilian life. Chiara, Chiara, and Ashlock explain the differences between the military process of learning and the learning process in academia to identify potential challenges and sources of frustration for veterans. Ultimately, based on their lived experiences with over 48 years of military affiliation, Chiara, Chiara, and Ashlock provide readers with tangible strategies and recommendations for better understanding and serving veterans in higher education.

Chapter 4: Supporting Military Veterans in the Graduate Classroom: A Case Study of Political Beliefs and Discussion

Howe provides readers with a unique perspective of veterans' political beliefs. Often assumed to be Republicans, Howe's research suggests that veterans' political identities are complex and multifaceted. Impacted by both the Uniform Code of Military Justice and their own personal lived experiences, how veterans may voice their opinions within the higher education classroom may not actually coincide with how veterans identify politically or with how veterans are perceived by non-MALs. After explaining this juxtaposition of assumed vs. actual political beliefs, Howe provides instructors with ways to effectively engage veterans within graduate classrooms in controversial or political discussions.

Chapter 5: Challenges for Military Partners Pursuing College and Resilience-Based Strategies to Support Their Academic Success

While many may think that MALs solely consist of active-duty service members and veterans, the MAL title encompasses everyone affected by military service to include spouses, children and even parents and family members of service members. As such, Rossetto and Owlett provide unique insights into the lived experiences of military partners pursuing college (MPPCs). Although a majority of research focuses on the lived experiences of active-duty service members and veterans, this firsthand look into the barriers faced by MPPCs identifies three overarching challenges partners encounter with their academic success. Through the identification of these barriers, Rossetto and Owlett conclude with strategies for resilience to improve instructors' capacity for empathy and support for this student population within higher education.

Educator and Support Staff Experiences with MALs

Chapter 6: Speaking Chicana/o Secrets to Military Students: An Autoethnography of Teaching Senior Military Noncommissioned Officers

Pérez provides the untold account of Mexican Americans' role in military history. Often neglected in historical accounts, these brave troops have played an important role in the success of the U.S. military with no recognition of their sacrifices and valor. Through a reflection of his relatives' roles within the U.S. military, as well as his experience teaching MALs for ten years, Pérez provides an exploration of the pedagogical imperative of speaking secrets within the classroom. Additionally, Pérez provides educators with ideas and strategies for speaking secrets within their own classrooms.

Chapter 7: Nepantla, Liminality, and Teaching and Learning in the Space Between: A Borderlands Approach to Military-Affiliated Learner Pedagogy

Understanding the changing demographic landscape of the military is crucial for understanding the nuanced identity characteristics, assets, and challenges that diverse MALs can bring into the classroom. Hernández, De Los Santos Upton, & Aguirre apply a borderlands pedagogical approach to explore how to cultivate a space of awareness, inclusion, and equity in the higher education classroom. From the innovative understanding of the intersection of diverse MAL identities and a borderlands pedagogy approach, Hernández, De Los Santos Upton, & Aguirre provide readers with practical case studies for developing inclusive spaces for MAL educational success using Chicana feminist and borderlands approaches to nepantla, liminality, and honoring students' lived experiences.

Chapter 8: Supporting the Military-Affiliated Learner's Medical School Aspirations

Applying to medical school can be a daunting process for many MALs. Moullet and Belding recount their experiences as professionals in the field who have helped and mentored MALs for years to reach their goals of entering medical school. Moullet and Belding explore the knowledge gaps MALs may encounter when working through the medical school application process from picking an undergraduate institution to preparing for a medical school interview. Moullet and Belding provide in-depth insights into successfully preparing and setting MALs' expectations for medical school.

Chapter 9: Nuclear War! Theorizing the Negative Impacts of Policy Debate's Tropological Mistreatment of War for Military-Affiliated Learners

Sciullo takes a comprehensive look at the exclusive language associated with policy debate. As debates often devolve into "nuclear war," Sciullo argues that current policy debate practices make light of war, and ignore the experiences of those impacted by war by jumping drastically from international tensions to war at almost every debate tournament. Providing recommendations for change, Sciullo invites readers to consider the polarizing language used when describing war and invites the debate community to reform how the topic of war is approached to become more inclusive of MALs, as well as all those impacted by war more broadly.

Chapter 10: The Hard Sell: Supporting MALs through Resource Centers

Although campus resource centers were developed to help students succeed in higher education, they have been and remain immensely underused by college students. However, those who do use campus resource centers are more likely to attain higher grades, experience increased rates of retention, and feel more connected to their campus community. Velez, McDermott, and Marlin examine how to better support MALs through the use of campus resource centers. Through an autoethnographic account of their experiences managing and working in campus resource centers, the authors provide strategies for resource centers and administrators to use to encourage MALs to all of the resources they have at their disposal.

Building Inclusive Learning Communities Virtually and Beyond

Chapter 11: Best Practices in Supporting Military-Affiliated Learners: Suggestions from Communication and Psychological Sciences

Hernández and Belding highlight how the rising population of MALs in higher education necessitates instructors' need to effectively engage this diverse student population. Outlining the contributions MALs bring to the classroom through a variety of frameworks, Hernández and Belding provide readers with tangible pedagogical strategies for easing MALs' cultural shift from the military to academic settings, enhancing students' learning through mastery and growth mindsets, promoting increased academic self-efficacy through effective study habits, and mitigating the challenges associated with geographic mobility. Ultimately, Hernández and Belding challenge readers to rethink the level of engagement and support provided within the higher education classroom.

Chapter 12: Two-Way Communication as a Constructivist Pedagogy Tool for Enhancing Meaning Making in Digital Learning Classrooms with Military-Affiliated Learners

Presenting an overview of constructivist pedagogy for online learning, Cobos shares the importance of understanding the unique characteristics of the

MAL population for more successful distance learning classrooms. Through a review of literature surrounding the challenges MALs may face when attempting to succeed in online classes, Cobos outlines the importance of engaging in constructivist pedagogy and provides options for implementing constructivist strategies into remote learning environments. With an overview of digital platforms that are useful for initiating two-way communication, Cobos provides an overview of technologies and strategies instructors can implement to increase online engagement and communication with MALs, as well as with their peers.

Chapter 13: "So We Hit It with a Hammer, and It Worked": How Active-Duty Marines Communicate, Collaborate, and Forge Their Own Learning Communities in Social Media

With the ubiquity of social media, McGregor provides an in-depth analysis of the importance of Marines' social media communities. Through a mixed-methods study using interviews and survey responses, findings indicate that Marines engage in self-directed resource-seeking behaviors, demonstrate creative cross-generational collaboration, and reflect a culture of trust and support through shared purpose and communication practices within social media communities. McGregor's findings provide campus administrators and faculty with recommendations to create and foster MAL mentorship communities in higher education. To enhance learning and increase MALs' success and retention rates, McGregor suggests developing virtual MAL communities to help socialize, mentor, and support MALs as they transition from military to academic life.

Chapter 14: Collaborative Apprenticeship Approaches to Writing Instruction in the Military

While the research to date has explored a deficit model for MALs' literacy competency, Ludewig approaches literacy from the perspective of an asset-based approach. Using in-depth interview data from previously enlisted MALs, Ludewig uncovers a literacy model used in the military and how their previously learned literacy skills can be transferred to the higher education classroom. Although previous research has assumed that MALs' literacy skills are done in lock-step fashion, Ludewig uncovers the rhetorical complexities of military writing and argues that the connection between literacy instruction and practices in the military and those of academia are stronger than previously thought. Identifying potential areas of frustration for MALs, Ludewig provides tangible strategies for higher education instructors to develop and improve literacy relationships with their MALs.

Chapter 15: Adapt and Overcome: Military-Affiliated Learners and Transfer of Learning

St. Pierre challenges readers to reject the deficit model of learning, focusing on the skills MALs may not have to embrace an asset frame of mind in the classroom. Through this change in perspective, St. Pierre argues that instructors can use MALs' knowledge and skills from their military experience to enhance learning in higher education. Through a transfer of learning approach, St. Pierre offers instructors strategies for cultivating MALs' already-developed educational skills to avoid frustration among in academia.

From Application to Graduation: Collective Recommendations for Better Serving Those Who Serve

In a final review of recommendations, May, McDermott, and Hernández offer final thoughts for administrators, policymakers, educators, and support staff to better understand MALs' identities and experiences in higher education. Summarizing key findings from the edited volume, the authors offer final recommendations for better serving MALs throughout their educational careers in higher education from beginning their undergraduate journey to completing a doctorate degree.

CONCLUSION

Ultimately, military culture impacts and influences opportunities and access to education. Presently, MALs are not being provided with the adequate tools to succeed in the transition from the military to higher education, with barriers evident in mental health, information navigation, and access contexts (Burnett & Segoria, 2009; Griffin & Gilbert, 2015; Hernández & Belding, 2019). Student veterans, for example, face information barriers distinctive from other student populations (McBain, Kim, Cook, & Snead, 2012), leading scholars to conclude that veteran students require academic system streamlining (Burnett & Segoria, 2009; McBain, Kim, Cook, & Snead, 2012; Vance & Miller, 2009) and intra-institutional collaboration to attend to the unique needs of veteran students. Active-duty service members and reservists may struggle with structural and cultural barriers to success. Spouses may experience a variety of barriers from a lack of social support to a loss of identity.

To better serve the military community, it is imperative to understand MALs' lived experiences and sense-making processes to better support their success within the higher education classroom. If you are an instructor, a MAL, an administrator, a policymaker, an instructional designer, etc., we hope this volume provides you with the insight to recognize the challenges and barriers those affiliated with the military may face when attempting to succeed in academia. Although the obstacles they face may be great, MALs will continue to be resilient, determined, and committed in the classroom. As such, it is our role to support their aspirations by better serving those who serve in higher education.

NOTES

1. "Military-affiliated learner" is the term used to signify our broader focus on and inclusion of the experiences of active duty service members, reservists, veterans, spouses/significant others, and adult children who are transitioning to higher education.

2. Data reflect funding between 2009 and 2018 for 2.3 million beneficiaries (CBO, 2019).

3. We acknowledge the dual-edged nature of the linguistic sword, namely that some MALs may identify with the D in PTSD because of the nature of the diagnosis; however, we also seek to destigmatize negative perceptions of MALs who may have PTS and other affiliated diagnoses.

REFERENCES

Alschuler, M., & Yarab, J. (2016). Preventing student veteran attrition: What more can we do? *Journal of College Student Retention: Research, Theory & Practice, 20*(1), 47–66. doi:10.1177/1521025116646382.

American Council on Education. (2015, Spring). *By the numbers: Undergraduate student veterans*. https://www.acenet.edu/the-presidency/columns-and-features/Pages/By-the-Numbers-Undergraduate-Student-Veterans.aspx.

Artino, A. R. (2009). Think, feel, act: Motivational and emotional influences on military students' online academic success. *Journal of Computer Higher Education, 21*, 146–166.

Barry, A. E., Whiteman, S., MacDermid Wadsworth, S., & Hitt, S. (2012). The alcohol use and associated mental health problems of student service members/veterans in higher education. *Drugs: Education, Prevention, and Policy, 19*(5), 415–425. doi:10.3109/09687637.2011.647123.

Burnett, S. E., & Segoria, J. (2009). Collaboration for military transition students from combat to college: It takes a community. *Journal of Postsecondary Education and Disability, 22*(1), 53–58.

Congressional Budget Office [CBO]. (2019, May). *The Post-9/11 GI Bill: Beneficiaries, choices, and cost.* https://www.cbo.gov/system/files/2019-05/55179-Post911GIBill.pdf.

Defense Manpower Data Center. (2018). *DoD Personnel, Workforce Reports & Publications*. https://www.dmdc.osd.mil/appj/dwp/dwp_reports.jsp.

Drummet, A. R., Coleman, M., & Cable, S. (2003). Military families under stress: Implications for family life education. *Family Relations, 52*(3), 279–287.

Eagan Chamberlin, S. M. (2012). Emasculated by trauma: A social history of post-traumatic stress disorder, stigma, and masculinity. *The Journal of American Culture, 35*(4), 358–365.

Green, S., Nurius, P. S., & Lester, P. (2013). Spouse psychological well-being: A keystone to military family health. *Journal of Human Behavior in the Social Environment, 23*(6), 753–768.

Griffin, K., & Gilbert, C. K. (2012). *Easing the transition from combat to classroom: Preserving America's investment in higher education for military veterans through institutional assessment*. https://www.americanprogress.org/wp-content/uploads/issues/2012/04/pdf/student_veterans.pdf.

Griffin, K. A., & Gilbert, C. K. (2015). Better transitions for troops: An application of Schlossberg's transition framework to analyses of barriers and institutional support structures for student veterans. *The Journal of Higher Education, 86*(1), 71–97.

Hart, D. A., & Thompson, R. (2016). Veterans in the writing classroom: Three programmatic approaches to facilitate the transition from the military to higher education. *College Composition and Communication*, *68*, 345–371.

Hernández, L. H. (2019). "We get it done because we have to:" Military spouses with advanced degrees, career and educational experiences, and grit amongst uncertainty. In L. H. Hernández & J. Belding (Eds.), *Military spouses with graduate degrees: Interdisciplinary approaches to thriving amidst uncertainty* (pp. 239–258). Lexington Books.

Hernández, L. H., & Belding, J. (Eds.) (2019). *Military spouses with graduate degrees: Interdisciplinary approaches to thriving amidst uncertainty.* Lexington Books.

Hernández, L. H., & Belding, J. (2019). Introduction: Military spouses with advanced degrees: Current findings and pathways moving forward. In L. H. Hernández & J. Belding (Eds.), *Military spouses with graduate degrees: Interdisciplinary approaches to thriving amidst uncertainty* (pp. 1–14). Lexington Books.

Levin, A., & Dean, D. R. (2013, November/December). 5 ways today's students are radically changing our colleges. *The Association of Governing Boards of Universities and Colleges, 21*(6). https://agb.org/trusteeship-article/5-ways-todays-students-are-radically-changing-our-colleges/.

Lim, N., & Schulker, D. (2015). *Measuring underemployment among military spouses*. RAND Corporation. http://www.rand.org/pubs/monographs/MG918.

Livingston, W. G., Havice, P. A., Cawthon, T. W., & Fleming, D. S. (2011). Coming home: Student veterans' articulation of college re-enrollment. *Journal of Student Affairs Research & Practice, 48*(3), 315–331.

May, A., & McDermott, V. (2019). Cultural collision: A review of the literature. In L. H. Hernández & J. Belding (Eds.), *Military spouses with graduate degrees:*

Interdisciplinary approaches to thriving amidst uncertainty (pp. 59–85). Lexington Books.

McBain, L., Kim, Y. M., Cook, B. J., & Snead, K. M. (2012). *From soldier to student II: Assessing campus programs for veterans and service members.* American Council on Education. https://www.acenet.edu/Documents/From-Soldier-to-Student-II-Assessing-Campus-Programs.pdf.

Mitchell, M., Leachman, M., & Masterson, K. (2017, August 23). *A lost decade in higher education funding: State cuts have driven up tuition and reduced quality.* Center on Budget and Policy Priorities. https://www.cbpp.org/research/state-budget-and-tax/a-lost-decade-in-higher-education-funding.

Olsen, T., Badger, K., & McCuddy, M. D. (2014). Understanding the student veterans' college experience: An exploratory study. *The United States Army Medical Department Journal, 2014*, 101–108.

Padden, D. L., Connors, R. A., & Agazio, J. G. (2011). Stress, coping, and well-being in military spouses during deployment separation. *Western Journal of Nursing Research, 33*(2), 247–267.

Rumann, C. B., & Hamrick, F. A. (2009). Supporting student veterans in transition. *New Directions for Student Services, 126*, 25–34. doi:10.1002/ss.313.

Segal, M. W. (1986). The military and the family as greedy institutions. *Armed Forces & Society, 13*(1), 9–38.

Smith, R. T., & Whooley, O. (2015). Dropping the disorder in PTSD. *Context, 14*(4), 38–43.

Sollitto, M., & Cole, K. (2019). Military spouses' uncertainty management: Navigating academic goals and military needs. In L. H. Hernández & J. Belding (Eds.), *Military spouses with graduate degrees: Interdisciplinary approaches to thriving amidst uncertainty* (pp. 125–142). Lexington Books.

Vance, M. L., & Miller, W. K. (2009). Serving wounded warriors: Current practices in postsecondary education. *Journal of Postsecondary Education and Disability, 22*(1), 18–35.

Vieira, K. (2020, January 27). *PTS vs PTSD: Dropping the "D"*. The Recovery Village. https://www.therecoveryvillage.com/mental-health/ptsd/related/pts-vs-ptsd/.

Weiss, E. L. (2015). Building community in a diverse college classroom. In J. E. Coll & E. L. Weiss (Eds.), *Supporting veterans in higher education: A primer for administrators, faculty, and advisors* (pp. 104–129). Lyceum.

Part I

EXPLORING MAL IDENTITIES

Chapter 1

Setting the Stage

MALs in Higher Education

Victoria McDermott, Margaret C. Stewart, Carol M. Walker, Amy R. May, and Leandra Hinojosa Hernández

Although they may be considered a minoritized group in higher education, Military-Affiliated Learners (MALs) have been on campus and in the higher education classroom for decades. Throughout their history in academia, MALs have frequently been met with opposition and skepticism as educators, administrators, and students alike have expressed reluctance and apprehension toward the idea of integrating battle buddies into the classroom (Olson, 1973; Wilkes, 2017). Historically, it has been difficult to convince the public that including combat veterans with traditional students could yield any benefits or positive outcomes, beginning with the troops returning home from World War II (Olson, 1973). Currently, research shows this hesitancy is still being perceived today (in this edited volume, see May & McDermott, chapter 2; Sciullo, chapter 9; Hernández & Belding, chapter 11). However, higher education has been revolutionized through the inclusion of MALs, changing the rules and regulations for who is and isn't allowed in college classrooms (Humes, 2006; Olson, 1973). As such, this chapter provides an overview of the history of MALs in higher education beginning from the end of World War II to the present, illustrating the role they have played in increasing the accessibility of academia for all students.

THE G.I. BILL'S ROLE IN FOSTERING GROWTH OF MALS

Veterans and active duty members of the Armed Forces began joining college campuses after World War II, due to the implementation of the original

G.I. Bill. Created in 1944, the original G.I. Bill was developed to avert mass unemployment for millions of veterans returning from war. Formally entitled the Servicemen's Readjustment Act of 1944, this controversial bill was passed under the administration of then-president Franklin D. Roosevelt to provide educational assistance and unemployment insurance to returning war veterans (Humes, 2006; Olson, 1973).

The passing of the G.I. Bill, however, was not without much debate. The educational benefits proposed in this bill from the government were unprecedented, which led to extensive deliberation in Congress. The opposed provisions including the educational grants of $500 for veterans who served 90 days or more of active duty and to those discharged because of a service-related disability, as well as a living stipend of $65 per month for single veterans and $90 month for married veterans, with an additional $20 a month for unemployed veterans (Bound & Turner, 2002; Olson, 1973). Benefits extended from a minimum of one year to a maximum of four years, depending on age and length of service. Opponents of the bill argued that it would minimize the veterans' incentive to look for work, while other legislators questioned the pragmatism of sending battle-hardened veterans to colleges and universities, an opportunity that was largely viewed as being reserved only for the privileged few (Olson, 1973).

With the passing of the G.I. Bill in 1944, 2.2 million veterans received an education upon return from World War II (Olson, 1973). Although initially designed to aid the 16 million individuals who served the nation during World War II, the G.I. Bill of 1944 has been credited with democratizing education by affording less "elite" populations access to higher education (Humes, 2006). Prior to the bill's passing, higher education was composed almost exclusively of society's existing upper class (Field, 2008; Humes, 2006). By the time the original G.I. Bill ended on July 25, 1956, 7.8 million World War II veterans had participated in some sort of education or training program with these benefits (Olson, 1973), and between 1940–1950, the number of U.S. college and university degree holders more than doubled (DoD, 2019). Humes (2006) credits the G.I. Bill and the educational opportunities it provided with building modern America, creating a thriving middle and upper class among individuals who were raised during Depression-era America. Previous research estimates that 20% of the veterans who received an education through the G.I. Bill would not have attended college without this financial support (Bound & Turner, 2002; Turner & Bound, 2003).

Ultimately, the G.I. Bill removed financial barriers to education for veterans and lowered the barriers to access for others. After the bill was enacted, the first signs of a non-traditional student population began emerging (Field, 2008; Olson, 1973). Overarchingly, the bill showed that a large portion of the nation's youth had the ability to succeed in, but not the money to enter into, higher education.

Benefits of the G.I. Bill & MALs

The G.I. Bill was initially criticized by educators, fearing that the quality of education would be compromised by the attendance of war veterans. Certain scholars frowned upon the idea of being required to educate a population of people that they believed did not traditionally possess college aptitude. To the surprise of many, the G.I. Bill proved to be extremely successful. The anticipated friction between veteran and nonveteran students was almost nonexistent (Olson, 1973). Among the original critics were some skeptical professors who subsequently acknowledged the unforeseen maturity, commitment, and overall academic achievement by the veteran-student population (Olson, 1973).

Franklyn Snyder, former President of Northwestern University, called the G.I. Bill the "greatest experiment in democratic education that the world has ever seen" (Olson, 1973, p. 606). In November of 1947, *The New York Times* published a study on veterans in higher education. Editor Benjamin Fine revealed that the veterans were making Dean's Lists, earning top marks, and filling honor roll lists. He declared veterans an asset to higher education, a far cry from the hardened problem child that some were anticipating (Olson, 1973). Additionally, in 1949, *Fortune* published the findings of its study on the class of 1949 reporting, "to most of those who have worked closely with it [class of 1949] from the professors and the placement directors to the recruiters for industry '49 is the best class the country has ever produced" (Olson, 1973, p. 596). However, it is important to note that in some cases veterans were "tolerated, not embraced" in their transition to higher education (Olson, 1973, p. 609).

Overall, the G.I. Bill led to a variety of long-term effects both in and out of the academic world. Within the academic realm, enrollment numbers skyrocketed. In addition to the rise in enrollment numbers, the knowledge gaps veterans faced in the classroom led to a reformation of degree paths. Institutions began providing refresher courses for veterans. Institutions also began providing credits for certain military training and experience, which is still in place today. Moreover, since many veterans were married, they normalized the idea of married students in higher education, an idea that was once a sufficient cause of dismissal (Brambaugh, 1944; Olson, 1973; Waller, 1945). Overall, the presence of veterans in higher education permitted students to combine marriage, children, and college, which meant that a family no longer meant the end to a college degree. In opposition to the initial fear of veterans' academic performance, the best students were veterans who had been away "from school the longest, those who were married, and those who were older" (Olson, 1973, p. 606).

Outside of academia, the G.I. Bill positively impacted the economy. The G.I. Bill led to an increase in economic and personal knowledge capital stock (i.e., habits, social, and personality attributes embodied in the ability to perform labor so to produce economic value; Zhang, 2017). Through this increase in college-educated workers, the job market became more competitive. Furthermore, by diverting some veterans' career paths from flooding the job market upon their return from World War II, mass unemployment and high job demand was avoided. Researchers suggest that every dollar spent on the G.I. Bill was multiplied many times in benefits to the post-war economy (Quindlen, 2008). As Americans earned higher degrees and wages, they were able to pump more money into the economy by buying homes and consumer goods (Mettler, 2005).

While the G.I. Bill afforded many veterans increased access to higher education and a pathway to post-service careers, there were and remain challenges, most notably related to racial/ethnic and gender disparities. Despite being touted as the first "race-neutral" legislation, Black veterans faced significant discrimination related to segregation especially in the South (Turner & Bound, 2002, p. 7).

At the conclusion of World War II, Black veterans wanting to attend college in the South were

> restricted in their choices to about 100 public and private institutions delineated in the Office of Education publications as "Colleges for Negroes," as segregation in public higher education remained a legal mandate in many southern states. The public institutions for blacks were founded largely under the Second Morrill Act in 1890, which specifically prohibited the distribution of federal funds to states that did not provide separate accommodation for blacks if the primary state institution denied admission to blacks. (Turner & Bound, 2002, p. 4)

Due to limitations of available data, driven in large part by the lack of race-related tracking at predominantly white institutions (PWIs) (Turner & Bound, 2002), there is a lot we do not know about how "transformative" (DoD, 2019) the G.I. Bill was for marginalized and disenfranchised populations, including the 500,000 Hispanic or Latino soldiers, 13,499 Chinese American soldiers, and 25,000 Native American soldiers who served in WWII alone (Department of Veteran Affairs, 2017).

Similarly, Mexican Americans who served in World War II encountered several barriers and challenges when trying to utilize the G.I. Bill's provisions. Mexican American World War II veterans expected to have full access to the G.I. Bill and America's post-war reintegration promises, yet access to the Bill's provisions and services was not always straightforward (Rosales, 2011). Such barriers included access issues, uneven implementation as a result of logistical downfalls, and personal battles with post-war trauma, all of which "occurred within a larger social, political, and racial environment that sought to maintain the discriminatory practices that existed prior to World War II" (Rosales, 2011, p. 627). Rosales (2011) illustrates, however, that as a result of Mexican American veterans' racial/ethnic identity in the middle of the American white/black racial binary, Mexican American veterans had a "nuanced" experience with the G.I. Bill overall, illustrating the "complicated matrix" of experiences veterans of color had (p. 599). For example:

> Training received under the bill, and one's status as a veteran, did facilitate a degree of social mobility for some while also fostering activism in the pursuit of social equality. The bill's entitlements had the capacity to serve as a "bridging environment"—a term that refers to the many potential opportunities that have benefited veterans as compared to their non-veteran counterparts, such as preferential hiring practices, military training, and exposure to mainstream bureaucratic organizations, as well as the GI Bill. (Rosales, 2011, p. 599)

Ultimately, Rosales (2011) contends that the use of a deficit frame to view Mexican American veterans' experiences with the G.I. Bill should not be the primary frame utilized from a historical memory perspective, yet the real impacts of racism and discrimination for Mexican American veterans, Black veterans, and other veterans of color remain.

In addition to racial disparities, gender disparities also characterized early access to the G.I. Bill in the post-World War II era. During World War II, women acquired both "new responsibilities and unprecedented opportunities" to contribute to wartime efforts (Hartmann, 1995, para. 1), although such efforts were not without their challenges. In occupying this unique role in military efforts, "Americans acknowledged the critical need for women's contributions to the war effort, but they also cherished conventional gender roles and worried about how women's new activities outside the home would affect male-female relations and family life" (Hartmann, 1995, para. 2). To advocate for women's rights, leaders of women's organizations asserted that women's right to serve in the military was a fundamental component of their citizenship, as was the right to benefit from such policies as the G.I. Bill. While women's college enrollments grew after war, they declined in relation to men's enrollments (Hartmann, 1995). Given that women comprised a smaller percentage of veterans than men (less than three percent of all military personnel at this time), the vast majority of student-veterans were men (Hartmann, 1995). Hartmann (1995) highlights the gender disparity that characterized women's use of the G.I. Bill in the post-war years:

> Women, in fact, found it more difficult to gain admission to crowded colleges that gave preference to veterans. In some respects the G.I. Bill was a forerunner of affirmative action, though in this case men—and their families—benefited.

> As World War II veterans completed their education, women's share of college degrees rose, but even by 1960, women claimed just one-third of all degrees, still less than their share in 1940. The great postwar expansion of higher education eventually benefited women, but not until the 1960s and 1970s. (para. 19)

Ultimately, while the G.I. Bill should be celebrated, given the fact that it was "one of the most far-reaching pieces of social policy legislation in the second half of the twentieth century," one of its major limitations is that even though it was created for all service members, it was mostly available to only white, middle-class men (Canaday, 2003, p. 935).

The G.I. Bill Today

Due to the success of the first G.I. Bill, it has been extended multiple times, helping over 10.3 million veterans access college education after the Korean and Vietnam wars (DoD, 2019). In 2008, the Post-9/11 G.I. Bill was passed. Since its implementation in August of 2009, 800,000 veterans and their families have received educational benefits and opportunities (DoD, 2019). Earlier versions of the bill paid up to the highest public in-state undergraduate tuition and fees and initiated the Yellow Ribbon program to designate institutions that were considered to be military-friendly or that had programs in place to support MALs during their transition to higher education (Fields, 2008). The most recent version passed, the "Forever G.I. Bill," offers additional benefits and funding for specific degrees, such as science, technology, engineering, and math (Harry W. Colmery Veterans Educational Assistance Act, 2020).

Since the Post 9/11 G.I. Bill, dependents of veterans have been eligible for G.I. Bill benefits through the Transfer of Entitlement provisions (Department of Veterans Affairs, 2010). Moreover, in alignment with the G.I. Bill, more benefits have been developed for both active duty service members and dependents spouses and children of active duty service members or veterans. These financial aid programs include Tuition's Assistance, Military Spouse Career Advancement Accounts (MyCAA), Dependents Education Assistance Program (DEA), and a variety of branch specific scholarships and grants (MilitaryOneSource, n.d.).

CONCLUSION

In an interview featured in *La Salle Magazine* (2009), Dr. Earl Goldberg, a former U.S. Army Reservist and veteran of Operation Desert Storm, described veterans as a "welcome asset" to any institution or program, acknowledging that veterans tend to be motivated, disciplined, and goal-oriented. He indicates that they are a group that understands teamwork and are willing to reach out to their peers, seek cohesion, and promote collaboration. Similarly, research indicates that military spouses are resilient and dedicated to their education and careers, even in the midst of military-related unemployment and underemployment when compared to their civilian counterparts (see Hernández & Belding, 2019). However, this favorable shift in attitude toward MALs is not always the case in institutions of higher learning. Christinger (2016) shares:

> For those who haven't ever served the gap between the military and civilians can be easily ignored. For those who have served, the military-civilian divide is everywhere and it affects how Veterans are able to connect with their families, their classmates, their coworkers, and their communities. (p. 6)

The tension between MALs and non-MALs in the classroom continues to occur. The everyday lives and lived experiences of MALs tend to differ from the lives of traditional undergraduate college students, which impacts their role and place in higher education (Field, 2008). Overarchingly, the G.I. Bill has had a lasting effect within the United States. Although originally controversial, the G.I. Bill has resulted in significantly beneficial impacts, both inside and outside of academia from stimulating the economy (Olson, 1973; Griffin & Gilbert, 2012) to increasing accessibility for all students to obtain a college degree. With the implementation of the G.I. Bill, U.S. Americans were no longer restricted from career advancement because of limited access to higher education, a trend that continues today as thousands of MALs are able to overcome the financial barriers of higher education to successfully earn a college degree.

REFERENCES

Bound, J., & Turner, S. (2002). Going to war and going to college: Did World War II and the G.I. Bill educational attainment for returning veterans? *Journal of Labor Economics, 20*, 784–815.

Brambaugh, A. J. (1944). Planning education for returning members of the Armed Forces. *American Association of Teachers Colleges, 23*, 53–54.

Canaday, M. (2003). Building a straight state: Sexuality and social citizenship under the 1944 GI Bill. *The Journal of American History, 90*(3), 935–957.

Chrisinger, D. (2016). *See me for who I am: Student veterans' stories of war and coming home.* Whitman Excelsior College Press.

Department of Veteran's Affairs. (2010, December). *Education and training*. https://benefits.va.gov/gibill/.

Department of Veteran's Affairs. (2017). *Minority veterans report: Military service history and VA benefit utilization statistics*. https://www.va.gov/vetdata/docs/SpecialReports/Minority_Veterans_Report.pdf.

Department of Veteran's Affairs. (2019). *Federal benefits for veterans, dependents, and spouses*. https://www.va.gov/opa/publications/benefits_book/2014_Federal_Benefits_for_Veteran_English.pdf.

DoD. (2019, January 9). *75 years of the G.I. Bill: How transformative it's been*. https://www.defense.gov/Explore/Features/story/Article/1727086/75-years-of-the-gi-bill-how-transformative-its-been/.

Field, K. (2008). *Cost, convenience drive veterans' college choices. The Chronicle of Higher Education*. https://www.chronicle.com/article/cost-convenience-drive-veterans-college-choices/.

Griffin, K., & Gilbert, C. (2012). *Easing the transition from combat to classroom: Preserving America's investment in higher education for military veterans through Institutional Assessment*. Center for American Progress. https://www.americanprogress.org/wp-content/uploads/issues/2012/04/pdf/student_veterans.pdf.

Harry, W. (2017). *Colmery Veterans Educational Assistance Act of 2017*. 3218 U.S.C. § 115c.

Hartmann, S. M. (1995, September). Women, war, and the limits of change. *National Forum, 75*(4), 15–20.

Hernández, L. H., & Belding, J. N. (2019). *Military spouses with graduate degrees: Interdisciplinary approaches to thriving amidst uncertainty*. Lexington Books.

Humes, E. (2006). *Over here: How the G.I. Bill transformed the American Dream*. Harcourt, Inc.

Mettler, S. (2005). *Soldier to citizens: The GI Bill and the making of the greatest generation*. Oxford Press.

MilitaryOneSource. (n.d.). *For spouses-benefits*. https://www.militaryonesource.mil/education-employment/for-spouses/for-spouses-benefits.

Olson, K. W. (1973). The G.I. Bill and higher education: Success and surprise. *American Quarterly, 25*(5), 596–610.

Quindlen, A. (2008, March 31). Because it's right. *Newsweek*. https://www.newsweek.com/quindlen-because-its-right-84315.

Rosales, S. (2011). Fighting the peace at home: Mexican American veterans and the 1944 GI Bill of Rights. *Pacific Historical Review, 80*(4), 597–627.

Turner, S., & Bound, J. (2002). *Closing the gap or widening the divide: The effects of the G.I. Bill and World War II on the educational outcomes of Black Americans*. https://www.nber.org/papers/w9044.

Waller, W. (1945). Which veterans should go to college. *Ladies' Home Journal, 62*, 142–169.

Wilkes, L. (2017). Framing a veterans syllabus statement for faculty members. *Journal of Veterans Studies, 2*(1), 113–116.

Zhang, L. (2017). Veterans going to college: Evaluating the impact of the Post-9/11 GI Bill on college enrollment. *Educational Evaluation and Policy Analysis, 40*(1), 82–102.

Chapter 2

Understanding Dual Identity Formation

Identifying Commonalities and Resolving "Differentness" for Military-Affiliated Learners

Amy R. May and Victoria McDermott

Making the decision to join the military is life changing. Whether an active-duty service member, reservist, veteran, spouse, or child, "it's not just a job" (U.S. Naval Academy, 2019, para. 9). Service members and their families join a group defined by

> service, military customs and courtesies, proper wear of the uniform, military bearing, military values and ethics, and other information that is critical to their success in the service, including how to listen to and follow orders and how to function within the military chain of command. (DHHS, 2010, p. 10)

Through their group identification, service members and their families adopt the identity of "military service member," defining who they are, in part, by their position within the Armed Forces and commitment to the values, beliefs, and customs instilled by military culture (Koehler, 2014). Ultimately, service members (and potentially their families) may adopt multiple identities as they navigate service and transition between military and civilian groups; however, they "will continue to identify themselves as soldiers" and interact with others based on this identity (Koehler, 2014, p. 2). When competing identities are congruent with the military, group membership may be perceived as beneficial, facilitating engagement, commitment, and connectedness (Tajfel & Turner, 1986). When competing identities are in conflict with what it means to be a "service member," an individual may withdraw to avoid identity conflict or bias in favor of the dominant identity, for example, the military (Tajfel & Turner, 1986). Within the context of higher education, a military-affiliated learner (MAL) is tasked with building the identity of "college student," which is oftentimes shaped and formed by the framework of their military identity. If MALs and their selected colleges and universities are able to identify and promote commonalities, dual identity formation is supported (i.e., service member and student), and a reconciliation of the student and MAL identity occurs (Dovidio, Gaertner, & Saguy, 2007). As a result, academic success for the MAL college student is facilitated. On the other hand, if MALs are unable to connect or engage with peers, faculty, and/or administration due to perceived "differentness," retreat is likely to occur and dual identity formation may be stalled. Research shows that MALs struggle with persistence and degree completion in higher education (Mentzer, Black, & Spohn, 2015), suggesting some service members and/or their dependents (e.g., spouses or children) may struggle with dual identity formation, highlighting the need for further investigation and setting the agenda for this research project.

While previous research explicates military identity formation (DHHS, 2010; Koehler, 2014; Meiners, 2019), little is known about how the military identity impacts experiences in higher education, specifically the formation of the student identity. To address this deficiency in the literature, this chapter defines the service member identity framed by social identity theory, explores differences between the military and academy, and elucidates the results of a qualitative investigation designed to learn more about how dual identity is both supported, negated, and impacted by a MAL's experience in higher education. The chapter ends with recommendations for colleges and universities to make commonalities salient to increase rates of retention and degree completion.

SOCIAL IDENTITY THEORY AND THE SERVICE MEMBER IDENTITY

"A soldier for life" underscores the commitment between a service member and the Armed Forces (Koehler, 2014, p. 1). Since its inception in 1789 (NCC, 2019), the military has defined its culture with specific jargon, rituals, and structure (Halvorson et al., 2010; May & McDermott, 2019), establishing strong group identification for members through social integration. As framed by social identity theory (SIT), people formulate their social identity based on group identification (Tajfel, 1978). "I am a member of the military" represents a social identity: "that part of an individual's self-concept which derives from knowledge of membership in a social group (or groups) together with the value or emotional significance attached to that membership" (Tajfel, 1978, p. 63). For service members and their families, the strength of the "service member identity" is forged over time through cultural assimilation, living the military life, and communing with other military families. Members of the military community are seen as "family" or in-group participants, sharing a sense of solidarity and camaraderie as military group members share a rich culture, characterized by strong values, beliefs, and norms.[1] For military families living on bases or installations, the connection may be even stronger as "it is possible to live life on a base or post and never interact with the 'outside world'" (DHHS, 2010, p. 3).

For many service members, being part of the military represents a call to serve the greater good (Helmus et al., 2018), and their experiences affirm their sense of self-worth, resulting in positive group identification and continued membership (Tajfel & Turner, 1986). Being part of the military group and embodying the culture is critical for building and maintaining service members' identities and commitment to the calls of the military (i.e., long deployments, hazardous situations, and time away from family). As these aspects of culture are ingrained in members' identities, they shape and frame how a service member perceives and responds to out-groups (Tajfel & Turner, 1986), such as the academy. As such, these aspects of culture offer a starting point for identifying commonalities and points of contention which may facilitate or hinder dual identity formation.

Defining the Military Group

Military culture has a specific set of rituals and roles prescribed for all members. Particularly, the military places a high value on cultivating a member's sense of identity through rank and adherence to a strict moral code.

Ranks

A defining feature of military groups is the system of hierarchy defined by the rank structure. Through this system, each group member knows their role, level of authority, and purpose across tasks, jobs, and locations (Devries et al., 2012). This structure is vital in maintaining order, respect, and formality within the group and defines membership and specific benefits for service members and by extension their families. The rank system includes three overarching categories: enlisted, warrant officer, and commissioned officer (Devries et al., 2012).

The enlisted ranks are considered the "blue-collar" workers of the military and comprise 82 percent of the Armed Forces (CRS, 2019). Enlisted service members do much of the hands-on work that keeps the military running. Enlisted ranks are divided E-1 through E-4, which are considered junior enlisted. E-5 and E-6 are junior non-commissioned officer (NCO) positions, and E-7 through E-9 service members are senior NCOs; these service members are expected to take on leadership roles. As enlisted service members make up a majority of the military, they are considered the body of the force. Warrant officers are technical specialists, resident experts, and teachers who hold a unique position as the highest enlisted and lowest commissioned officers. Warrant officers do not take on the leadership roles and tasks that commissioned officers and NCOs are expected to assume. Conversely, commissioned officers can be likened to the "white-collar" worker (Devries et al., 2012). Commissioned officers fill leadership and staff roles. Commissioned officers make up a smaller portion of the military than their enlisted service member counterparts at less than 20 percent (CRS, 2019). Overall, service members of all ranks work together in various configurations to support military operations.

Much of a service member's identity is impacted by their rank. For example, a commissioned officer must have a college degree. Many commissioned officers enter into higher education with the intent to pursue a military career through either a military academy or ROTC program at a four-year institution (ASVAB, 2019). Enlisted members, however, may enter into the military first and use their G.I. Bill second, after years of service, to obtain a college degree (ASVAB, 2019). As commissioned officers step back into the classroom, they bring with them successful previous experiences with higher education, potentially supporting the reignition of the student identity. Furthermore, ranks impact the jobs (military occupational specialty; MOS) and opportunities available to service members while they serve. For example, officers are considered the "managers of the Military, acting in leadership roles that require planning, directing operations and making critical decisions" (DoD, 2020, para 4). Conversely, enlisted careers are said to include mostly "hands-on training for mechanical, transportation, human service or office fields" (DoD, 2020, para 2).

Moreover, service members' rank and MOS can impact the skills they learn while in the service (Parker et al., 2019). Ultimately these skills and experiences are extremely important when service members transition back to civilian life. For example, a majority of enlisted members felt their military skills and experiences provided them with little to no marketable experience when entering into the civilian world (Harkins, 2019). Additionally, a recent study found that while 78 percent of commissioned officers said the military was very or fairly useful in giving them the skills needed for a job outside of the military, only 54 percent of enlisted personnel felt the same way (Parker et al., 2019). These feelings of nontransferable skills have real-world implications, as Parker and colleagues (2019) found that commissioned officers are twice as likely to have a civilian job lined up upon leaving the service.

Transferring these skills and experiences to the classroom represents a potential source of contention as service members enter the classroom. For commissioned officers who may already have a degree and time in service, the transition to college for a graduate degree may impact their "student" identity differently than an enlisted member who has been out of school and serving in uniform for an extended period of time. Furthermore, since ranks in the military establish people's roles and expectations across locations, posts, and jobs, there can be some confusion for service members when entering into academia with a lack of visible rank structure. For example, while a rank system exists for staff, faculty, and administrators, there isn't typically a ranking structure for students. While classes are divided based on professor and student role, class status (freshmen, sophomore, junior, senior) doesn't impact students' interactions with one another. As such, a service member's rank has no bearing or weight within the academy and affords them no privileges in terms of class hierarchy.

Moral Code

In addition to the use of ranks, the military holds its members to a strict moral code to maintain order and uniformity of behavior. The military tends to perpetuate conservative[2] and moral ideologies (Taber, 2009; May & McDermott, 2019). These conservative and traditional mindset ideals are reinforced through ethical guidelines and honor codes such as the *Uniform Code of Military Justice* (UCMJ) (UCMJ, 2018). Violations of ethical behavior or immoral actions can lead a service member to being punished and/or discharged from service (UCMJ, 2018). The many rules and customs are in place to train service members to be effective at following orders and defending one's country, while heavily reinforcing the culture the military wants to portray (Turchik & Wilson, 2010). Since the main function of the military is combat, these unwavering standards allude to the military being composed of citizens of the highest moral and patriotic value (Dunivin, 1994).

Much of the military culture is dictated by the values it encourages. While each branch may have different core values, such as the Navy and Marines values of honor, courage, and commitment or the Coast Guard's values of honor, respect, and devotion (DHHS, 2010), overall, they emphasize the responsibilities to the military above all else. Service members must be willing to leave behind the safety and security of home, sacrificing all that is valuable to them for the greater cause (Dervies et al., 2012). Structure, order, and precision permeate the military environment and are considered critical for executing tasks.

The moral code represents both a potential source of commonality and point of conflict as it relates to both level for formality and consequence.

Students in higher education are expected to adhere to a code of conduct. While each institution defines its own rules and expectations, most institutions have set guidelines for ensuring academic integrity, citizenship, and shared value systems. Moreover, when students violate the student code of conduct, there are formal processes for evaluating the severity of the infraction, assigning blame, and punishing the offender. Although the military and academia may have moral codes, the level of commitment to these values can vary in academia, whereas in the military they cannot. For example, students can attend a faith-based university without being a member of that religion. In this instance, a student would not be formally reprimanded for this "violation" of the code of conduct. Additionally, for students, their potential spouses or children are not required to follow the university's moral code if they are not on campus. For the military, however, their commitment to military values cannot waiver without consequence. The military regulates service members and their dependents' actions both in and outside of uniform. This all-encompassing commitment ingrains military code and values deep into its members, forever impacting their ability to form new and alternative identities (Kramer & Dailey, 2019; Shpeer & Howe, 2020). While the student identity can allow for people to build additional identities, the dominance and devotion required by the military identity make it difficult for any additional identity formation (Williams et al., 2018).

Formation of the Student Identity

Service members and their families choose to join the academy for a variety of different reasons, ranging from career advancement within the military to increased marketability after service. MALs encounter and respond to many of the same issues and concerns as their civilian counterparts, with added challenges from military-specific experiences that impact their transition to college, their preparedness to learn (Drummet et al. 2017) and ultimately the formation of their student identity.

For Service Members

Education is part of the military experience. Service members have access to educational opportunities to develop specialized skills needed to perform their assigned occupational roles and duties defined by rank. However, some service members and their families may choose to join the academy to further their military careers or prepare for civilian life. Enlisted service members often cite the promise of paid education as a primary motivator for selected service (Helmus et al., 2018). Moreover, enlisted active-duty members need a college degree to continue moving up rank and to transition from enlisted to commissioned officer. For example, a four-year or specialized degree is required for Officer Candidate School or Direct Commission[3] (DOD, 2020). However, some military occupations may not leave service members with skills applicable outside of the military, requiring additional education to be competitive in the job market. Upward of 65 percent of U.S. job openings will require postsecondary education (Carnevale, Smith, & Strohl, 2019), thus earning a degree provides a pathway to increased marketability post-service.

For Spouses

Military spouses may feel their educational opportunities are negatively impacted by the military lifestyle (Orthner, 2005). Frequent relocations, for example, present a significant challenge as military spouses are fourteen times more likely than their civilian counterparts to move across state lines (Marshall, 2014). With frequent moves, spouses may struggle to transfer credits, complete degrees, and obtaining licensure (RAND, 2001). This can lead to a lack of opportunities for many military spouses to start or advance in their chosen career paths (Marshall, 2014). While earning a degree, many military spouses may be dealing with a deployment or long training exercises in which they must take on additional household responsibilities, moving coursework lower on the priority list. Ultimately, the military lifestyle makes it difficult for military spouses to pursue and complete a college degree or certificate (Hernández & Belding, 2019).

In-Group vs. Out-Group

As MALs transition to group membership in the academy, they bring with them a strong sense of connection to their military group. Following a SIT model of group identification, the more MALs are able to connect their service member identity with the student identity, the more likely they are to succeed in attaining their educational goals. Research indicates that military-affiliated students sensemake their experiences in higher education based on military service (Hernández & Belding, 2019). Current research associates the MAL identity with characteristics such as persistence, driven or mission-focused, and resilient (Ford & Vignare, 2015). These values are highlighted and reinforced in the military, as well as the values of a strict hierarchy, order-based decision-making, and the emphasis on acting in favor of the greater good (Halvorson et al., 2010). All of these values and experiences help MALs develop transferable skills that support academic success (May & McDermott, 2019) and the adoption of the student identity. Furthermore, MALs are likely to possess "rich experiential history with exposure to organizational culture, leadership, decision-making, different management styles, and other managerial involvement and organizational engagement" (Starr-Glass, 2011, p. 150) that have the potential to support academic achievement and further define their student identity. However, there are challenges that exist for MALs as they attempt to fully integrate.

According to the psychological underpinnings of SIT, service members and their families will engage in a three-step process as they evaluate the value of membership in the academy (Tajfel & Turner, 1979): social categorization, social comparison, and social identity. Step 1 involves social categorization, initially dividing the social world of higher education and their place in it into one of three groups or categories: student, non-student, or people not interested in the academy. Next, MALs engage in social comparison, comparing their identity as military-affiliated students to the traditional student identity to assess commonalities and points of divergence (Hogg, 2000). Through the process of social comparison, which may play out over the course of many semesters, MALs may determine strong points of similarity and adopt dual identities—service member and student—setting the stage for academic success. Equally likely, the process of social comparison and their experiences in higher education may result in rejection of the student identity. This is particularly salient for MALs who feel their service member identity is devalued or the demands associated with the student identity are too much in conflict with the demands of military group membership. While many factors may contribute to feelings of devaluation or conflict for service members and their dependents, the deficit model and culture clash are of particular interest to this investigation.

Deficit Model

Higher education has a rich history of viewing students who fall outside of the "traditional" student model as lacking or deficit in the needed skills to be successful in the classroom (Valencia, 1997). MALs fall outside of the "traditional" student spectrum in terms of age, marital status, and outside employment (Postsecondary National Policy Institute, 2019); moreover, MALs bring with them multiple identities. For example, MALs are often first-generation students, and up to 51 percent of MALs have reported a disability to the VA (Student Veterans of America, 2017a), adding additional layers of complexity to their academic membership. If the student identity is not supported and developed in conjunction with competing identities, bias in favor of the dominant identity, for example, the military-affiliated identity, may occur. An asset-based model, on the other hand, focuses on the strengths a learner brings with them to the learning community. For MALs, these strengths include discipline, leadership, teamwork, adaptability, and self-discipline (Student Veterans of America, 2017b).

As MALs engage in social comparison, failure to identify the assets they bring to the classroom may create a divide and promote in-group vs. out-group identification. For military spouses or children, deployments and frequent duty station relocations create added stress and uncertainty, differentiating them from their civilian peers which may be perceived as a deficit (Weinstein & White, 1997). If members of the academy fail to recognize and validate the unique lived experiences of MALs, they may determine that membership in the academy is not sustainable, weakening the value of the student identity.

A Cultural Gap

Research suggests that MALs may struggle to connect with peers (Kim & Cole, 2013), develop study skills in a way that supports learning in the college classroom (Ford & Vignare, 2015; Hernández & Belding chapter 11 in this edited volume), and struggle to integrate back into civilian life (Naphan & Elliot, 2015; Zogas, 2017). Moreover, when transitioning to academia, MALs may experience a culture shock. Previous research suggests that military and academic cultures highlight different values (May & McDermott, 2019). The military promotes uniformity, cohesion, and loyalty, whereas academia promotes freedom of thought and questioning the status quo (May & McDermott, 2019). Service members are often given checklists, guides, and routines to ensure accuracy and precision. Troops are encouraged to check and recheck gear constantly, with supervisors or commanders coming by to check gear again. While this highly structured system may make sense when considering the safety of troops, it may be difficult to transition from a highly structured military environment to the unstructured environment of the academy, further challenging the student identity. In college, it can be difficult to know what classes to take, how to complete assignments, and who to contact when you have questions. In academia, there are a variety of "unwritten" rules to success that MALs may not be aware of. Overall, when these cultures clash, MALs can experience alienation and confusion, leading to challenges navigating higher education effectively (Barry et al., 2014; Steele et al., 2011), ultimately negatively impacting their ability to cohesively connect their military and student identities.

Collectively, MALs may struggle to form dual identities supporting both the formation of the student identity and validation of membership in the military community. The dearth of understanding regarding MAL identities in higher education, specifically through an asset-based framework, led to the research question:

RQ: *How does military identity impact development of MAL student identity in higher education?*

METHODS

To answer the research question, the researchers used qualitative methods to collect data from participants. Online questionnaires were distributed to participants who self-identified as MALs (i.e., active duty, reservists, veterans, spouses, dependents). Qualitative content analysis of the open-ended questionnaires responses was used to answer the research question to further understand the experiences of MALs in higher education.

Population and Sampling

To collect the experiences of MALs in higher education, the researchers distributed an anonymous questionnaire. All MALs over the age of eighteen currently enrolled in an institution of higher learning were eligible for participation in the study. To obtain a diverse sample of respondents, the researchers used a variety of recruitment strategies, such as asking faculty to share the questionnaire with their students and posting the questionnaire link on social media networks (i.e., Facebook and Reddit). The researchers also used personal networks to distribute the questionnaire. No incentives were offered for participation.

Instrument

Participants began the questionnaire by reading through the informed consent document. Following the informed consent, respondents were invited to share demographic information. The demographic data collected from respondents included gender identity, race & ethnicity,[4] branch affiliation, officer vs. enlisted status, level of education, and age. The researchers did not ask participants to disclose if they had children or their marital status. The researchers also collected information on the MAL identity: active-duty member, reservist, veteran, spouse, children, or other. Respondents were able to click more than one MAL status if applicable. After completing the demographic information, participants began the open-ended and likert scale questions. These questions were broken down into the following variables to understand how the military impacts the student identity formation of MALs in higher education: decision-making process, connectedness, resources, and faculty. Decision-making process refers to factors influencing a MAL's decision to enroll in college along with how they selected their current or former institution(s). Connectedness explored a MAL's sense of belonging and included multiple dimensions, such as connectedness to faculty, peers, and the institution. Resources were explored by asking MALs to characterize the type of resources available along with utilization rates, as well as resources they would like to have available. Finally, specific questions were asked to assess relationships with faculty and their impact on teaching and learning outcomes.

Qualitative Analysis

The online questionnaire responses were analyzed using qualitative analysis to determine thematic patterns in responses (Vaughn & Tuner, 2015). Using open-ended questionnaire responses, the researchers were able to obtain a diverse and rich set of answers regarding MAL identity experiences in higher education (Reja et al., 2003). The researchers read through the dataset three times. The first time, the researchers read through the questionnaire responses without taking notes. After the first read, the researchers read through the responses again, this time taking notes and highlighting keywords or similar conceptualizations to determine the similarity between the responses provided by participants (Gibbs, 2018). The researchers looked for repetition, frequency, and forcefulness to find common patterns among questionnaire responses (Vaughn & Tuner, 2015; Gibbs, 2018). During the first two readings, the researchers read separately. Finally, in the third read-through, the researchers came together to compare results looking for places of similarity and difference. Once the results were compared, the researchers organized the emerging patterns into categories to make meaning and answer the research question (Elliot, 2018; Vaughn & Tuner, 2015). During this read-through, the researchers identified and recorded representative quotations to exemplify each theme.

RESULTS

The results of the data analysis identified the following themes: Get In/Get Out, Barriers to Student Identity Formation, Benefits of Dual Identity, and What MALs Want You to Know. First, the researchers will share the demographic information for the sample population followed by a discussion of each theme. Twenty-nine respondents completed the online questionnaire. The sample population is represented by table 2.1.

Table 2.1 Participant Sample

	Study Population	*2018 Demographics (Council on Foreign Relations, 2020; DoD, 2019; Vespa, 2020)*
Branches		
Army	45%	35%
Navy	14%	24%
Airforce	14%	24%
Marines	17%	14%
Coast Guard	3%	3%
Gender		
Male	34%	84%
Female	66%	16%
Race and Ethnicity		
White	72%	69%
African American/ Black	7%	16.2%
Latinx	7%	16.1%
Asian	3%	4.2%
2 or more ethnicities	10%	1.7%
Rank		
Officer	48%	18%
Enlisted	52%	82%
Military Identity[1]		
Active Duty	5	1,304,418
Reservists	6	1,039,308
Veterans	11	18,000,000
Spouses	14	966,604
ROTC	1	Data not available

[1]Respondents were given the option to select multiple identities, generating a total greater than 29.

Get In/Get Out

Get-in/Get-out describes the mission-oriented way of approaching the student identity. MALs entered into higher education and adopted the student identity with a specific goal in mind, to get an education and a degree to further their service member identity. This was true for both enlisted and commissioned respondents. No matter the level of degree, undergraduate or graduate, college was a stepping stone for military identity success later on. For example, one student stated they expected "to obtain the knowledge and skills to perform my job duties," while another student stated they expected to "to gain enough knoedge [*sic*] to practice prudently in my field." As one student stated, their desire to enter into higher education was to get "a degree that would be looked st [*sic*] as creditable." Moreover, MALs stated clear expectations regarding their experience in higher education. They were ready to get their degree as soon as possible. MALs were mission-focused on getting a degree, as opposed to traditional college students who might have additional goals for college, such as exploring new disciplines or making lifelong friends (Bean & Metzner, 1985; Southwell et al., 2018). For example, one participant stated "To earn a degree. Be treated as an adult." Other respondents provided similar answers, such as "To obtain my degree in the specified time frame" or "Earning my degree." College is seen as a stepping stone for career succession for MALs (i.e., active duty, veterans, spouses), affirming the student identity serves to further the service member identity.

For a small group of respondents, self-directed learning experiences and growing their own skills and knowledge were part of the mission. Although getting a degree was the main objective, students were interested in learning from a "great education that [teaches] me skills" relevant to their future goals. In addition to learning, MALs were excited for the challenge associated with higher education. For example, one participant shared the following sentiment regarding college: "a challenge, exposure to more open-minded enviroment [*sic*], & opportunity to excel." Other participants stated that they were excited about "significant opportunities to experience new things." For these MALs, higher education and the student identity support goals both inside and outside of furthering their service member identity.

Barriers to Student Identity Formation

Even though MALs were interested and excited to pursue a college degree, they did express some concern with their ability to successfully navigate higher education. These barriers to student identity formation included the culture gap, their ability to work with others, work-life balance, cost, the difficulty of leaving the military culture behind, whether or not they should disclose their military affiliation, and the difficulty of being a spouse.

Cultural Gap

A major concern for all MALs was the potential gap in culture. The culture of academia and the lack of understanding regarding the military were both causes of concern for MALs and present a challenge for dual identity formation. For example, one participant mentioned, "teachers won't understand my situations." Of the MALs who participated in this study, 85 percent of respondents had either never had a MAL instructor or only had a MAL instructor once or twice in their college career. MALs rarely saw themselves in academia due to the lack of role models. Additionally, because most instructors did not have military affiliation, students were worried they would not be understood by faculty. For example, MALs were concerned that instructors would not understand the impact that field training can have on an active-duty student's ability to complete work. Participants also mentioned that they felt looked-down on for missing or turning in late assignments due to a military exercise. As one participant explained:

> They don't understand the demand of the job and that we as military students come up on orders that we can't get out of and that missed classes for training etc is going to happen. it's not that we are trying to get out of doing the work we just need more time than a student that is not military to complete it.

In addition to a lack of understanding of the military functioning, MALs felt that universities were not always military friendly. One participant stated that they felt like, "university just wants the VAs money they are actually anti-military." MALs also felt judged by both instructors and their peers, suggesting members of the academy stereotype or see members of the military as "out-group" members. When asked how instructors and peers communicate support for MALs, one participant responded "they dont they are either Liberal and think I killed babies or Conservative and want to idolize me." Often MALs felt unsupported by academia and their institution stating things such as, "they don't" communicate support or being a "military affiliated learner hasn't afforded me very much support except for qualifying for in-state tuition rates." These data suggest when faced with threats to their service member or military identity, MALs feel devalued and may struggle to see value in academic membership.

Working with Peers

In addition to the cultural divide, MALs struggled to work with other students. One participant stated, "Not feeling accepted by other classmates, could be a distancing issue with maturity." MALs mentioned that they felt mentally older than peers and due to their cultivation of critical thinking skills and enhanced intercultural competence developed through international travel/duty station assignments, they felt detached from peers. Being older than their traditional student counterparts, MALs perceived a distance between themselves and their classmates. A participant stated they struggled with "being around people much younger than I am but at the same level." Another participant described their struggle with collaborating as, "I'm not 'young' anymore. A lot more life experience than a lot of others. Be a little understanding with me." Another participant stated that their biggest struggle was, "working in groups, talking in front of other students." Furthermore, MALs felt as though their communication with peers was difficult. One participant stated, "I am too vulgar and direct at times. I come off as being a know it all when I'm just trying to help or figure out how to 'fit in'." During the social comparison phase of identity formation, MALs may see more differences than similarities, negatively impacting their development of the student identity.

Work-Life Balance

Another concern shared by participants is work-life balance. For students who may be active duty or spouses of active-duty service members, balancing school, work, and home life represents a potential barrier for maintenance of the student identity. For example, one participant stated, "my biggest concerns involve the amount of time I need to dedicate to coursework while maintaining a full time enrollment pace and a family/home balance." Another student stated they were concerned over the "time required to do the courses after work." Not only was time management a cause for concern, if a MAL has been out of school for a while, participants also noted that they were worried about being able to develop strong study habits and absorb the material. These findings are not unique to the MAL experience but may represent a confounding variable impacting dual identity formation.

When transitioning from the military to academia, MALs were worried about time management. For those who were still active duty or married to an active-duty member, the military can make it difficult to find time to attend classes. For example, one spouse stated, "My biggest struggles involve scheduling classes while my young daughter is in school to avoid childcare costs." This struggle of time management was reinforced by several military spouses who shared their experiences. "Being able to find time while working and then having a newborn, all while my active-duty husband was gone for training for six months" represents a reality for MALs that does not exist for most traditional civilian learners. For spouses, the student identity must compete with both the military and motherhood identities. Another spouse noted, "My biggest struggle was finding balance btween [*sic*] school and family, especially while my spouse was on TDY often," reinforcing the challenges associated with military group membership for MALs experiencing identity conflict. Finally, for participants who identified as both a reservist and a military spouse, the work-life balance could be even more difficult to juggle. For example, one participant noted the hardest part of their transition into higher education was "Work life balance, Reserve duties, spouse away as active duty." The role of a reservist represents complex competing identities: service member, military spouse, parent, and student.

Cost

Of significant concern to military spouses was cost, a concern shared by civilians and other MALs. Although there are a variety of tuition assistance programs for active duty, reservists, and veterans, there is limited funding and scholarships for military spouses. In addition to the cost of tuition, in order to take classes, spouses mentioned the need for finding and paying for childcare options if their active-duty spouse was away for training or on deployment. Furthermore, due to frequent relocations, spouses were worried about their credits transferring to different institutions; when credits do not transfer, the time invested in the course in addition to the financial investment is considered a "waste." Spouses were concerned about having to retake classes, spending more time and money because previous courses from another institution may not transfer. Lack of funding for spouses represents a unique challenge that may impact alignment with both military and student identities. For military spouses, life is heavily dictated by the chain of command and often fragmented, broken down into two to three year "chunks" as duty station assignments come and go. "Temporary" often defines the relationship between military families and a given community, limiting opportunities for employment and education for spouses.

Leaving the Military Culture

Leaving military culture behind was noted as difficult for MALs in both positive and negative ways. As one participant stated "adjusting from a very low IQ setting (Army) to an academic setting and having to shift gears" was difficult. This service member is representing a negative aspect of service member group identity, characterizing the military as "very low IQ." SIT suggests negative attributes associated with group identity may result in conflict and generate negative feelings of self-worth (Tajfel & Turner, 1986). As previously mentioned, service members, especially enlisted members, may view their military identities negatively believing they did not gain any useful skills during their military service. For this MAL, feeling unprepared to perform as a result of one group identity has the potential to impact the development of new identity formation, such as that of the student identity.

Some MALs also identified one aspect of group membership in the academy as problematic. For some MALs, the strict and regimented nature of the military was valued in opposition to what was perceived as a lack of structure or discipline in the academy. One participant expressed this sentiment by noting, "having too much freedom without any strict left and right limits to abide by/dictate focus" was difficult. Another student mentioned that they felt they had to engage in a "complete disposal of discipline carried into and developed within the Army" when entering into higher education.

Finally, MALs struggled to let go of the roles and responsibilities defined by rank as they transitioned into the classroom. For example, after having leadership assigned responsibilities within the military, MALs were frustrated with the lack of faith instructors had in their ability to complete basic course work. Another student stated they experienced, "being treated and talked to as if I'm a child." The privilege of rank and recognition of their lived experience are valued by the service member identity and may not be reflected in some learning communities, leading to identity conflict with the student identity.

To Disclose or Not Disclose

Overall, students struggled with the decision to disclose their military identity. Most participants stated that unless directly asked, they would not disclose their military affiliation. Participants explained that they had had negative experiences in the past after disclosing their MAL status, suggesting a protection mechanism was in place to counteract stereotyping directed at their service member identity. For example, one participant stated that they "Always get asked about Trump" because of their military affiliation. Another student mentioned that previously they were willing to disclose their MAL status; however, they became hesitant after negative experiences. The participant went on to explain:

> I am hesitant now since I've oftentimes been overlooked for lab/research positions because faculty thought there was no point in my having those types of academic enrichment experiences since they assumed I would be out of the work force, not live up to my potential, etc as a military spouse.

Other participants mentioned that after disclosing they received "animosity towards the military from certain members of society," which caused them to withhold their military-affiliated status. Another student mentioned that people might be able to guess based on their clothes and beard, but their military-affiliated status is not something they volunteered. Additional participants mentioned that they "don't flaunt it," but openly admit it if necessary. This suggests that the military identity equates to a specific stereotype that may or may not be accepted and approved by others. To avoid potential conflict with others, as well as identity conflicts, MALs chose to withhold a vital piece of their identity from others.

Spouses Left without Support

Finally, spouses, in particular, struggled with the transition to higher education. Spouses frequently mentioned the lack of support they received. Many spouses felt completely alone in trying to pursue a degree. For example, as mentioned previously spouses receive little to no financial support. Although active-duty members can transfer their G.I. Bill to dependents after a specific amount of time served, many families try to save that money for their children. Furthermore, having to take on additional duties while an active-duty spouse is training or deployed limits a spouse's ability to commit the time necessary for all their coursework. For example, one spouse stated being a military spouse "is not a part of my identity, it is something I have to get through to be able to reach my goals. The military doesn't [*sic*] support transitional careers for spouses."

On top of the added responsibilities, the locations of military bases or posts can greatly impact spouses' ability and success in higher education. One spouse stated, "Just limits my options about where I can go." Additionally, a participant mentioned that their military-affiliation made it "difficult to commit to a local college for a graduate degree. This fact has led me to put off my higher education for over 10 years." To overcome this, some participants mentioned that they had to live away from their spouses for extended periods of time in order to complete a degree. Another spouse mentioned wanting to pursue a Ph.D. but was afraid that the closest university to their next duty station would not provide enough prestige for a useful degree in the competitive academic job market.

In addition to the military barriers, spouses noted feeling left out of the campus community, impacting their ability to form the student identity. One student mentioned, "It was difficult to socialize and integrate woth [*sic*] my peers, most of whom were unmarried and/or unfriendly/disrespectful of military service. They assumed I was an extremely politicaly [*sic*] conservative person, which I'm not." Spouses had a hard time feeling like a part of the campus community. On the flip side, however, spouses also received backlash from the military community, "many of the spouses of my husband's colleagues also ostracized me bc [*sic*] they felt I was being selfish by attending a brick-and-mortar school." When in the academy, military spouses felt like they could not be proud of their military spouse identity. One participant explained, "being a military spouse was something that I felt I shouldn't be proud of while I was in graduate school, which was kind of hard for me." When in the military culture, spouses felt like they could not be proud of their academic achievements. Overall, military spouses feel unsupported both by the military and academia in their pursuit of higher education, resulting in identity conflict.

Benefits of the MAL Identity

When asked how the military shaped their identity, MALs felt as though the military prepared them for college. For example, one participant mentioned that they felt, "more organized. More on time. More mature and willing and wanting to learn and grow." Other participants mentioned not getting enough credit for being a military member. The military provided participants with "worldly experience from travel and duties" that they felt could not compare with other civilian jobs. Another student mentioned that the "real world life experience" can provide different perspectives than a traditional student might hold. Participants mentioned that the military provided "significant life experience and soft skills/personal development." In addition to the maturity and experience, MALs felt like the military prepared them with the resilience they needed to succeed in higher education. Participants stated that the combination of "stress management" and military experience as a member or dependent gave "a lot of resiliency and gave me a more open world view."

What MALs Want You to Know

They Want Help

MALs may not be good at asking for help, but they need it. As one participant stated, "Many veterans have chips on their shoulders but not all veterans are dysfunctional and antisocial." This sentiment was echoed by another participant who stated, "Just reach out. Check the ones for disabilities." Another participant stated,

> Help communicating. Feeling like I'm not alone. Psychologist. Counseling. Just knowing someone is there looking out. I'm sure adult degree students feel this way as well, but you're going from a tight bond community of what essentially you consider family, to young kids, who are living off their mom and dad who have no idea what the other side of life is when it comes to military.

The military identity poses a unique challenge for students returning to school. Military service members may be reluctant to ask for help when struggling in part because of lingering stigma regarding mental health and disability that defined their military service (Acosta et al., 2014). MALs want more support from peers, instructors, and administrators. For example, one participant stated, "Don't overlook military students (vets, AD, or spouses) based on personal assumptions that we won't be able to hold a job due to future moves/PCSes." Other students mentioned needing support through tutoring assistance and compassion, "some Veterans have been out of school for quite some time and it takes them time to adjust back into being dedicated to schoolwork and doing things such as remembering good study habits." Other MALs mentioned wanting, "greater tolerance overall for those students who are military-affiliated" and "compassion for difficult and out of the box situations."

Flexibility

MALs need instructors, staff, and administrators to be flexible. MALs don't make their own schedules; they are at the mercy of the military at all times. For those that have transitioned out of the military, they may still be beholden to military schedules for things such as Veterans' Affairs (VA) Health Care appointments. The VA Health Care system has been known to be difficult to access with long wait times and inadequate healthcare (GAO, 2017;

Urbi, 2018). Oftentimes, MALs may need to visit a VA Health Care center hours away in a larger city, multiple times to get the healthcare they require. Understanding the schedule of a MAL student is a vital step in further supporting their success in higher education. As one participant mentioned, "they need to know about the hardships of moving, changing dates, deployments etc." Another stated, "It is important to know that we are a highly mobile family and it is important that I am able to work at a pace that best suits my needs." Instructors should recognize that MALs require flexibility that might not be needed by their traditional student counterparts. As one participant stated, "be patient."

DISCUSSION

The findings from this study reinforce previous research (Drummet et al., 2003; Starr-Glass, 2011) while highlighting new information about how the military impacts the identity of MALs in higher education. Previous research has focused heavily on the veteran experience within the higher education classroom (Livingston et al., 2011; Howe & Shpeer, 2019). Additionally, previous research has explored the struggle military spouses have faced attaining and living with graduate degrees (Hernández & Belding, 2019). However, a dearth of research to date has explored how the military impacts MALs student identity formation in the higher education classroom. Exploring the experiences of active-duty service members, reservists, veterans, and spouses, this study sought to understand how the military culture impacts MALs higher education identity.

Findings from this questionnaire indicate that MALs take with them into the classroom a very mission-focused mindset driven by their service member identity (Ford & Vignare, 2015). From the get-go, MALs are focused on getting a degree to get a job, refining the skills they need to excel at their current job, or to progressing through the rank system. While some more traditional students may go to college to explore different disciplines and career paths (Bean & Metzner, 1985; Southwell et al., 2018), MALs feel pressured to complete their degree as quickly as possible. For example, military spouses were always afraid of PCSing without being able to finish their degree, and active-duty service members were trying to strategically time classes between trainings and deployments. Commissioned officers were focused on completing their graduate degrees to move to the next rank.

In addition to supporting previous research on MALs about the struggles they may encounter in higher education (work-life balance, time management), the data illustrate additional challenges MALs may encounter when building the student identity. For example, while most research on MALs characterizes these learners as independent and self-sufficient (Ford & Vignare, 2015), the findings in this study suggest that MALs need help and have trouble reaching out for help. Unlike the structured and regulation-driven military group, the answer can't always be found "in the regs" to ensure military-affiliated students have the information they need. Even though MALs are older than their traditional student counterparts, thus expected to be able to handle the stressors of college better, MALs still need help and empathy from peers, faculty, and administrators. MALs may not be aware of the resources they have available to them or know how to access them. Ultimately, this can leave MALs feeling helpless and alone attempting to navigate college. Additionally, MALs may find it difficult to disclose or choose not to disclose their military affiliation. MALs noted that they often experienced negative reactions from peers, faculty, and staff after disclosing their affiliation. If MALs are uncomfortable identifying as such, they may not be willing to use resources that have been developed specifically for them or resist seeking support from those they perceive do not value their military identity. Moreover, long-term identity conflict can have negative impacts on self-esteem and self-efficacy (Hirsh & Kang, 2016; Pratt & Corley, 2007). Previous research suggests that identity conflict can produce high levels of anxiety and stress, negatively impacting people's mental health (Hirsh & Kang, 2016). Specifically, in regard to the military identity, previous research has found that the contrast between military and civilian identities is distressing to many active-duty and veteran participants causing adverse mental health effects[5] (Smith & True, 2014). If MALs are not able to reconcile the differences between their military and student identities, the long-term negative impacts on mental and emotional health can lead to missed assignments, class failure, and even dropping out of higher education.

Finally, findings of this investigation further support the challenging experiences military spouses face when trying to obtain a degree (Hernández & Belding, 2019; May & McDermott, 2019). Despite decades worth of research on the military spouse and their experiences in higher education, colleges and universities are still struggling to support them. Although barriers such as relocations and deployments have already been identified as challenges to completing a degree (Ott et al., 2018; Runge et al., 2014), this study found that a lack of support for military spouses on college campuses, costs, and feeling isolated from the military culture negatively impacted spouses' ability and desire to complete their degree. While spouses expected a lack of support from academia, they did not expect to be isolated from other military spouses as a result of their student identity. Spouses getting a degree received backlash from other spouses denoting them as selfish and unsupportive, suggesting additional cultural barriers within the military community may complicate degree completion. The negative repercussions from others regarding their dual identities added to the stress and difficulties military spouses already faced trying to harmonize the differentiating military and academic identities.

Even though many previous studies have identified the challenges MALs may face while pursuing a college degree, this study sought to illustrate the assets military identities bring to the table. As such, findings from this study indicate that MALs bring critical thinking and worldly experiences (Starr-Glass, 2011), organization and maturity (Halvorson et al., 2010), and resiliency and adaptability into the classroom. With their previous experiences in the military or moving around as a military spouse, MALs bring with them different lived experiences than their traditional student counterparts (Starr-Glass, 2011). Thus, promoting and supporting MALs in accommodating the needs of both identities, military and student, will both bolster MALs success in the classroom and enrich and enhance the learning environment for all.

LIMITATIONS AND FUTURE RESEARCH

Although this research provides valuable insights into the lived experiences of MALs transition to higher education, there were some limitations to this study. The sample size of participants was small, as only twenty-nine participants responded to the questionnaire. Although the demographics of the questionnaire sample do not reflect the current makeup of the military, not all military members pursue a college degree before or after service. Findings indicate that only 47 percent of veterans enroll in school after leaving the military (Parker et al., 2019). These numbers are not broken down into demographic specifics such as age or race & ethnicity. Even though there were limited responses, due to the questionnaire including a variety of open-ended questions, the responses provided valuable insight into the lived experiences of MALs transitions to higher education. Additionally, the authors identify as military spouses, potentially adding a level of bias in the data analysis and interpretation of the results. To address this limitation, the researchers first analyzed the responses separately and then came together to review common similarities and differences in the analysis. However, bias remains a concern. The researchers also relied heavily on research affirming "service members for life" way of thinking, positing the military or service member identity as dominant. However, service members and their families have different levels of identification with their service, which was not assessed in this research investigation.

As researchers and administrators look to better understand MALs, there are several possible directions for future research. First, data from this investigation suggest spouses may experience backlash from within the military community while pursuing their educational goals. Identifying in-group barriers that complicate degree completion may provide a more complicated picture of how dual identity formation is challenged. Learning more about how active-duty and veteran students would like to be engaged in the classroom to employ the use of their military skills and experience would help to create more military-specific pedagogy for faculty. Additionally, understanding how children of military families view their identities as military children and students would provide significant insights into dual identity formation.

CLOSING REFLECTIONS AND RECOMMENDATIONS

The present chapter explored how the military impacts the identity of MALs in higher education. From commonalities that promote dual identity formation to barriers that challenge academic success, MALs have unique experiences in higher education that provide learning opportunities for students, faculty, and administrators. From the information gathered through qualitative questionnaire analysis, the researchers have developed recommendations for better supporting MALs in higher education. These recommendations come directly from the MALs who participated in this study, paired with relevant research and literature to further support the MAL as they pursue their educational goals.

Reach Out

MALs want help. Students, faculty, and administrators need to recognize three important dimensions of the military identity that may influence how resources are utilized and received. First, service members receive guided support based on their current rank and career trajectory. Respondents noted that there are so many resources thrown at them once they enter college, it is difficult to know which ones "match" their current development needs. For example, one participant stated, "the largest issue is always finding them [relevant resources] within the sea of information and resources." They want more targeted resources and assistance that reflects where they are in their academic journey. Second, vital information flows through the chain of command. Mission-critical information flows top-down and in formation, many times through face-to-face meetings where information is communicated from the top military member in charge. Universities need to consider how information is strategically communicated to MALs as a result. As noted by another participant, "the resources can be better advertised. Many Veterans don't know these resources exist." Finally, there is a "hidden curriculum" on many campuses that define unspoken rules of engagement. Likewise, the military is defined by unspoken norms that impact a service member's ability or willingness to seek out support. As such, a more active approach may be required vs. "here's a flyer." As one participant explained:

> Help. Reach out. We were molded to shut up and color and do what we're told, reaching out for help is hard. Mentor us on how to interact with others. Just because I don't show emotion doesn't mean I don't feel it. If a vet has ptsd or depression . . . watch them, if there are any signs at all, reach out, even if it's just a little. Try and understand, we are not the same as anyone else. When we find other vets, it's like we found our family we haven't seen in years. When we feel supported or understood, it's an entirely different environment. I don't want thank you for my service (though it's appreciated), I want fucking genuine help and understanding.

This participant references Post Traumatic Stress (PTS) and other mental health concerns for veterans. While the military is committed to providing resources to support the mental health of service members and veterans, stigma still exists regarding the use of these resources, impacting disclosure and willingness to seek out support (Acosta et al., 2014). For some MALs, the stigma may carry over to their student identity and impact their willingness to seek out these services and even additional resources such as writing centers or math labs on college campuses. Partnering with veteran organizations in the local community and inviting them to join campus communities may promote more open and honest discussions about mental health and encourage help-seeking behaviors with MALs.

Onboarding

With these additional, in your face resources, MALs also want a more structured onboarding process when entering college. For many MALs, the freshmen orientation or new student orientation does not provide enough structure or information for successfully navigating higher education (Wardley, Belanger, & Leonard, 2013); moreover, veteran or military-specific information may not be included (e.g., Veterans Center location, ROTC information, using the G.I. Bill). In the military, onboarding is extremely structured from the application process to the process of leaving the military. There are forms, resources, classes, and mentorship every step of the way. Since MALs are used to this form of onboarding, providing this level of support within universities will help MALs feel more connected and supported while establishing and building their student identity (Southwell et al., 2018).

VA Centers

Additionally, MALs want more military groups on campus to help navigate the system. While Veterans Affairs (VA) centers provide a good starting point, not all VA centers are the same. Well-funded and supported VA Centers may be based in central locations on campus, making it easier to find and access; however, not all VA Centers are easy to find or even fully staffed. In many cases, VA Centers are staffed by one or two people who must support all the MALs within a university. Furthermore, while VA Centers may provide emotional support for active-duty and veteran service members, they do not always specialize in helping students figure out how to logistically access their tuition funding (Steele et al., 2011). For example, one participant stated, "the biggest issue I've had is not having a centralized office for working with military education benefits. It has always been 4 or 5 different people in different locations to figure out what I need to do to utilize my benefits." VA Centers need more funding to meet three fundamental needs: (1) centralized support for logistics, to include enrollment and financial aid; (2) community building spaces where MALs are invited to commune and socialize; and (3) tutoring services for MALs by MALs (peer-to-peer tutoring programs).

MAL-Specific Advising

With the implementation of more VA Center administrator support, MALs need support in identifying the right classes and career paths to take. While all students may have an academic adviser, these advisers may not know much about the MAL experience (DiRamio et al., 2008; Steele et al., 2011). Thus, these advisers may not have the skills and knowledge to fully support MALs in their desired career paths. For example, an adviser may not know what the best degree for an enlisted service member interested in becoming an officer may be or what jobs a military spouse may be able to get easily with frequent relocations (see Moullet & Belding chapter 8 in this edited volume). One participant stated, "Our career search needs are extremely nuanced and different compared to civilian counterparts, and we need more help getting connected with alumni or employers." As such, training advisers, or having specific military-affiliated academic advisers, would be a step to close the cultural knowledge gaps MALs may have when it comes to academia. Additionally, providing MALs with priority registration is critical for supporting MALs academic success; MALs don't have the luxury of playing class roulette semester after semester when they often have limited time at a duty station.

Furthermore, MAL-knowledgeable advisers may also be able to help strategically plan degree completion around deployments and PCSing. For example, advising spouses to take all the face-to-face classes needed before

PCSing to the next duty station and moving to online learning or taking a difficult class in the fall before a spouse deploys in the spring may offer a realistic pathway to degree completion. MAL-knowledgeable advisers need to help MALs with their immediate educational goals based on how long they will be in the area before PCSing. Talking at length about graduation is pointless when a MAL will only be attending an institution for a short period of time; directing these learners to transferable general education classes is vital. If a MAL has already transitioned out of the military, they may need advising help to know how to get their military service to count for college credit or which degree paths they may already have partially completed. For example, with their military experience, MALs may be further along in a homeland security or criminal justice degree than a physics degree. Moreover, MALs may need assistance learning what career paths might be available that allow them to maintain a relationship with the military (i.e., contracting or working for the DoD). Finally, a MAL-knowledgeable adviser may be able to build relationships that promote disclosure when they are able to speak the language that connects with the service member identity.

Community Building Spaces

VA centers are challenged to reimagine their spaces to invite military spouses and address their unique needs, especially during deployment and field training exercises. As previously mentioned, VA Centers play a role in supporting MALs; however, VA Centers may not feel like a welcome place for military spouses. The name alone excludes military spouses who are not veterans from feeling invited or comfortable in that space. Since moving can make it difficult for spouses to finish their degrees, having a supportive community space would assist military spouses in being successful in higher education.

Tutoring Services

As noted by the MALs in our study, MALs may not know how to jump back into their education after time in the service. Even just four years away from school can impact how MALs come into the classroom, absorb class material, and engage in studying. Thus, MALs need additional tutoring services to assist them in their educational success. Although there may be a variety of services already available on college campuses, there are barriers MALs face to utilizing these services (in this volume see chapter 10 by Velez, McDermott, & Marlin). To encourage students to use these services, the researchers recommend implementing both the Peer Advisor for Veteran Education (PAVE) program and inviting campus tutors to sit in the VA Center. The PAVE program pairs successful veterans on campus with incoming veterans (see Howe, chapter 4 in this edited volume). Once paired with a current student veteran, MALs may find it is easier to ask for tutoring, mentoring, and help from these peers (Koenig et al., 2014). Additionally, the researchers have had success building stronger community connections and rapport by meeting minoritized campus groups "halfway." To better engage with the MAL population, VA Centers can set study hall hours once or twice a week in which they have tutors from all campus resource centers come to the VA Center. This way, MALs are receiving the help and assistance they need, in a space in which they feel comfortable (in this volume see chapter 10 by Velez, McDermott, & Marlin).

Spousal Specific Support

As mentioned above, military spouses need more support to succeed in their higher education endeavors. For example, access to more funding and tuition support would greatly assist military spouses in completing their degrees. While there are some systems in place (e.g., MyCAA), the limited funding and stipulations of its use make it difficult for spouses to access and use these funds (Runge et al., 2014). As one participant mentioned, "If officer spouses had more support financially for re-training or higher education, that would enable us to make less sacrifices in terms of institution and location." Additionally, funds are needed to assist with childcare when an active-duty spouse is away.

Moreover, after obtaining a degree, spouses need more help with career counseling. For example, when an active-duty service member transitions out of the military, they have mandatory classes and briefs for preparing resumes and entering into the civilian job market. These are not services military spouses have access to. If a spouse has been taking care of the children or frequently switching jobs due to permanent change of station (PCS), they may not have the knowledge or expertise to excel in the civilian job market. More career counseling and support is needed to make sure that military spouses are able to find opportunities to keep progressing their careers.

For military spouses living away from their active-duty spouse, more emotional and mental support is needed. As findings from King and colleagues (2020) illustrate, widows of veterans reported statistically lower levels of loneliness compared to widows of nonveterans due to their perceived level of social support from friends. Ultimately, their findings suggest widows of veterans may have more resilient social support structures (King, Carr, & Taylor, 2020). However, the participants in our study noted that they felt left out of both the military and higher education because of their desire to get a degree. These findings have been supported by previous research by Hernández and Belding (2019) and May and McDermott (2019). Military spouses need social and emotional support from other military spouses, as well as from academics to increase their academic resilience. Collaborating with student support services, mental health resources, and VA Centers, universities can try to develop specific support programming for military spouses to better ensure their academic success.

Military Groups: Beyond the VA

Many MALs noted the lack of representation of MAL-affiliated faculty during their tenure in higher education. Colleges and universities are encouraged to consider establishing informal networks of faculty and staff who are military-affiliated/military-friendly and available to provide additional support. Moreover, Green Zone, a program guided by the goal of "training a cadre of individuals throughout the institution who [are] supportive and knowledgeable about issues facing our military students and the various resources available to assist them" (Nichols-Casebolt, 2012, p. 29), may be an option for providing additional educational and networking opportunities for faculty. Originally launched at Virginia Commonwealth University (VCU) in 2010, this program now serves colleges and universities across the globe through a standardized curriculum (Nichols-Casebolt, 2012).

MALs Are a Form of Diversity

As such, participants in our study wanted MALs to be considered a diverse group, with a participant stating, "educate people about how veterans are a form of diversity." One participant stated "that diversity is based on differing cultural experiences. I may be more diverse as a white male veteran" than others may think. As with other diverse cultures, MALs want others to know and learn more about military culture (Barry et al., 2014; Steele et al., 2011). For example, one participant stated, "tutors, good advisors, and dedicated professors are needed the most to support academic success" while another noted, "educated teachers and other students more on the demand of the military job." A better understanding of the military culture and how the military impacts both active and former members will help higher education better support MALs throughout their educational careers, as well as form their student identity.

CONCLUSION

As we head into an uncertain future in the academy, faculty, staff, and administrators are challenged to meet the needs of an increasingly diverse population of learners. MALs, as underscored by one participant in this study, have rich diversity of experiences, requiring access to genuine and informed resources to better support academic achievement and degree attainment. MALs join our campus communities shaped and defined, in part, by their membership and connection to their military service and may struggle to build and connect with their student identity. Ultimately, MALs are here to learn. MALs are excited to be in the classroom, entering into higher education because they want to learn more, improve their skills, and get a degree. Although they might not always be in class, MALs are committed to their education. One participant noted, "we may need more time to complete the work due to missing the class not due to not wanting to get the work done." By partnering with VA Centers and working directly with faculty, staff, and educators to build military competence, institutions of higher learning can establish a framework and network of support to help MALs strengthen the level of perceived connectedness and reduce perceived difference. Supporting MALs in higher education can be the difference between never finishing a degree and developing a stronger network of lifelong learners.

NOTES

1. The authors note that service members and their spouses and children may have varied levels of "connectedness" with the military culture and identity. Factors that can impact connectedness to the military identity and community may include but are not limited to living on or off a military base or post, the desire to be part of the military community, and a person's support networks inside and outside of the military. The authors note that this is an area for future inquiry and research.

2. The authors acknowledge that the term "conservative" has a political connotation; however, the authors characterize "conservative" as withholden to traditional social institutions (e.g., traditional gender roles) and valuing hierarchy and authority (Heywood, 2012).

3. Officer Candidate School (OCS) is the "U.S. Army's main training academy for prospective Army Officers. The school is generally open to qualified enlisted Noncommissioned Officers, along with civilians who hold at least a four-year college degree" (U.S. Army, 2018, para. 1). Direct commission is available for civilians with at least an undergraduate degree and experience in high demand fields, such as medicine and legal services (U.S. Army, 2020).

4. The authors recognize the complexities associated with defining race & ethnicity. The researchers used categories consistent with the Military Leadership Diversity Council (2009): White non-Hispanic, Black non-Hispanic, Asian non-Hispanic, Other non-Hispanic (includes American Indians, Pacific Islanders, Alaska natives, and "more than one race."

5. The authors of this study did not specify the exact "mental health effects" they were referencing.

REFERENCES

Acosta, J. D., Becker, A., Cerully, J. L., Fisher, M. P., Martin, L. T., Vardavas, R., Slaughter, M. E., & Schell, T. L. (2014). *Mental health stigma in the military*. https://www.rand.org/content/dam/rand/pubs/research_reports/RR400/RR426/RAND_RR426.pdf.

ASVAB. (2020). *What's the difference?: Enlisted vs officer*. https://www.asvabprogram.com/media-center-article/66.

Barry, A. E., Whiteman, S. D., & MacDermid Wadsworth, S. (2014). Student servicemembers/veterans in higher education: A systematic review. *Journal of Student Affairs Research and Practice, 51*(1), 30–42. doi:10.1515/jsarp-2014-0003.

Bean, J. P., & Metzner, B. S. (1985). A conceptual model of nontraditional undergraduate student attrition. *Review of Educational Research, 55*, 485–540.

Binkin, M., & Bach, S. S. (1977). *Women and the military*. The Brookings Institution.

Carnevale, A.P., Smith, N., & Strohl, J. (2019). *Job growth and education requirements through 2020*. https://cew.georgetown.edu/wp-content/uploads/2014/11/Recovery2020.ES_Web_pdf.

Clever, M., & Segal, D. R. (2013). The demographics of military children and families. The *Future of Children, 23*(2), 13–39.

Congressional Research Service [CRS]. (2019, December 26). *Defense Primer: Military enlisted personnel*. https://fas.org/sgp/crs/natsec/IF10684.pdf.

Council on Foreign Relations. (2020, July 13). *Demographics of the U.S. Military*. https://www.cfr.org/backgrounder/demographics-us-military.

Department of Defense [DoD]. (2019). *2018 demographics profile of the military community*. https://download.militaryonesource.mil/12038/MOS/Reports/2018-demographics-report.pdf.

Department of Defense [DOD]. (2020). *Becoming an officer*. https://www.todaysmilitary.com/joining-eligibility/becoming-military-officer#jump-officer-candidate-school.

Department of Health and Human Services [DHHS]. (2010). *Understanding the military: The institution, the culture, and the people*. https://www.samhsa.gov/sites/default/files/military_white_paper_final.pdf.

Devries, M. R., Hughes, H. K., Watson, H., & Moore, B. A. (2012). Understanding the military culture. In B. A. Moore (Ed.), *Handbook of counseling military couples* (pp. 7–18). Routledge.

DiRamio, D., Ackerman, R., & Mitchell, R. L. (2008). From combat to campus: Voices of student-veterans. *Journal of Student Affairs Research and Practice, 45*, 73–102.

DoD. (2020). *Ways to serve: Enlisted and officer paths*. https://www.todaysmilitary.com/ways-to-serve/enlisted-officer-paths.

DoD, ODASD, & MC&FP. (2015). *2015 Demographics profile of the military community*. http://download.militaryonesource.mil/12038/MOS/Reports/2015-Demographics-Report Pdf.

Doenges, T. J. (2011). *Calling and meaningful work among student military veterans: Impact on well-being and experiences on campus* [Doctoral sissertation, Colorado State University]. Researchgate.

Dovidio, J. F., Gaertner, S. L., & Saguy, T. (2007). Another view of "we": Majority and minority group perspectives on a common ingroup identity. *European Review of Social Psychology, 18*, 296–330. doi:10.1080/10463280701726132.

Drummet, A. R., Coleman, M., & Cable, S. (2003). Military families under stress: Implications for family life education. *Family Relations, 52*(3), 279–287.

Dunivin, K. O. (1994). Military culture: Change and continuity. *Armed Forces & Society, 20*(4), 531–547.

Elliot, V. (2018). Thinking about the coding process in qualitative data analysis. *The Qualitative Report, 23*(11), 2850–2861.

Enloe, C. (1983). *Does khaki become you?* South End Press.

Ford, K., & Vignare, K. (2015). The evolving military learner population: A review of the literature. *Online Learning, 19*(1), 7–30.

Gibbs, G. R. (2018). *Analyzing qualitative data* (2nd ed.). Uwe Flick.

Government Accountability Office. (2006). *Military personnel: Reporting additional servicemember gemographics could enhance Congressional oversight.* U.S. Government Accountability Office. http://www.gao.gov/new.items/d05952.pdf.

Government Accountability Office. (2017). *VA health care: Improvements needed in data and monitoring of clinical productivity and efficiency.* https://www.gao.gov/assets/690/684869.pdf.

Halvorson, A., Whitter, M., & Taitt, S. B. (2010). *Understanding the military: The institution, the culture and the people.* https://www.samhsa.gov/sites/default/files/military_white_paper_final.pdf.

Harkins, G. (2019, September 6). Marine grunts say their skills won't get them civilian jobs. The Corps wants to change that. *Military.com.* https://www.military.com/daily-news/2019/09/06/marine-grunts-say-their-skills-wont-get-them-civilian-jobs-corps-wants-change.html.

Helmus, T. C., Zimmerman, S. R., Posard, M. N., Wheeler, J. L., Ogletree, C., Stroud, Q., & Harrell, M. C. (2018). *Life as a private: A study of the motivations and experiences of junior enlisted personnel in the U.S. Army.* https://www.rand.org/content/dam/rand/pubs/research_reports/RR2200/RR2252/RAND_RR2252.pdf.

Hernández, L. H., & Belding, J. N. (Eds.) (2019). *Military spouses with graduate degrees: Interdisciplinary approaches to thriving amidst uncertainty.* Lexington Books.

Hernández, L. H., & Belding, J. N. (2020). Best practices in supporting military-affiliated learners: Suggestions from Communication and Psychological Sciences. In V. McDermott, L. H. Hernández, & A. May (Eds.), *Supporting the Military-Affiliated Learner: Communication Approaches to Military Pedagogy & Education* (pp. 231–256). Lexington Books.

Hirsh, J. B., & Kang, S. K. (2016). Mechanisms of identity conflict: Uncertainty, anxiety, and the behavioral inhibition system. *Personality and Social Psychology, 20*(3), 223–244.

Hogg, M. A. (2000). Social identity and social comparison. In J. Suls & L. Wheeler (Eds.), *Handbook of social comparison.* Springer.

Howe, W. T. (2020). Supporting military veterans in the graduate classroom: A case study of political beliefs and discussion. In V. McDermott, L. H. Hernández, & A.

May (Eds.), *Supporting the military-affiliated learner: Communication approaches to military pedagogy & education* (pp. 67–86). Lexington Books.

Hunter, M. (2007). *Honor betrayed: Sexual abuse in America's military*. Barricade Books.

Jenner, B. M. (2017). Student veterans and the transition to higher education: Integrating existing literature. *Journal of Veterans Studies, 2*(2), 26–44.

Kanter, R. M. (1993). *Men and women of the corporation*. Basic Books.

Keats, P. A. (2010). Soldiers working internationally: Impacts of masculinity, military culture, and operational stress on cross-cultural adaptation. *International Journal for the Advancement of Counselling, 32*(4), 290–303.

Kelty, R., Kleykamp, M., & Segal, D. R. (2010). The military and the transition to adulthood. *Future Child, 20*(1), 181–207.

Kim, Y. M., & Cole, J. S. (2013). *Student veterans/service members' engagement in college and university life and education*. American Council on Education. https://vtechworks.lib.vt.edu/bitstream/handle/10919/97817/StudentVeteransEngagement.pdf?sequence=1.

King, B. M., Carr, D. C., & Taylor, M. G. (2020). Loneliness following widowhood: The role of the military and social support. *The Journal of Gerontology: Series B*. doi:10.1093/geronb/gbz164.

Koehler, C. H. (2014). *Soldier for life: Professional soldier identity in the U.S. Army*. https://publications.armywarcollege.edu/pubs/159.pdf.

Kramer, M. W., & Dailey, S. L. (2019). Socialization and organizational culture. In J. McDonald & R. Mitra (Eds.), *Movements in organizational communication research* (pp. 96–115). Routledge.

Marshall, S. L. (2014). Helping military spouses address employment challenges through career communications. *Career Planning & Adult Development Journal, 30*(2), 85–90.

May, A., & McDermott, V. (2019). Cultural collision: A review of literature. In L. H. Hernández & J. N. Belding (Eds.), *Military spouses with graduate degrees: Interdisciplinary approaches to thriving amidst uncertainty* (pp. 59–85). Lexington Books.

Meiners, E. B. (2019). Student veterans' assimilation in higher education: The role of identity complexity. *Ohio Communication Journal, 57*, 47–57.

Mentzer, B., Black, E. L., & Spohn, T. (2015). An analysis of supports for persistence for the military student population. *Online Learning, 19*(1), 31–47.

Military Leadership Diversity Council. (2009). *How we define race and ethnicity categories for MLDC research*. https://diversity.defense.gov/Portals/51/Documents/Resources/Commission/docs/Issue%20Papers/Paper%2001%20-%20Race%20and%20Ethnicity%20Categories.pdf.

Morris, M. (1996). By force of arms: Rape, war, and military culture. *Duke Law Journal, 45*, 651–781.

Moullet, Z., & Belding, J. N. (2020). Supporting the military-affiliated learner's medical school aspirations. In V. McDermott, L. H. Hernández, & A. May (Eds.), *Supporting the military-affiliated learner: Communication approaches to military pedagogy & education* (pp. 159–180). Lexington Books.

Naphan, D. E., & Elliot, M. (2015). Role exit from the military: Student veterans' perceptions of transitioning from the U.S. military to higher education. *The Qualitative Report, 20*(2), 36–48.

National Constitution Center. (2019, September 29). *On this day: Congress officially creates the U.S. Army*. https://constitutioncenter.org/blog/on-this-day-congress-officially-creates-the-u-s-army/.

Nichols-Casebolt, A. (2012). The Green Zone: A program to support military students on campus. *About Campus*, 26–29. https://www.naspa.org/images/uploads/kcs/Nichols_The_Green_Zone_A_program_to_Support_Military_Students_on_Campus_2012.pdf.

Orthner, D. K. (2002). Deployment and separation adjustment among Army civilian spouses. *Survey of Army Families [SAF IV] Survey Report*. http://www.armymwr.com/corporate/operations/ planning/surveys.asp.

Ott, L. E., Morgan, J. K., & Akroyd, H. D. (2018). Impact of military lifestyle on military spouses' educational and career goals. *Journal of Research in Education, 28*(1), 30–61.

Parker, K., Igielnik, R., Barroso, A., & Cilluffo, A. (2019, September 9). *The Transition to Post-Military Employment*. Pew Research Center. https://www.pewsocialtrends.org/2019/09/09/the-transition-to-post-military-employment/.

Pincus, S. H., House, R., Christenson, J., & Adler, L. E. (2001). The emotional cycle of deployment: A military family perspective. *Journal of the Army Medical Department, 2*, 15–23.

Postsecondary National Policy Institute. (2019, November 9). *Veterans in higher education*. https://pnpi.org/veterans-in-higher-education/.

Pratt, M. G., & Corley, K. G. (2007). Managing multiple organizational identities: On identity ambiguity, identity conflict, and members' reactions. In C. Bartel, S. L. Blader, & A. Wrzesniewki (Eds.), *Identity and the modern organization* (pp. 99–118). Psychology Press.

Proudford, K. L., & Smith, K. (2003). Group membership salience and the movement of conflict: Reconceptualizing the interaction among race, gender and hierarchy. *Group and Organizational Management, 28*(1), 18–44.

RAND Corporation. (2013). *Lengthy military deployments increase divorce risk for U.S. enlisted service members*. http://www.rand.org/news/press/2013/09/03.html.

Reja, U., Manfreda, K. L., Hlbec, V., & Vehovar, V. (2003). Open-ended vs. close-ended questions in web questionnaires. *Developments in Applied Statistics, 19*, 159–177.

Runge, C. E., MacKenzie, A., & McGuire, A. C. L. (2014). Spouses of military members' experiences and insights: Qualitative responses to an open-ended question in a survey of health and wellbeing. *PLoS One, 10*(2), e0118433. doi:10.1371/journal.pone.0118433.

Shpeer, M., & Howe, W. T. (2020). Socialization, face negotiation, identity, and the United States military. *International Journal of Communication, 14*, 726–744.

Smith, R. T., & True, G. (2014). Warring identities. *Society and Mental Health, 4*(2), 147–161. doi:10.1177/2156869313512212.

Soeters, J. L., Winslow, D. J., & Weibull, A. (2006). Military culture. In *Handbook of sociology and social research* (pp. 237–254). Springer Science and Business Media, LLC.

Southwell, K. H., Whiteman, S. D., MacDermid Wadsworth, K. M., & Barry, A. E. (2018). The use of university services and student retention: Differential links for student service members or veterans and civilian students. *Journal of College Student Retention: Research, Theory, & Practice, 19*(4), 394–412. doi:10.1177/1521025116636133.

Steele, J. L., Salcedo, N., & Coley, J. (2011). *Service members in school.* Rand National Defense Research Institute. https://www.rand.org/content/dam/rand/pubs/monographs/2011/RAND_MG1083.pdf.

Student Veterans of America. (2017a). *Research brief: Profile of the contemporary student Veteran.* https://nvest.studentveterans.org/wp-content/uploads/2017/04/Profiles-of-a-Contemporary-Student-Veteran.pdf.

Student Veterans of America. (2017b). *Student veterans: A valuable asset to higher education.* https://ivmf.syracuse.edu/wp-content/uploads/2018/03/Student-Veterans_A-VAluable-Asset-to-Higher-Education.pdf.

Taber, N. (2009). Gender in children's books written for military families: the gendered portrayal of women and men, mothers and fathers in the Canadian military. *Journal of Integrated Social Sciences, 1*(1), 120–140.

Tajfel, H. (1978). Social categorization, social identity and social comparisons. In H. Tajfel (Ed.), *Differentiation between social groups* (pp. 61–76). Academic Press.

Tajfel, H., & Turner, J. C. (1986). The social identity theory of inter-group behavior. In S. Worchel & L. W. Austin (Eds.), *Psychology of intergroup relations.* Nelson-Hall.

Turchik, J. A., & Wilson, S. M. (2010). Sexual assault in the U.S. military: A review of the literature and recommendations for the future. *Aggression and Violent Behavior, 15*, 267–277.

Uniform Code of Military Justice, 1951. (2018). 109 U.S.C. UMCJ §§ 801–946.

Urbi, J. (2018, May 28). The VA's history of setbacks and missteps. *CNBC.* https://www.cnbc.com/2018/05/28/va-veterans-affairs-history-setbacks-missteps.html.

U.S. Army. (2018). *Careers & jobs.* https://www.goarmy.com/careers-and-jobs/current-and-prior-service/advance-your-career/officer-candidate-school.html.

U.S. Army. (2020). *Direct commission officers.* https://www.goarmy.com/careers-and-jobs/become-an-officer/how-to-become-an-officer-in-the-army/direct-commission.html.

U.S. Naval Academy. (2019, April 11). *Naval ads through the years.* https://go.navyonline.com/blog/naval-ads-part-1.

Vacchi, D. T. (2012). Considering student veterans on the twenty-first-century college campus. *About Campus, 17*(2), 15–21. doi:10.1002/abc.21075.

Valencia, R. R. (1997). *The evolution deficit thinking: Educational thought and practice.* Taylor & Francis.

Vaughn, P., & Turner, C. (2015). Decoding via coding: Analyzing qualitative text data through thematic coding and survey methodologies. *Journal of Library Administration, 56*(1), 41–51.

Velez, M., McDermott, V., & Marlin, K. W. (2020). The hard sell: Supporting MALs through resource centers. In V. McDermott, L. H. Hernández, & A. May (Eds.), *Supporting the military-affiliated learner: Communication approaches to military pedagogy & education* (pp. 201–228). Lexington Books.

Vespa, J. E. (2020, June). *Those who served: America's veterans from World War II to the War on Terror.* https://www.census.gov/content/dam/Census/library/publications/2020/demo/acs-43.pdf.

Wardley, L. J., Belanger, C. H., & Leonard, V. M. (2013). Institutional commitment of traditional and non-traditional-aged students: A potential brand measurement? *Journal of Marketing for Higher Education, 23*, 90–112. doi:10.1080/08841241.2013.810691.

Weinstein, L. L., & White, C. C. (1997). *Wives and warriors: Women and the military in the United States and Canada.* Bergin & Garvey.

Williams, R. K., Allen-Collinson, J., Hockey, J., & Evans, A. B. (2018). 'You're just chopped off at the end': Retired servicemen's identity work struggles in the military to civilian transition. *Sociological Research Online, 23*(4), 812–829.

Zogas, A. (2017). US military veterans' difficult transition back to civilian life and the VA's response. *Costs of War.* https://watson.brown.edu/costsofwar/files/cow/imce/papers/2017/Zogas_Veterans'%20Transitions_CoW_2.1.17.pdf.

Chapter 3

What is Impacting the Military-Affiliated Learner Experience in Higher Education?

Nancy A. Chiara, Michael Chiara, and Mary Z. Ashlock

Military service is a life-changing experience. People who have served in the military experience situations that civilians are not exposed to and that are impossible to replicate. The military asks individuals to place their personal desires and expectations secondary to the needs of the service. It does not matter how long one serves; the impact of the military lifestyle—and it is a lifestyle, not a job—is so immersive that it is impossible to remain unchanged by the experience. This lifestyle is so encompassing that military members do not have to experience extensive traumatic situations, such as death or extreme physical or emotional shock, to be affected. The very nature of military service is sufficient to permanently affect someone's identity and perspective. These perspectives impact the behaviors and responses of those who serve and those affiliated with the military throughout their lives on all levels.

The authors have all had some form of military affiliation, understanding first-hand how the military has impacted their lived experiences, perspectives, and identities. Michael Chiara served on active duty in the U.S. Army for twenty years and continues his relationship with the military working as a civilian on a military installation. He was an Army instructor, and when he retired at age thirty-nine, he went back to school to earn his bachelor's degree. Nancy Chiara served for ten years on active duty with the U.S. Army and twelve years as a reserve component soldier with the Army Reserves and Army National Guard. Both Michael and Nancy were enlisted and retired as senior noncommissioned officers (NCO). Upon returning to school, they met Mary Z. Ashlock, an Associate Professor at the University of Louisville. Mary is one of four children of a U.S. Army military police officer (MP) and later married a U.S. Air Force Officer. Her son is currently a cadet at the U.S. Air Force Academy, and her daughter is a national Air Force ROTC scholarship recipient at a four-year university. Mary has previously taught military personnel in evening college courses and served as an instructor for the Transition Assistance Program for those separating from the Air Force. Given the authors' experiences, with forty-eight years of combined military affiliation, we understand first-hand from both a military and an academic perspective the military-civilian-academic divide.

Ultimately, the all-encompassing culture the military requires can create a divide for military-affiliated learners (MALs) when transitioning back to civilian life (May & McDermott, 2019). Although MALs bring worthy attributes into the classroom, such as a sense of duty and commitment, the military lifestyle may also pose some challenges, as MALs can have trouble adapting to a less regimented lifestyle after leaving the service (Howe & Shpeer, 2019; May & McDermott, 2019). In order for educators to effectively design instructional and educational activities, they must understand the complexities and unique lived experiences of MALs. For example, an area that has not seen a great deal of attention is the experience of female veterans[1] and how their situation may differ from male veterans. Additionally, there is a dearth of research seeking to understand veterans' educational history and learning styles; this is a critical area for future research, as the process of acquiring new knowledge and skills in the military is fundamentally different from the process of learning in a civilian context (U.S. Army, 2017). As such, the present chapter explores the nuances that can impact veteran identities in the academic classroom, as well as some of the dichotomies between military education/training and traditional higher education environments. We also discuss effective communication strategies and techniques that can be used to mitigate some of the obstacles for both educators and MALs in higher education.

MALS: A NUANCED POPULATION OF LEARNERS

While every student is an individual, certain groups have similar characteristics unique to that group (Tajfel & Turner, 1979). MALs, specifically veterans, share many common experiences that lend to the similar characteristics they may showcase both inside and outside the classroom. As such, it is important to define this specific demographic of learners. A 2018 report by the Department of Defense reported that the total number of troops serving in the four branches (Army, Navy, Marine Corps, Air Force) was 1.3 million (DoD, 2018, p. 3). The largest branch was the U.S. Army with nearly 472,000 service members, and the smallest was the Marine Corps with over 185,000

(DoD, 2018, p. 4). Furthermore, the report stated that 60 percent of service members were under the age of thirty (DoD, 2018, p. 8) and that those enlisting in the military typically come from middle-class households (Reynolds & Shendruk, 2018, para. 4).

As these individuals leave military service, they find themselves facing adjustment to life in higher education (Borsari et al., 2017; Napan & Elliot, 2015; Castro, 2018). The U.S. Department of Education National Center for Education Statistics reports that veteran students' average age was thirty-four compared to an average age of twenty for civilian students (Walton-Radford, Bentz, Dekker, & Paslov, 2016). As such MALs, while proud of their service, may hesitate to share their experiences with their classmates because of a lack of common ground with younger peers or familiarity with the MAL experience (Zoli, Maury & Fay, 2015). Additionally, with their lived experience ranging anywhere from four to twenty-plus years of military service, they may enter the higher education classroom with a variety of different responsibilities their younger counterparts may not have, such as family and financial commitments. Ultimately, this can lead to a variety of perspectives and attitudes that exacerbate the perceived generational divide between MALs and traditional college students (Zoli et al., 2015). MALs may feel distanced from their classmates because they may feel they are at a different point in their life and the non-veteran students are "just kids" (Borsari et al., 2017, para. 8). In addition to the perceived age impacts on integration, transitioning from the life of a military member to that of a student in higher education can be intimidating for those who may have no college experience or who have only attended "on post" classes offered at their duty stations. The subsequent sections briefly overview some of the nuances that can impact veteran learners within the higher education classroom including misconceptions of the veteran stereotype and the effect of gender.

Misconceptions of the Veteran Stereotype

Due to the complex nature of MALs' backgrounds, it may not be possible to recognize MALs by their outward appearance when among their civilian counterparts in higher education and the workforce. Even if someone discloses their military service or affiliation, it is still presumptuous to assume one knows exactly how that MAL thinks and responds, specifically as it relates to political affiliation and mental health status. Pew Research (2019) surveyed 1,284 veterans and found that while the majority of service members (59%) identify as Republican, almost 40 percent identify themselves as Democrats (Igielnik, Parker, & Cilluffo, 2019). While those with little military interaction may assume all military members are staunch Republicans, the data show otherwise (Igielnik et al., 2019). Thus, assuming a veteran's political stance within the classroom can stereotype and isolate veteran students (Howe, 2019; Kranke, Weiss & Brown, 2017).

In addition to political affiliation and viewpoints, one cannot assume to understand a veteran's mental health status. Research indicates that civilians overestimate the instance of Post-Traumatic Stress Disorder (PTSD) within the veteran population and more heavily associate veterans with mental health issues (Schreger & Kimble, 2017; Veteran Stereotypes, 2014). Due to the potential stereotyping of MALs that can occur, MALs may not feel comfortable identifying as such when entering into higher education. They may be especially hesitant if they predict potential conflicts in viewpoints with instructors, peers, and administrators.

While these negative stereotypes can be harmful to MALs, one must also consider the individual veteran's self-perceptions that may also hinder their success in a non-military environment. Several studies indicate that veterans may be reluctant to seek help or report a need because the military often stigmatizes help seeking as what can be perceived as weakness. Often the expectation in the military is for the service member to "pull oneself up by one's bootstraps" (Kranke et al., 2017; Shackelford, 2009; Selber, 2015). As described earlier in the chapter, I (Mick) came to college later in life—as a non-traditional student and veteran I was, for the most part, twenty years older, or more, than my fellow students. In many of my classes, I was older than the professor. Being older than the professor, however, did not present me with any real challenges. I still saw the professor as the person to look to for information, advice, and guidance; they were the person of authority within the classroom. However, if I was having difficulty with an assignment, or if I wanted to get more in-depth with some of the material presented in class, I would not go to the professor. I was concerned that the professor was "busy" and I should not bother her or him with my questions or concerns. I was used to resolving issues at the lowest possible level, and that usually meant "do not bother your superiors." I really did not know what office hours were or what they were for, and for the most part why they were needed. I expected that I had to be self-sufficient, and if I could not get the questions I had answered in class, then I missed my opportunity.

Gender

There are many variables that determine the experience of service members and MALs. These might include military occupational specialty (MOS), duty stations, number of deployments, age when enlisting, officer or enlisted, and number of years in service. While the overall veteran experience can be broadly generalized, there is research that indicates there are important differences between male and female experiences in and out of the military. There are currently two million women with military experience, which represents approximately 10% of the total number of veterans, and women make up the fastest-growing subgroup in both the active duty and veteran populations (Women's Task Force, 2020).

Although the population of active duty and veteran women service members continues to grow, research indicates that male and female veterans often reintegrate into society in different manners (Burkhart & Hogan, 2015; Castro & Kintzle, 2017; Culver, 2013; Diramio, Jarvis, Iverson, Seher, & Anderson, 2015; Goldstein, 2019; Iverson, Seher, DiRamio, Jarvis & Anderson, 2016; Perez, 2019; Thomas & Hunter, 2019; Women Veterans Task Force, 2020). Thus, the term "Invisible Veteran" is frequently applied to women veterans (Thomas & Hunter, 2019; Perez, 2019). The "Invisible Veteran" term is appropriate as it can be difficult to identify female veterans, given that they are much less likely to disclose their veteran status to a group of strangers or even coworkers or classmates (Diramio et al., 2015; Goldstein, 2019). Feeling that their service is not valued or recognized, female veterans are less likely to self-identify as MALs or use veterans' resources due to the negative responses they receive from others (Sander, 2012). These negative responses can vary from undermining the type of job they may have held while in service (i.e., assuming they were a secretary) to dismissing their service entirely. This idea is supported by a 2016 survey of 1,200 participants, including 1,000 female veterans, conducted by the Service Women's Action Network (2016) that found "when asked if their service is recognized and valued by the general public, three-quarters of respondents (74%) said 'No'" (para. 12).

Not only has research shown the lack of support for female MALs, I (Nancy) have also experienced negative or demeaning reactions after disclosing my veteran status. For example, several years ago, I was called to the office of a former civilian boss. He said, "I know you were in the Army, but not really, you're a woman . . . you weren't in Special Forces or Delta Force. Do you know any 'real' soldiers?" Another female veteran, Terry, with whom I served, experienced some uncomfortable moments when she arrived at a middle school to attend a Veteran's Day ceremony in 2019. She had been invited by a student who knew her from church. When Terry arrived at the school, she parked in the section reserved for veterans. She was told by the parking lot attendant she had to move her car. When she took a seat in the section roped off for veterans, she was told by the event organizer she needed to sit somewhere else (T.C., personal communication, November 16, 2019). Terry is in her sixties and retired after twenty-six years of service. She presents no outward sign of her military service, so it was assumed she could not be a veteran. While these situations did not take place on a university campus, they do illustrate the fact that it is nearly impossible to determine if someone is a veteran simply by looking at them, as it is a diverse population. It also illustrates the potentially negative reactions women veterans encounter when disclosing their prior service.

In addition to the lack of support for female veterans, I (Mary) have experienced the gendered stereotype as a spouse. During my first four years of marriage, I worked as a university administrator and professor near the Air Force base where my husband was stationed. However, as a military spouse, I was expected to attend social events on a regular basis at night, including baby showers and monthly squadron wives' meetings along with formal squadron events. In fact, I served as a "wife" for one of my husband's friends (at his request) as he needed help entertaining. This was in stark contrast to my university role where I served in a leadership position. In fact, ironically enough, several wives of Air Force colonels and generals (who supervised my husband) would sometimes visit me throughout the years at the university asking for assistance in planning for their return to college.

While in the military, many women receive support from their peers and other military women. Their identity is tied to their military experience and the challenges they encountered. Military transition theory (Castro, 2018) postulates that service members are more vulnerable during life transition periods because of the added stress, changes in relationships, differences in social settings, and impact on personal identity. Although male veterans experience the same transitions from service member life to civilian life, many women veterans struggle even more because the traditional gender norms of military service and the "veteran identity" do not coincide with also identifying as a woman (Bruggeman & Rosen, 2019). As such, these feelings of invisibility and isolation may contribute to the lack of specific services designed for the female veteran (Thomas & Hunter, 2019). Military-affiliated social groups, such as the Veterans of Foreign Wars (VFW) and American Veterans (AmVets), are still heavily male-dominated and may not offer the same sense of comradery and support to female veterans. In 2013, the VFW had 1.6 million members, but only 12,130 were women (Regan, 2013).

Understanding these differences in the female veteran experience, some universities have developed veterans' groups specifically for female MALs. For example, Temple University has developed a Military and Veteran Service Center. This center is supported by a campus Veteran Taskforce that meets monthly and plans events and activities designed to support the university's military-affiliated population. Additionally, the Taskforce recognized the unique needs of the female veteran and intentionally initiated events and strategies for them, including an annual Women Veteran Forum (Elfman, 2018).

In order to address the needs of female MALs on campus, it is important to understand that every veteran and MAL has a unique story, just as every student is an individual. Programs and policies that serve one set of students may not fill the needs of others. It is not enough just to develop programs or strategies for MALs. It is necessary to consider the wide range of MAL identities and the differing variables that may effectively impact their learning success, including retention and graduation. The number of MALs leaving the service and pursuing higher education is on the rise, and the number of those MALs with combat experience is increasing (DiRamio, 2017). Thus, it is vital to explore the differences between the military process of learning and learning in higher education to illustrate how educators and administrators can implement military learning within higher education classrooms to support the nuanced needs of all MALs and specifically women MALs.

CULTURAL CLASH: MILITARY PROCESS OF LEARNING VERSUS THE ACADEMY

Much like many institutions, both the military and the academy have their own specific norms for learning processes. As stated by May and McDermott (2019), the academy encourages "constant questioning of the status quo" and promotes liberal ways of thinking, academic freedom, and bringing about social change (p. 65). Alternatively, military learning is more prescriptive, with specific systems for maintaining standards across a variety of branches, bases, instructors, learners, and more. As such, we explain the military learning process to explore how military learning and training may differ from the learning processes used within the higher education classroom.

Instructional Methods

In order to maintain standards and regulations, the military requires all members to go through a specific educational process. This military education system is a heavily regulated training environment, especially for enlisted ranks. However, military assessments of understanding differ greatly from assessments in formal education settings. For example, military learning occurs both in training and education; however, military members "often see training as a field or unit task-based activity, while education brings to mind a specific learning institution or structured self-development activity" (U.S. Army, 2017, p. 8). While professors can create their own syllabi and present their material to their classes in their preferred method, military instructors, for the most part, are constrained in how they present. The branches of the armed services typically use Programs of Instruction, or POIs, to teach standardized material to service members.

While instructional techniques have changed over the years, POIs are still at the heart of how information is taught within the military. POIs ensure the homogeneity of the instructional material so that the information one instructor presents is the same as what a different instructor presents. Although each branch of the service has their own unique format, the topics are usually presented as: *a task*, which is the subject matter or what is to be completed; *the conditions*, which are the conditions under which the task is to be completed; and *the standard*, which is the expected outcome and how basic understanding of the information is demonstrated. This instructional method allows large groups of service members to be trained rapidly to a specified standard. Service members are taught to approach a topic in a certain way, to "stay within the box." They follow specified steps—Standard Operating Procedures (SOPs)—a systematic approach to just about every activity in which the military participates to arrive at the desired outcome (Roost & Roost, 2014). To receive a "go" (pass), service members must complete each of the tasks in a specific order or risk receiving a "no-go" (fail). This impacts the learner as information is communicated in a linear system. The learner follows the specified steps without questioning the method of instruction.

Seat Hours versus Credit Hours

The MAL experience as a linear learner is in stark contrast to sitting in a college classroom. Professors continue their preferred methods of instruction; however, interestingly enough, they adhere to the strict norm of making sure students have enough "face time." For example, many professors tell their graduate students and new instructors, "Do NOT let class out early." This may stem from an institution's policies on credit hours guided by national and regional accreditation standards (Hegji, 2017). For example, a student may be enrolled in a three-credit-hour course requiring professors to have three hours of facetime weekly with students in the classroom (or online). The accreditation standards would then require professors to teach, for example, a fifteen-week semester course three hours a week for a total of forty-five contact hours. Whew! This can be aggravating for MALs if the facetime isn't being used effectively or for MALs who are used to competency-based models of teaching.

The saying "train to standard not to time" (Leaders Guide to Unit Training, 2013) is something most service members are familiar with, and this concept may be perceived in direct opposition to the concept of required seat hours. Creating a classroom environment that provides learning experiences that build on students' experiences, provide opportunities for student agency, and incorporate project-based instructional strategies can shift the perceived focus from "sit n git" instruction based on time to a focus on relevant learning experiences connected to outcomes regardless of how long it takes to achieve mastery. While there are models that increase student agency and mastery learning such as standards-based learning and competency-based education that do not rely on seat time, these models need careful facilitation, preparation, and support.

Classroom Culture

Military acculturation can impact how well, or how poorly, a student veteran adapts to classroom life. MALs may feel that the college classroom lacks the structure to which they are accustomed. MALs tend to respect authority figures, such as their professors, and may feel that their younger classmates do not (Roost & Roost, 2014). MALs will usually "follow orders" as long as they are not illegal, immoral, or unethical, and will respect their superiors—cornerstones of the military lifestyle. This is not to say that the military only wants non-thinking, order-following automatons. The Chairman of the Joint Chiefs of Staff, General Mark Milley, said in 2017 while he was the Army Chief of Staff, "We're the military, so you're supposed to say, 'Obey Your Orders,' that's kind of fundamental to being in the military. We want to keep doing that" (Lopez, 2017, para. 10), but also that subordinates know they are allowed to think on their own, what Gen. Milley called "Disciplined Disobedience" (Lopez, 2017). This is not to suggest that student MALs will "disobey" their professors—just that while order-following is part of the daily culture of military service, student MALs can and do think independently.

Communication Styles

Additionally, most MALs experience a very direct type of communication style during their time in service and may adopt that same style while communicating even after departing military service. Many MALs use short, concise statements to get their points across, a trait that is preferred and necessary, especially when in a combat situation (Citroën, 2018). Professors and fellow students could perceive this direct communication style as rude, forceful, aggressive, or condescending (Carter, 2011; Easton, 2016; Joyce, 2012). Academic writing may also frustrate MALs—a student veteran may ask, "Why write a five-page paper when one page conveys all of the required information?" Army Regulation 25-50, *Preparing and Managing Correspondence*, specifies that "writing will be clear, concise, and effective . . . aid effective communication . . . [and] the reader must be able to understand the writer's ideas in a single reading" (AR 25-50, 2013, para. 1–10). Family members surrounded by the military conventions may also adapt this communication style resulting in similar experiences to that of the service member when in higher education.

Mentorship

Mentorship in the Army is defined in Army Regulation 600-100, *Army Profession and Leadership Policy*, as "the voluntary developmental relationship that exists between a person of greater experience and a person of lesser experience that is characterized by mutual trust and respect" (2017, p. 32, para. 8). During a periodic counseling, some mentoring can occur, but if the person being counseled does not agree with or trust the individual performing the counseling, the counseled may only do what is minimally expected to meet the standard to receive a positive evaluation. Mentoring does not automatically come from the position of authority—it comes from being a positive leader who someone younger or less experienced wishes to emulate. Sometimes this is a voluntary, informal relationship where a seasoned leader takes someone "under their wing" to professionally develop.

In the academy, it may be hard for a MAL, specifically a veteran, to receive the type of mentoring they may want or need. When I (Mick) went back to school after I retired to finish my undergraduate degree, I really felt like a fish out of water. I was not sure whom to turn to, who to trust, or who to believe. I looked to the persons who I believed were in positions of authority to help guide me. I was forty years old and had successfully completed a twenty-plus year career in the Army. While I was not new to the college environment, all of my classes up to that point were on military bases with professors who were familiar with the military, and in classes filled with my fellow service members and some military family members. I was generally successful throughout the years of taking classes on base. I planned on finishing a degree in business; imagine my complete shock when I was told by the registrar that I did not meet the acceptance requirements for the School of Business. It was at that point that I recalled my first abandoned try at college twenty-two years prior where I spent two semesters essentially goofing off, which is what prompted me to join the Army and "get my act together."

I share that story because at that time I did not have a mentor to help me navigate through the academic roadblocks that were now put up in front of me. Discounting that one bad year many years ago, I did not think of myself as a bad student, and I was genuinely surprised that the plan I had formed over the last two years in my Army career was now completely thrown out. After the registrar shared the bad news with me, I sat silent for a few moments until I asked him, "Well, what do I do now?" He did his best to help me and directed me to a department that did not have the same strict entrance requirements. Ultimately, that was the Communication department, where I met the person who would be my undergraduate academic advisor, mentor for the ten semesters I had remaining at the university, and who remains a close colleague and mentor today.

RECOMMENDATIONS

In the previous sections, we have explored the nuances that may impact a veteran's identity, as well as how the military process of learning and the academic process of learning differ. While having an understanding of these similarities and differences is a strong starting point for instructors and administrators, we have developed recommendations for addressing the potential challenges and gaps in understanding MALs may encounter when entering or re-entering the higher education classroom. The subsequent section provides tangible strategies that both instructors and administrators can implement to better support MALs' success in higher education.

Competency-Based Education Model

Competency-based education (CBE) can be an effective bridge between the training environment military members are familiar with and traditional instructional practices typically found in higher education degree programs. CBE is based on a set of frameworks that outline outcome competencies that a student must demonstrate, but the process of acquiring the skills and understanding necessary for this demonstration are personalized and adapted by the student's individual experiences, skills, and abilities (Colby, 2018). This is very similar to the military's mode of training and would provide not only a familiar format but also recognize the skills and experiences MALs bring to the classroom. CBE programming is trending across the United States in K-12 education, and based on a recent survey conducted by American Institutes for Research, "of the 602 respondents, 51% reported being in the process of adopting CBE, 23% reported being interested in CBE but hadn't started" with equal implementation at two- and four-year institutions (2019).

I (Nancy) am currently pursuing a graduate certificate through an online program with a state university. This program is competency-based and for someone with a traditional education background, it was somewhat disconcerting to adapt to a program that focuses on culminating projects rather than regularly scheduled formative assignments. In this program, the only required assignment is the culminating task which is a collaborative project graded as "met" or "not yet," indicating the student has demonstrated competency or needs to provide more evidence of their mastery learning. There are assignments included that cover the material one needs to master to complete the culminating task, but students only need to submit those if they want/need feedback. My group decided to focus solely on the culminating task which, resulted in building a complete website that included a variety of products from essays we had written, research results of surveys we conducted, recordings of interviews we had hosted, and presentations we built that explained components of the overall topic. This website was substantive and functional and was comparable to what I produced as a capstone for a graduate program. We had not only thoroughly learned the content, but we also were able to enhance our workplace communication skills such as our ability to collaborate and prepare and facilitate presentations.

Active Learning

Traditional lecture-style learning may not be the most effective form of learning for MALs. As previously mentioned, MALs are often used to hands-on training with a prescribed process for learning, a well-defined task, understanding of the conditions, and a specific standard to be met. Katie Beck shares this research in her article, *From GI Joe to College Joe: Bridging the Gap Between Military and College Life* (2020):

> If institutions are serious about creating a campus community that welcomes student veterans, they must look beyond student services and emphasize a change in a more practical aspect of attending college: classroom interaction. Implementing curriculum and materials using different modalities demonstrate that faculty members are willing to adapt courses to the various challenges student veterans face within the classroom, thus creating solutions for the challenges that student veterans face in the classroom. (Barnard-Brak et al., 2011; González & Elliot, 2016, p. 49)

To better implement this style of learning in the classroom, instructors can use techniques such as exploring in-depth case studies and working through simulations based on events or problems relevant to students. This way, students are able to connect the material and their lived experiences to complete a task while building new knowledge. These strategies allow students to develop a tangible product or skill-building, to avoid potential frustrations among MALs who may not see the connections between coursework and their ultimate goals post university. As I (Mick) mentioned earlier, "Why write a 5-page paper when I can say it efficiently in 1 page"—not feeling as though the assignments they are doing or the course work they are learning is relevant to their needs can lead to discontent among MALs.

Second, professors can use a writing prompt at the end of class to have students reflect on what they learned that day possibly incorporating the After

Action Review model described in the *Leader's Guide to After Action Reviews* (2013). Writing prompts can also be used as a time for personal reflection and to connect the material with students' lived experiences. Additionally, Kagan (1994) recommends using cooperative learning strategies such as "Think, pair and share" where students are given challenging questions from the lecture material in class to enhance coursework engagement. Students first think individually about the question and then pair up with someone else to solve the answer and then share their findings with the class.

Finally, using interactive games to get the class up and moving can replicate the active, hands-on training MALs are used to. For example, some games are set up with multiple choice questions where students move to the appropriate corner of the room to assess their learning of a new concept. Another technique involves students moving from one side of the room to another to show whether they agree or disagree with an issue (Ray, 2004).

Several of my (Mary's) MALs recently stayed after class to talk with me about similar experiences within university classrooms. They agreed that getting a chance to become involved in the learning environment and discuss their backgrounds with other students makes it much more comfortable to learn and share with one another. We remind instructors to consider the variety of alternatives to lectures that enhance active student engagement within the classroom.

Mentorship

As professors, we inherently know how important it is to connect with our college students. Felten and Lambert (2020) describe in their book, *Relationship-rich Education: How Human Connections Drive Success in College*, how professors and staff have such a profound impact on students even through mentoring moments that occur randomly throughout the semester, the academic year, or longer. Mentoring involves a single, well-timed conversation that can make a huge difference in a student's life, particularly in an inclusive environment with MALs. Mentoring can be peer-based from one MAL to another and/or mentee-mentor relationships with professors and staff. Additionally, given the importance of mentorships with college students and MALs, in particular, we recommend a published "directory" of MAL-affiliated faculty. This would open the door for MALs to feel comfortable in connecting with professors through informal or monthly scheduled mentee conversations.

Expectation Management

MALs need clear guidelines on what their individual degree paths will look like. They need to know what college looks like and what they need to do to succeed. Professors and administrators have the know-how to provide important basic information such as transfer credit evaluations, placement testing, degree requirements, and military tuition assistance that MALs need to know just to enter into higher education. Further developing programs, such as the American Council on Education (ACE) Military Guide, into a more robust resource would provide more assistance to MALs in the admissions, transfer, and advising processes. Furthermore, programs like the Warrior Scholar Program (WSP) which offers free, immersive, academic boot camp programs provide veterans with an opportunity to experience college before they enroll. Overall, just as in the classroom, MALs and all students need to feel as though they are part of an inclusive and structured learning environment geared for student success (Sathy & Hogan, 2019).

Administrator Consideration

Syverud (2019) points out that many universities "designed their infrastructure around 18-year-olds coming directly from high school or a gap year. This includes recruitment and admissions processes, advising and class registration, and transfer of credit" (para. 7). These infrastructure and policy requirements can be especially frustrating for MALs. For example, not granting credit for military educational experiences can delay a MALs degree progression. Additionally, many universities may have a list of requirements for those considered "freshmen" including that freshmen must live on campus or buy a meal plan. For older MALs with families and financial obligations, these policies are not practical or helpful. They may actually hinder a MALs ability to complete their desired degree. As such, policies should be altered to accommodate the unique needs of non-traditional students and MALs.

We are fortunate to have such a diverse group of MALs from various backgrounds on our college campuses. With this in mind, administrators need to acknowledge the lived experience of each of their students. Campus administrators can work to increase the visibility of MALs on campus. They can create opportunities to invite MALs to present at such events as annual campus teaching and learning conferences or online instructional design workshops for faculty. The results can be powerful insights and specific teaching strategies for faculty and administrators for campus communities. MALs can also serve as campus resources for incoming peers with similar military experiences. Moreover, colleges and universities have the ability to connect with local bases and installations which in turn provide strong community connections. We encourage administrators to be creative going so far as to provide a "military basic training" opportunity for all faculty, creating a dual perspective on what their MAL students have experienced within the military. Each of these ideas is within the range of possibilities for administrator consideration.

CONCLUSION

Professors usually seek out advice from other colleagues as to which teaching strategies will be most effective in the classroom. Julie Ray provides clear examples in her publication *Effective Teaching Strategies in Higher Education* (2004) that the goal of teaching is to actively involve students in new concepts, particularly where in-depth critical thinking needs to occur. This goal of actively involving students becomes even more important with all MALs, specifically older MALs and female MALs, who may struggle to transition back to civilian life because of societal norms such as being older than the "traditional" college student or traditional gender roles (Bruggeman & Rosen, 2019).

Overall, just as every civilian student is a distinct and diverse individual, there is not a one-size-fits-all student veteran. Like their civilian counterparts, each student veteran has their own personality, quirks, and life events that have informed their worldview and impacted how they act in the classroom and adapt to academia. Some MALs may thrive where others may need additional support. To better serve their MALs, instructors need to know and understand the unique stressors and challenges that many student MALs face. By ensuring that MALs understand that they are a valued part of the class, and by being aware of any deeper physical or psychological injuries MALs may be facing with their transition to higher education, instructors and administrators can increase MALs potential for success in academia.

NOTE

1. While the researchers acknowledge the differences between biological sex and gender, the military guidelines DTM 19-004 determine gender based on biological sex unless a waiver is approved for transgender members. As such, we refer to gender as biological sex throughout this chapter.

REFERENCES

American Council on Education. (2020). *Military guide online*. https://www.acenet.edu/Programs-Services/Pages/Credit-Transcripts/Military-Guide-Online.aspx.

American Institutes for Research. (2019). *State of the field: Findings from the 2019 national survey of postsecondary competency-based education.* https://www.air.org/sites/default/files/National-Survey-of-Postsecondary-CBE-Lumina-October-2019-rev.pdf.

Barnard-Brak, L., Bagby, J. H., Jones, N., & Sulak, T. (2011). Teaching post 9/11 student veterans with symptoms of PTSD: The influence of faculty perceptions and self-efficacy. *Journal of Vocational Rehabilitation, 35*, 29–36.

Beck, K. E. (2020). From GI Joe to college Joe: Bridging the gap between military and college life. *Journal of Student Affairs at New York University, 16*, 15–22. https://steinhardt.nyu.edu/sites/default/files/2020-07/JoSA-Volume-XVI-2020-Journal.pdf.

Borsari, B., Murphy, J. G., McDevitt-Murphy, M. E., Darcy, M. G., Yurasek, A., Miller, M. B., Martens, M. P., & Carey, K. B. (2017). Student service members/veterans on campus: Challenges for reintegration. *American Journal of Orthopsychiatry, 87*(2), 166–175. https://doi-org.msu.idm.oclc.org/10.1037/ort0000199.

Bruggeman, M. B., & Rosen, G. (2019). The way forward: Successful outreach to women veterans. In K. H. Thomas & K. Hunter (Eds.), *Invisible veterans: What happens when military women become civilians again* (pp. 203–213). Praeger.

Burkhart, L., & Hogan, N. (2015) Being a female veteran: A grounded theory of coping with transitions. *Social Work in Mental Health, 13*(2), 108–127. doi:10.1080/15332985.2013.870102.

Carter, S. B. (2011, April 27). Are we talking the same language? How communication styles can affect relationships. *Psychology Today*. https://www.psychologytoday.com/us/blog/high-octane-women/201104/are-we-talking-the-same-language-how-communication-styles-can-affect.

Castro, C. A. (2018, March 5). *Military transition, identity and health*. http://www.cimvhr.ca/documents/Military%20Transition%20Identity%20Health.pdf.

Castro, C. A., & Kintzle, S. (2017). *The State of the American Veteran: The San Francisco Veterans Study*. USC CIR.

Citroën, L. (2018). Don't communicate like you're still in the military: 5 tips for transitioning veterans. *Military.com.* https://www.military.com/veteran-jobs/career-advice/military-transition/dont-communicate-youre-still-military-5-tips-transitioning-veterans.html.

Culver, V. (2013). Woman-warrior: Gender identity development of women in the American military. *Journal of the Student Personnel Association at Indiana University*, 64–74. https://scholarworks.iu.edu/journals/index.php/jiuspa/article/view/3674.

DiRamio, D. (2017). *What's next for student veterans?: Moving from transition to academic success*. Stylus Publishing, LLC.

DiRamio, D., & Jarvis, K. (2011). Women warriors: Supporting female student veterans. *ASHE Higher Education Report, 37*(3), 69–80.

Diramio, D., Jarvis, K., Iverson, S., Seher, C., & Anderson, R. (2015). Out from the shadows: Female student veterans and help-seeking. *College Student Journal, 49*(1), 49–68.

DoD. (2018). *2018 Demographics*. https://download.militaryonesource.mil/12038/MOS/Reports/2018-demographics-report.pdf.

Easton, D. (2016, June 14). *What is the impact of your communication style on others?* Kent State University. https://www.kent.edu/yourtrainingpartner/what-impact-your-communication-style-others.

Elfman, L. (2018). Temple University takes comprehensive approach to serving military-connected students. *Diverse: Issues in Higher Education, 35*(9), 11–12.

Felten, P., & Lambert, L. M. (2020). *Relationship-rich education: How human connections drive success in college*. Johns Hopkins University Press.

Goldstein, A. N. (2019). No one understands us: Mapping experiences of women in the U.S. military. In K. H. Thomas & K. Hunter (Eds.), *Invisible veterans: What happens when military women become civilians again* (pp. 3–21). Praeger.

González, K. P. (2002). Campus culture and the experiences of Chicano students in a predominantly white university. *Urban Education, 37*(2), 193–218. doi:10.1177/0042085902372003.

Hegji, A. (2017). *An overview of accreditation of higher education in the United States. CRS Report R43826*. Updated. Congressional Research Service.

Howe, W. T., & Shpeer, M. (2019). From military member to student: An examination of the communicative challenges of veterans to perform communication accommodation in the university. *Journal of Intercultural Communication Research, 48*(3), 203–220.

Igielnik, R., Parker, K., & Cilluffo, A. (2019). *Trump draws stronger support from veterans than from the public on leadership of U.S. military*. Pew Research Center. https://www.pewsocialtrends.org/2019/07/10/trump-draws-stronger-support-from-veterans-than-from-the-public-on-leadership-of-u-s-military/.

Iverson, S. V., Seher, C. L., DiRamio, D., Jarvis, K., & Anderson, R. (2016). Walking a gender tightrope: A qualitative study of female student veterans' experiences within military and campus cultures. *NASPA Journal about Women in Higher Education, 9*(2), 152–168.

Joyce, C. (2012, November). The impact of direct and indirect communication. *The Independent Voice*.

Kagan, S., & Kagan, S. (1994). *Cooperative learning* (Vol. 2). Kagan Cooperative Learning.

Kranke, D. A., Weiss, E. L., & Brown, J. C. (2017). Student veterans with invisible disabilities: Accommodation-seeking in higher education. *Journal of Veterans Studies, 2*(2), 45–57.

The Leader's Guide to After-Action Reviews (AAR). (2013). Combined Arms Center – Training (CAC-T) Training Management Directorate (TMD) Fort Leavenworth, Kansas 66027. https://pinnacle-leaders.com/wp-content/uploads/2018/02/Leaders_Guide_to_AAR.pdf.

The Leader's Guide to Unit Training Management (UTM). (2013). Combined Arms Center – Training (CAC-T) Training Management Directorate (TMD) Fort Leavenworth, Kansas 66027. https://ssilrc.army.mil/wp-content/uploads/2019/05/Leaders_Guide_to_Unit_Training_Management.pdf.

Lopez, C. T. (2017). *Future warfare requires 'disciplined disobedience,' Army Chief says.* https://www.army.mil/article/187293/future_warfare_requires_disciplined_disobedience_army_chief_says.

May, A., & McDermott, V. (2019). Cultural collision: A review of literature. In L. H. Hernández & J. Belding (Eds.), *Military spouses with graduate degrees: Interdisciplinary approaches to thriving amidst uncertainty* (pp. 59–86). Lexington.

Office of the Deputy Secretary of Defense. (2020). *Military service by transgender persons and persons with gender dysphoria (DTM 19-004).* United States Department of Defense. https://www.esd.whs.mil/Portals/54/Documents/DD/issuances/dtm/DTM%2019-004.PDF?ver=2020-03-17-140438-090.

Perez, C. C. (2019). *Invisible women: Data bias in a world designed for men.* Abrams Press.

Radford, A. W., Bentz, A., Dekker, R., & Paslov, J. (2016). *After the Post-9/11 GI Bill: A profile of military service members and veterans enrolled in undergraduate and graduate education. Stats in brief. NCES 2016-435.* National Center for Education Statistics.

Ray, J. A. (2004, October). Effective teaching strategies in higher education. *Phi Kappa Phi Forum, 84*(4), 58.

Regan, M. D. (2013, February 23). Where the V.F.W. is both tough and feminine. *The New York Times.* https://www.nytimes.com/2013/02/23/nyregion/buffalo-v-f-w-post-is-both-tough-and-feminine.html.

Reynolds, G. M., & Shendruk, A. (2018, April 24). *Demographics of the U.S. military.* Council on Foreign Relations. https://www.cfr.org/article/demographics-us-military.

Sander, L. (2012). Female veterans on campuses can be hard to spot, and to help. *Chronicle of Higher Education*, A14–A15.

Sathy, V., & Hogan, K. A. (2019, July 22). Want to reach all of your students? Here's how to make your teaching more inclusive. *Chronicle of Higher Education.* https://www.chronicle.com/article/how-to-make-your-teaching-more-inclusive/?cid2=gen_login_refresh&cid=gen_sign_in

Schreger, C., & Kimble, M. (2017). Assessing civilian perceptions of combat veterans: An IAT study. *Psychological Trauma: Theory, Research, Practice, and Policy*, *9*(S1), 12.

Semer, C., & Harmening, D. S. (2015). Exploring significant factors that impact the academic success of student veterans in higher education. *Journal of Higher Education Theory & Practice*, *15*(7), 31–43.

Smith, J. G., Vilhauer, R. P., & Chafos, V. (2017). Do military veteran and civilian students function differently in college? *Journal of American College Health*, *65*(1), 76–79.

SWAN Releases 1st Annual Survey of Service Women & Women Veterans. (2017, October 26). *Service Women's Action Network.* https://www.servicewomen.org/press-releases/swan-releases-1st-annual-survey-of-service-women-women-veterans/.

Syverud, K. (2019). Commentary: Universities need to better serve military students and veterans. *US News.* https://www.usnews.com/news/education-news/article

s/2019-06-21/commentary-higher-education-must-better-serve-military-students-and-veterans.

Tajfel, H., & Turner, J. C. (1979). An integrative theory of inter-group conflict. In W. G. Austin & S. Worchel (Eds.), *The Social Psychology of Inter-Group Relations* (pp. 33–47). Brooks/Cole.

Thomas, K. H., & Hunter, K. (2019). *Invisible veterans: What happens when military women become civilians again*. ABC-CLIO.

The U.S. Army Learning Concept for Training and Education 2020–2040 (TRADOC Pamphlet 525-8-2). (2017). *U.S. Army Training and Doctrine Command.* http://fliphtml5.com/kbhsd/ulqa/basic/51-51.

U.S. Department of the Army. (2013). *Preparing and managing correspondence: Army regulation 25-50.* https://armypubs.army.mil/epubs/DR_pubs/DR_a/pdf/web/r25_50.pdf.

U.S. Department of the Army. (2017). *Army profession and leadership: Army regulation 600-100.* https://armypubs.army.mil/epubs/DR_pubs/DR_a/pdf/web/ARN3758_AR_600-100_FINAL_WEB_pdf.

U.S. Department of Education. (2018). *Fast facts: Back to school statistics.* Institute of Education Sciences, National Center for Education Statistics. https://nces.ed.gov/fastfacts/display.asp?id=372.

Ventrone, J., & Karczewski, P. (2015). *From the Army to college: Transitioning from the service to higher education.* Rowman & Littlefield.

Veteran Stereotypes: A Closer Look. (2014). Office of the Chairman of the Joint Chiefs of Staff Chairman's Office of Reintegration: Veterans/Families/Communities. https://www.jcs.mil/Portals/36/Documents/CORe/141024_veteran_stereotypes.pdf.

Warrior Scholar Program. (2019). *Warrior scholar program: From the battlefield to the classroom.* https://www.warrior-scholar.org/.

Women Veterans Task Force. (2020). The House Committee on Veterans' Affairs. https://veterans.house.gov/women-veterans-taskforce.

Zoli, C., Maury, R., & Fay, D. (2015). *Missing perspectives: Servicemembers' transition from service to civilian life.* https://surface.syr.edu/ivmf/7/.

Chapter 4

Supporting Military Veterans in the Graduate Classroom

A Case Study of Political Beliefs and Discussion

William T. Howe

A century ago, Agan (1920) described how military members learn differently than traditional students. Specifically, Agan found that "[e]verything points to military efficiency and preciseness" (p. 25). A century later, Howe (2020) found that military veterans still value these qualities and feel contempt toward civilian communication that is indirect. Military members learn to sacrifice themselves for the team, place the mission first, adopt military values, and communicate competently (Shpeer & Howe, 2020; Sørensen, 2011). However, these qualities may present challenges when entering higher education. Howe and Shpeer (2019) interviewed 20 undergraduate veterans attending a large university and found that these military veterans described their educational experience in mostly negative terms. They described a "culture clash" between what had been learned in the military and what they were now learning and experiencing on campus (see also May & McDermott, 2019).

An issue that veterans often voiced was a pressure to discuss politics in public, a practice the military forbids. Howe and Shpeer (2019) found that when these veterans experienced this clash of cultures, they often resorted to "self-silencing" because they perceived more risk in speaking up than the reward it could provide. Alternatively, veterans noted that they could only remain silent for a certain period of time before they expressed anger toward teacher and student communication. Some specific examples participants provided, regarding instructors, was a lack of classroom control and not standing up for themselves. Student activities that evoked this emotion were talking during lectures, coming late to class, and dressing too casually, to name a few. Recently Howe (2020) expanded on these findings and found that many veterans feel contempt toward civilian communication, although the amount of contempt felt ranged due to various prior experiences (e.g., combat deployment, months of combat). Scholars have proposed that anger is felt when an individual's autonomy is threatened, and contempt is felt when community is threatened (Rozin et al. 1999). Thus, when either the veteran themselves or veteran communal values are perceived as under attack, military veterans are likely to respond with these moral emotions.

The mainstream portrayal of military veterans (Rhidenour et al., 2017) may contribute to this inevitable eruption of emotions and further the social isolation or stigmatization of military veterans in higher education classrooms. For example, Parott et al. (2018) performed an extensive quantitative content analysis of messages regional news publications tweeted out across all fifty states. They found that the media framed veterans as a "charity case," a "hero," or a "victim" (p. 702). Although hero could be interpreted as a positive representation of military veterans, it was often used in the context of combat service—and not all veterans have served in combat. Parott et al. (2018) conclude that media representations of military veterans are very narrow, and "the results suggest the news media might neglect benefits of service and provide audience members a sentiment concerning service that is not shared by the men and women who serve in the Armed Forces." (p. 715). This media representation is especially important when considering the behaviors, values, and norms learned during military socialization are carried by individuals after they leave the military (Shpeer & Howe, 2020), and these issues are often central to the conceptualization of the military-civilian divide (Cooper et al., 2016). Furthermore, it has been found that both active duty military members and veterans feel a "widespread frustration with military-related reporting" (Schmidt, 2020, p. 16). Frustration may lead to issues, such as the culture clash between veteran and traditional students. This may be even more likely to occur in classrooms that have an open discussion format where students may hold opinions of the military based on the media representations that they have seen (Glasser et al., 2009) and especially if students have had little experience communicating with military veterans. Although many undergraduate instructors have made a move to encourage more open discussion, it is likely that the unique nature of graduate seminars (e.g., smaller class sizes, less lectures), combined with the propensity of graduate students to consume more news than undergraduate students (Noza et al., 2018), graduate programs and classrooms may be the most likely setting for this cultural clash to occur. Therefore, we may expect to see a wider military-civilian divide among graduate students than undergraduate students (see Wilson, 2020 for a review). As such, the present chapter focuses on understanding and supporting military-affiliated learners within the context of the graduate classroom.

Graduate courses are typically designed to go beyond learning the content of a textbook by asking students to theorize about the connections between theory and practical applications, such as current events portrayed in the news (Clements, 2021; Noza et al., 2018). Discussions in graduate seminars can become heated because there are multiple theoretical, methodological, axiological, and praxeological approaches students can adopt toward the topic at hand. For this reason, the firm and direct communication style learned in the military may be seen as offensive or dismissive when interacting with civilians who expect more indirect language use. Shpeer and colleagues (2020) provide examples of the clear and direct language used during basic training. For example, Drill Sergeants said, "I want you to write a letter home to your WH*RE MOTHER, and get her to start writing me CHILD SUPPORT CHECKS! Because I am TIRED of watching your stupid ass FOR FREE!" and "Private, if you ever f*cking embarrass me like that again I will make you eat that g*d-d*mned smart book. You are now Squad Leader and your people had better know their sh*t" (p. 732). Such clear and direct language might seem harsh outside of the context of basic training, but such communicative encounters illustrate how these communication norms and strategies are formed during military entry. The researchers also found that these norms were often carried after military exit. Therefore, the communication style of military veterans, when viewed through a civilian lens, may appear abrupt, uncaring, or insensitive. In this chapter, I will further describe the military socialization process and how it may have an impact on veterans' educational socialization. I will then examine a specific topic of interest, political discussions, to see how veteran responses to such discussions may not be rooted in political beliefs, but in inculcated communication norms. I will then conclude the chapter with practical steps that instructors can take to help ease classroom discussions from which veterans may normally withdraw.

SOCIALIZATION AND SOCIAL IDENTITY THEORY

Due to the unique experiences that military veterans encounter during basic training, a brief overview of the organizational socialization process is necessary. Scholars have conceptualized organizational socialization as having four phases (Jablin, 2001; Kramer, 2010). Anticipatory socialization occurs when potential members seek information and decide whether to join (Dailey, 2016). Encounter is when members enter the organization and learn its values and norms (Howe & Hinderaker, 2018). Metamorphosis occurs as a member internalizes the values and norms of an organization (Kramer et al., 2018). The traditional theorization of exit is the final stage of organizational socialization and occurs when a member officially leaves an organization

(McNamee & Gould, 2019). As will be discussed later, members of organizations that have high value commitments and deep identity requirements may experience a more dynamic, rather than static, exit process.

The military cultivates values and norms during the organizational socialization process, which are required to be performed both in and out of uniform, and undoubtedly change the identity of many military members (Howe & Hinderaker, 2018; Knight, 1990). The immersive nature of the United States Military (USM) has been classified by Howe and Hinderaker (2018) as a totalistic organization (see Hinderaker, 2015). Researchers have examined totalistic organizations for years, but only recently have communication scholars begun to investigate the experiences of members that belong or have belonged to these organizations. Hinderaker (2015) proposed that some organizations are totalistic or all-encompassing of the organizational member's speech, actions, dress, and relationships. Totalistic organizations differ from traditional organizations in that they are heavily reliant on values as a basis for organizational decisions. To date, most of the research on totalistic organizations has been about religious organizations (e.g., Garner & Peterson, 2019; Hinderaker & O'Connor, 2015).

The importance of recognizing the U.S. Military as a totalistic organization, for this project, is because of the documented difficulties of organizational members to exit. As Kramer (2010) notes, little is known about organizational exit from traditional organizations. However, in the context of totalistic organizations, more research has been conducted on this phase. Hinderaker and O'Connor (2015) document the difficulty of exiting a totalistic organization in an examination of the Church of Jesus Christ of Latter-day Saints, using microstoria narrative analysis (Boje, 2001). The researchers report that rather than a singular moment of exit, exit occurred over a period of time through microdisassociations or "quitting small parts of the faith that, to the member, represented either leaving behind or rebelling against core church values" (p. 522). Presumably the military commitment to values (Howe & Bisel, 2020) could lead military veterans to experience similar microdisassociations such as changes to dress, appearance, and actions which are controlled by the military during active duty (Howe & Hinderaker, 2018). Former members of the Church of Jesus Christ of Latter-day Saints found it difficult to leave the organization due to the heavy indoctrination of the church and the loss of relationships with primary ties (i.e., family and friends) who remained behind (Hinderaker & O'Connor, 2018). Howe and Shpeer (2019) found that military members described similar feelings about exiting the military. Therefore, it seems that members of totalistic organizations do not experience exit the same way members of traditional organizations do, and military exit may not fit the linear phasic model of organizational exit that scholars have proposed (e.g., Jablin, 2001). Hinderaker and O'Connor (2015) state:

> This exit process was described in the narratives of this study as lasting from 6 months to 20 years. Previous research has assumed a relatively short time period of preexit decision making that includes discussion of the decision with close friends, family, and/or coworkers. (p. 522)

This is pertinent to the study of military veterans because if veterans cannot cognitively process that they have exited the military, then they are likely to retain the values and norms instilled during their military service and may still identify with the military to a greater degree than their current civilian status (Howe, 2020). The U.S. Military instructs members that "Soldiers are required to live them [military values] daily in everything they do, whether on or off duty" (Philips, 2013, para. 3). Second, veterans report that even when off duty, they represent the military (Howe & Hinderaker, 2018). These instilled values create a distinct new culture for military members (see Philipsen, 1992). This new culture (Howe & Shpeer, 2019) carries a new morality (Haidt & Joseph, 2004). Furthermore, Fairhurst (2010) argues that the military provides a new frame for seeing the world.

When we consider the military's organizational culture as value-centric and all-encompassing in conjunction with the idea that it is natural for individuals to identify with the organization they serve (Burke, 1937, p. 264 as cited in Bisel, 2018), then perhaps part of the difficulty transitioning from the military to civilian life can be attributed to identity needs (Gade & Wilkins, 2012) and especially those needs proposed by social identity theory. Social identity theory proposes that individuals think about themselves in relation to the groups they belong to (Tajfel et al., 1979). Social identity, then, is derived primarily from group memberships and interaction with other members of a group (Brown, 2000; Haslam et al., 2010). In the context of this chapter, members of the military start thinking about themselves in the context of belonging to this group, assigning the characteristics of the military, and other members, to themselves. Entry into the military begins with taking an Oath of Enlistment. For non-commissioned officers, the Oath opens with an acknowledgment to the chain of command defined from the top down: "I will obey the orders of the President of the United States and the orders of the officers appointed over me, according to regulations and the Uniform Code of Military Justice" (U.S. Army Center of Military History, 2020, para. 1).

From the outset of military service, the military is political. However, the nature of those politics changes as the leader of the military changes. Due to the Uniform Code of Military Justice (UCMJ), which governs the actions of military members and creates both higher standards and harsher punishments for members, service members cannot speak out against the actions or statements of those in authority over them. According to Article 88 of the UCMJ:

> Any commissioned officer who uses contemptuous words against the President, the Vice President, Congress, the Secretary of Defense, the Secretary of a military department, the Secretary of Homeland Security, or the Governor or legislature of any State, Commonwealth, or possession in which he is on duty or present shall be punished as a court-martial may direct. (United States Congress, 2020, Article 88)

Article 89 extends such restrictions to include officers: "Any person subject to this chapter who behaves with disrespect toward his superior commissioned officer shall be punished as a court-martial may direct." (United States Congress, 2020, Article 88). Because the President of the United States is the Commander in Chief of the military, they are the highest-ranking officer and, therefore, no service member either enlisted or commissioned can "use contemptuous words" or "behave with disrespect" toward the President. Therefore, military members must either support or remain silent when asked about their political attitudes toward the President of the United States. After military exit, it may be easier for a veteran to claim the veteran group political stereotype, a conservative republican, and blend in with other veterans than to speak out about their personal political beliefs (Jetten et al., 2001). Understanding the socialization of military members may explain why there is conflicting evidence about the confidential professed political beliefs of military members and the publicly expressed beliefs of military veterans.

POLITICAL BELIEFS IN THE U.S. MILITARY

Both popular press (Newport, 2009; Thompson, 2012) and peer-reviewed journal articles (Bachman et al., 2000; Nail & McGregor, 2009) seem to agree that military members are more conservative than their civilian counterparts, although it is noteworthy that recent events have caused political division in the military (Shane, 2018). Feaver and Kohn (2001) posit that the reason military members, and by extant veterans, are more conservative are numerous and the result of a shift between military and civilian cultures (see also Foy & Restifo, 2018). Unpacking the various reasons why military members appear to be more conservative than their civilian counterparts may require a volume of books to even scratch the surface. However, there are some obvious differences that could be a good entry point into this conversation.

First, as previously mentioned, military members are duty bound to uphold the orders of the President of the United States, regardless of their personal feelings toward the policies of the President. Notably, this does not mean that a service member agrees with the actions that are being taken, but it does mean that service members have very little political agency unless such agency aligns with the political party of the commander in chief (Hasian, 2007; Milburn, 2010). Furthermore, the ability of an officer to engage in dissent may be more plausible than an enlisted member that has limited visibility. Imagine a soldier that enlisted under President Bush, continued service under President Obama, and exits service under President Trump. This soldier will have served under three presidents with remarkably different political ideologies. If that soldier is asked, while enlisted in military service, what their political beliefs are, it would not be surprising that they would mimic the official political stances of the President, the commander in chief. Even immediately after that soldier exits the military, they may not be willing to publicly express their view about political issues because of the years of trained responses. Therefore, it may be more likely that the soldier will engage in privacy management (Jetten et al., 2001; Wilson et al., 2019). Cox and Albright (2016) discuss in detail how military members must use privacy management or "motivation for disclosure, the possible need for privacy, and how this information is managed" (p. 13), even with spousal partners. The phrase "loose lips sink ships" may have begun as an awareness campaign in World War II, but it is still practiced in the military today (Cole, 2017). A second reason may be that veterans simply do not wish to talk about politics because people in general see risk in discussing politics (Chan, 2018) and, therefore, veterans may see greater risk than reward, as researchers have reported (Howe & Shpeer, 2019).

This begs the question, how exactly are the political beliefs of military veterans portrayed? News media consistently remind readers and viewers that military members support President Trump (Igielnik et al., 2019; Rassmussen Poll, 2018), although some recent reports have shown that this support is dwindling (Shane, 2019). Pictures of the President standing in front of service members saturate the Internet and are used as a rallying cry for his supporters. Somewhat lost in these reports is the fact that these same service members would have applauded, as many did, President Obama when he gave speeches because military members are both morally and legally obligated to do so. Yet headlines such as "Trump Draws Stronger Support From Veterans Than From the Public on Leadership of U.S. Military" (Igielnik et al., 2019) and "Military Veterans of All Ages Tend to Be More Republican" (Newport, 2009) paint a picture of a monolithic body of service members that are conservative republicans. Academic researchers, such as Klinger and Chatagnier (2015), have argued that this conservative leaning among military members is because of military values with conclusions such as:

> If veterans believe that duty is a paramount trait or virtue and they exhibit this trait to a degree that civilians do not, then veterans may be willing to utilize government intervention to make up for the perceived moral shortcomings. This

> alternative theory of how duty drives the political opinions of veteran citizens is consistent with our finding that veterans hold socially conservative ideological beliefs, which are in concordance with government policies that place a high priority on duty, discipline, and deference to authority or established norms. (p. 685–686)

Popular press articles about recently separated veterans argue that younger veterans lean more Republican than older veterans (Leal & Teigen, 2015; Ordway, 2018). According to Ordway (2018), veterans that turned eighteen before World War II were 55.4 percent Democrat, 38.3 percent Republican, and 6.2 percent Independent. However, Ordway (2018) also writes that veterans that turned eighteen between 1974 and 2016 were over half of veterans identified as Republican, a third as Democrat, and 16.7 percent as independent. Ordway (2018) argues that the end of the draft era and the rise of women in military ranks have contributed to the rise of conservatism. Shane (2018) offers conflicting evidence to this belief. Shane (2018) found that the election of President Trump led to more polarization in the political attitudes of the military. *The Military Times* (Shane, 2018) conducted a poll of military members to see what party they planned to support during the midterm elections; of the 829 responses that were received, 45 percent said that they planned to vote Republican, 28 percent Democratic, and 26 percent said that they did not support either party. When the percentages are split by gender an interesting trend can be seen, as only 28.1 percent of female self-identified military members said they would vote Republican and 53.5 percent said they would vote Democrat, 17.3 percent said None/Independent, and 0.9 percent said Libertarian. This stands in stark contrast to male self-identified service members of whom 48.7 percent said Republican, 23.6 percent said Democrat, 20.3 percent said None/Independent, and 4.4 percent said Libertarian. Due to the fact that there are more men than women in the military, and subsequently in the responses that were received, 45 percent of the overall sample said they planned to vote Republican. Although this is the largest percentage, it is not even a simple majority of service members that claim to be Republican. Yet news reports and academic findings promote the idea that military members and veterans tend to be more conservative and Republican than civilians (Leal & Teigen, 2015; Ordway, 2018).

Importantly, this representation of military members as mostly conservative Republicans can falsely shape the way that civilians, teachers, instructors, and even military members begin to view military members' political beliefs. If an instructor enters the classroom with the mental model that all veterans are conservative, they may approach veterans in a manner to which the veteran will have a negative response. One, because they may not support the Republican party and two, because they have been stereotyped. Yet, if the veteran also believes that it is the norm for military members, both current and former, to be conservative, they may self-silence and revert to the military group identity instead of verbalizing their own beliefs as some researchers predict (Jetten et al., 2001) and other researchers have shown (Howe & Shpeer, 2019). To test the veracity of this claim, a survey was designed to discover the political ideology of veterans and compare them to their civilian counterparts. In line with the previously cited findings of veterans being more conservative than civilians, it was predicted that:

H1: Veterans will self-report being more conservative than civilians.
H2: Veterans will score more conservative on a political ideology scale than civilians.

Additionally, because veterans were under the command of the President of the United States for so long, they may need to cognitively reconcile the President's beliefs as similar to their own. It was, therefore, predicted that:

H3: Veterans will rate President Obama significantly differently than civilians.
H4: Veterans will rate President Trump significantly differently than civilians.

The details of the method and the results of hypothesis testing follow.

METHOD

To test the claims that veterans are more conservative than their civilian counterparts, a one-time case study was conducted via cross-sectional data collection. After receiving Institutional Review Board (IRB) approval from the affiliated universities at the time of data collection, veterans and civilians were recruited for this study from the Amazon Mechanical Turk (MTurk) platform, a recruitment site that allows researchers to access participants and provide compensation. This project was the first step in a larger project to see how personal and presidential beliefs may interact regarding military veteran political participation. Funding for this project was provided by the Graduate Student Senate of the University of Oklahoma.

Data Collection

For civilian participants, the only screening requirement was that they be eighteen years of age or older. Current or former military status and eighteen years of age or older were required prerequisites for veteran participants.

Participants from both groups received \$2.00 for participation in a survey that took approximately ten minutes for an hourly rate of \$12.00 an hour. After participants provided online consent, acknowledgement of being eighteen or older, and acknowledgement of military service (if applicable), the survey began. Participants were asked to first provide demographic information before completing randomly ordered political identification scales for both their personal beliefs and their beliefs of President Obama and President Trump's political identification using the scale developed by Mehrabian (1996). Participants then answered open-ended questions about the office of the President of the United States, as well as communication competence in general.

Participants

Although 100 civilians and 100 veterans completed the survey, some responses failed attention verification questions, and, therefore, the final sample consisted of 174 responses (N = 174). Of these responses, 90 were civilians (n = 90) and 84 were veterans (n = 84). The average age of civilian participants was 48.16 (SD = 17.11), average income was \$52,905 (SD = 27,766), most were identified as men (65), and none were identified as non-binary. Regarding race and ethnicity, participants were identified as White (62), Black (13), Asian (12), Pacific Islander (2), and Latina/o/x (1). The highest education achieved by participants was High School Diploma (40), College Degree (35), and Advanced Degree (14). The average age of military veteran participants was 34.81 (SD = 9.04), average income was \$59,146 (SD = 34,909), most were identified as men (64), and none were identified as non-binary. Regarding race and ethnicity, participants were identified as White (75), Black (4), Asian (2), and Latina/o/x (3). The highest education achieved by participants was High School Diploma (37), College Degree (34), and Advanced Degree (13).

Participants were also asked to identify their political ideology on a scale from 1 = Extremely Liberal to 7 = Extremely Conservative. Table 4.1 shows these responses by population.

Measurements

The political ideology scale was used as both a self-report of the participants' beliefs and an other-report of President Donald Trump and President Barack Obama. The political ideology scale contains twelve statements such as "Stricter environmental laws and regulations cost too many jobs and hurt the economy" and "The best way to ensure peace is through military strength." Initial scale reliability analysis revealed that the self-report scale was reliable

Table 4.1 Self-Reported Political Ideology of Participants

Political Ideology	*Civilian*	*Veteran*
Extremely Liberal	8	4
Very Liberal	12	18
Liberal	13	20
Neutral	20	9
Conservative	20	28
Very Conservative	10	2
Extremely Conservative	7	3

with a Cronbach's alpha of .879. However, one item, "Marijuana should be legalized by the federal government," was negatively related with all the other scale items and was removed, which raised Cronbach's alpha to .912. Thus, this item was also removed from both the political ideology scale of President Donald Trump (α = .805) and President Barack Obama (α = .909). As all alpha scores were over .800, the scales were considered valid. No issues of skewness or kurtosis were noted. All data were analyzed using IBM's SPSS v. 25 software package.

RESULTS

First, a correlation analysis was run with all independent and dependent variables. These results are reported in table 4.2.

After the analysis of correlations was complete, a one-way multivariate analysis of variance (MANOVA) was conducted with the sample source, civilian or military, as the independent variable and political ideology, political ideology scale, President Obama political ideology scale, and President

Table 4.2 Correlations of Study Variables

Variable	*Age*	*Income*	*Political Ideology*	*Political Ideology Scale*	*President Obama PIS*
Income	-.11				
Political Ideology	.06	-.02			
Political Ideology Scale	.05	.02	.71***		
President Obama PIS	-.16*	-.03	.06	.03	
President Trump PIS	.11	.08	-.32***	-.06	-.28***

Note: Political Ideology is the single-item question answered, while Political Ideology Scale is the scale participants completed. * = $p < .05$, ** = $p < .01$, *** = $p < .05$.

Table 4.3 MANOVA between Civilians and Veterans on Dependent Variables

Variable	F	p	*Civilian M*	*Civilian* SD	*Veteran M*	*Veteran* SD
Political Ideology	1.78	.184	4.04	1.72	3.71	1.53
Political Ideology Scale	3.12	.079	43.92	24.19	37.92	19.67
President Obama PIS	0.09	.761	24.34	17.83	25.06	11.37
President Trump PIS	2.59	.109	70.03	15.03	65.82	18.35

Note: Political Ideology is the single-item question answered, while Political Ideology Scale is the scale participants completed. * = $p < .05$, ** = $p < .01$, *** = $p < .05$.

Trump political ideology scale as the dependent variables. The results revealed no significant differences between civilians and veterans, and complete results can be seen in table 4.3.

DISCUSSION

Politics is just one of many facets of military-civilian communication that could be examined. However, the belief that military members and veterans are conservative Republicans is pervasive and may provide an inaccurate mental model for those who have had little or no experience with military veterans. In this study, there was no evidence found that a random sample of military veterans differed from a random sample of civilians, to a statistically significant degree, in terms of political ideology. These findings are markedly different from the news reports that portray veterans as conservative (e.g., Newport, 2009; Thompson, 2012). Military veterans have described a desire to not discuss politics in the classroom (Howe & Shpeer, 2019) and according to the UCMJ active military members are not allowed to, unless their views are in alignment with the views of the current President. When political discussions, which are considered taboo in the military (Mohr et al., 2019), are coupled with the social identity needs to align self-presentation with leadership ideology, veterans may not be willing, or even able, to discuss political issues in the classroom. This may be especially apparent in a graduate classroom considering that graduate students are often more attune to political news and may, therefore, have a differential mental model of military veterans than undergraduate students (Noza et al., 2018). Additionally, the Department of Defense has sought to send more officers to attend graduate programs during their military service (Tilghman, 2015), which may heighten the chances of service members interacting with civilian students.

Furthermore, if veterans feel that they are already categorized by others as conservative, or that their group identity has been categorized in such a way, they may be unwilling to communicate any new information that may build upon or challenge that stereotype. Current members may report their views as aligning with the current leader of the military, the President of the United States, even if this is not their actual belief, due to the requirements of the UCMJ. Perhaps these issues have led to findings of the military leaning conservative in political beliefs (Bachman et al., 2000; Nail & McGregor, 2009). The representation of service members and military veterans as conservative, and often Republican, may also contribute to the unwillingness of veterans to speak in the classroom. If a veteran believes that others have stereotyped them as a conservative, but the veteran is actually more liberal, perhaps the choice is made to avoid engaging in political conversations because their personal self-categorization does not align with the mainstream framing of military members' political beliefs. Such a misalignment likely creates identity gaps (Jung & Hecht, 2004) in that the veteran cannot perform their true identity as it does not align with what is expected of them from civilians.

LIMITATIONS AND FUTURE RESEARCH

This study, like all studies, has limitations. The sample size is fair for a two-group comparison (Comrey & Lee, 1992), but perhaps a larger sample would find significant differences. G*Power 3.1 was used to calculate the number of participants that would be needed to see a significant effect and that number was 788, although such a prediction assumes that the answers of the next 588 participants would mirror those in this study. Additionally, it was thought that perhaps the homogeneity of the sample, mostly white men, may have contributed to non-significant findings. A previous study (Cooney, Segal, Segal, & Falk, 2003) found that White and African American veteran participants differed in only two areas of the eight studied, whereas civilians differed on all eight. Furthermore, with the two areas that were statistically different, percent married not separated and percent single never married, the mean difference was less than civilian participants.

Future research could continue to examine the unwillingness or inability of military veterans to discuss political issues. Researchers could also consider the influence that military service, especially for active duty military members, may have on how political views are expressed in public. Because it is illegal, according to the UCMJ, for military members to speak against the President of the United States and other elected federal officials, they are highly likely to remain silent or to provide answers that they do not believe to remain within military regulations even when they disagree. Military leadership can exert tremendous influence and pressure on military personnel, especially enlisted personnel; therefore, efforts should be taken to hear and report the voices of these individuals without the threat of repercussions from chain of command.

BEST PRACTICES FOR INSTRUCTORS AND CONCLUSION

As such, it is recommended that instructors be careful when approaching politically charged or related issues in the classroom, as well as to not assume that a veteran holds a conservative view on an issue. This practice is particularly relevant in graduate courses, such as seminars, that are more discussion oriented and often attended by students that are politically savvy (Noza et al., 2018). Assuming that veterans hold a certain view is detrimental not only to the veteran but also to the overall classroom environment as it alienates the veteran, discourages diversity in the classroom, and furthers stereotyping of veterans as conservatives. Conversations in classes should try to avoid designating or positioning veterans as conservative voices in the classroom, as they do not necessarily hold such beliefs. Politics, however, is only one of many areas where it is inappropriate to single out a veteran to answer a question simply because of their military service.

Instructors might consider asking veterans indirectly, rather than directly, to speak on behalf of the military or veteran populations by using innovative anonymous technology. Most instructors understand that it is inappropriate to call on students because of their heritage, nationality, race, or religion, yet many instructors are quick to call on veterans and ask them for the beliefs of the military (Howe & Shpeer, 2019; Osborne, 2014). Military veterans may feel singled out and pressured to answer questions when this practice occurs in the classroom. Instructors need a way to seek out these diverse opinions in a way that supports students rather than alienates them. New technology allows this to occur. Different student response systems, such as Mentimeter and PollAnywhere, allow students to respond to prompts anonymously and may, therefore, open up discussions in a classroom in a way that it was not open before. Additionally, the use of such a program could allow instructors to see a variety of students' beliefs, even if they differ from the instructor's beliefs, and allow for a more inclusive classroom climate to better support all students and their viewpoints.

At an institutional level, one program that has been found to help student veterans on campus is the Peer Advisor for Veteran Education (PAVE) founded by the University of Michigan (Kees et al., 2017). The idea of PAVE is relatively simple; veterans communicate best with other veterans (Koenig et al., 2014). Therefore, PAVE campuses find veterans that have been successful on campus and pair these veterans with incoming veterans to assist them in the transition to higher education. Such a move can also be accomplished by instructors on an individual level. Most instructors may have had, will have, or know of student veterans in their program. A personal introduction from an instructor to two veterans may aid a new veteran in transitioning to the program or class and help both veterans have an outlet, should they feel uncomfortable in the classroom.

Osborne (2014) provides a list of questions and comments that make veterans feel uncomfortable such as: "Did you seen anyone die?" and "This war is unnecessary" (p. 258). Osborne (2014) also provides several suggestions to help ease military veterans into the classroom such as: making contact prior to arrival, separate orientations for veterans, and communicating with veterans often. During data collection for Howe and Shpeer (2019), veterans described the difficulties of seeking help from an instructor, even when the instructor made it clear they had an open-door policy. Veterans often felt a high-power distance between the instructor and themselves and were less likely to initiate communication with the instructor, likely based on experiences of the strict hierarchical structure of the military. Therefore, one way that instructors can support veteran students, especially in graduate education, is to make it a point to ask veterans to come to office hours or go to lunch and begin a conversation. Once the conversation has begun and the veteran believes that the instructor is willing to listen, the veteran may open up. This also helps veterans engage in perspective taking.

Finally, instructors could encourage and support the varying beliefs of students by making it a point to express expectations of respect and civility for diverse viewpoints in the classroom. This is especially key for student veterans who may have gone directly from high school to military to college and without much civilian interactions as an adult. These veterans have likely normalized the military way of learning, which is to listen to the instructor and do exactly as they say. This approach may not be helpful in a higher education classroom and especially in classes that require higher order and individual cognition such as with graduate coursework. Additionally, traditional students, who may have never known life outside of the classroom context, might need assistance seeing a veteran's perspective. Such practices could also help any student that has a dissimilar view to the rest of the class feel more comfortable expressing their viewpoint and feeling supported by the instructor. There is no one approach to the instruction of military veterans, as they have unique experiences that may differ from the overall veteran population; for this reason, it may be more beneficial to approach veteran instruction from an individual rather than group level. Combat veterans have been found to report significant differences in loneliness, reintegration, and military identity than non-combat veterans, and these relationships have been found to increase as the number of deployments increase (Howe, 2020). Therefore, until an instructor knows the background and personal military experiences of each student veteran, it may be best to avoid general assumptions about this population and to focus on the individual needs of each veteran in the classroom.

REFERENCES

Agan, K. E. (1920). Public speaking in a military college. *Quarterly Journal of Speech, 6*, 25–32.

Bachman, J. G., Freedman, P. D., Segal, D. R., & O' Malley, P. M. (2000). Distinctive military attitudes among U.S. enlistees, 1976–1997: Self-selection versus socialization. *Armed Forces and Society, 26*(4), 561–585. doi:10.1177/0095327x0002600404.

Boje, D. M. (2001). *Narrative methods for organizational & communication research*. Sage.

Bisel, R. S. (2018). *Organizational moral learning: A communication approach.* Routledge.

Brown, R. (2000). Social identity theory: Past achievements, current problems and future challenges. *European Journal of Social Psychology, 30*(6), 745–778. https://doi.org/10.1002/1099-0992(200011/12)30:6<745::aid-ejsp24>3.0.co;2-o.

Chan, M. (2017). Reluctance to talk about politics in face-to-face and Facebook settings: Examining the impact of fear of isolation, willingness to self-censor, and peer network characteristics. *Mass Communication and Society, 21*(1), 1–23. doi:10.1080/15205436.2017.1358819.

Clements, G. (2021). Creating relevance and purpose in postgraduate education: Focus on India. In M. R. Gurubatham & G. A. Williams (Eds.), *Advancing innovation and sustainable outcomes in international graduate education* (pp. 70–88). IGI Global. Advanced Online Publication. doi:10.4018/978-1-7998-5514-9.ch005.

Cooney, R. T., Segal, M. W., Segal, D. R., & Falk, W. W. (2003). Racial differences in the impact of military service on the socioeconomic status of women veterans. *Armed Forces & Society, 30*(1), 53–85. doi:10.1177/0095327x0303000103.

Cooper, L., Caddick, N., Godier, L., Cooper, A., & Fossey, M. (2016). Transition from the military into civilian life. *Armed Forces & Society, 44*(1), 156–177. doi:10.1177/0095327x16675965.

Cole, I. C. (2017). *Loose lips sink relationships: Applying communication privacy management theory in military family support organizations.* Unpublished Master's Thesis, Texas A&M University – Corpus Christi. https://tamucc-ir.tdl.org/bitstream/handle/1969.6/24397/Cole_Catherine_thesis.pdf?sequence=1&isAllowed=y.

Cox, J. A., & Albright, D. L. (2016). Can you keep a secret? Examining military couples' disclosure through the lens of communication privacy management theory. *Journal of Military and Government Counseling, 4*(1), 12–21.

Dailey, S. L. (2016). What happens before full-time employment? Internships as a mechanism of anticipatory socialization. *Western Journal of Communication, 80*(4), 453–480. doi:10.1080/10570314.2016.1159727.

Fairhurst, G. T. (2010). *The power of framing: Creating the language of leadership* (Vol. 2). John Wiley & Sons.

Feaver, P. D., & Kohn, R. H. (2001). *Soldiers and civilians: The civil-military gap and American national security.* MIT Press.

Foy, S. L., & Restifo, S. J. (2018). Reliably Republican? Shifts in U.S. Veterans' political party affiliation from 1974 to 2016. *Sociological Spectrum, 38*(3), 174–193. doi:10.1080/02732173.2018.1469445.

Gade, D. M., & Wilkins, V. M. (2012). Where did you serve? Veteran identity, representative bureaucracy, and vocational rehabilitation. *Journal of Public Administration Research and Theory, 23*(2), 267–288. doi:10.1093/jopart/mus030.

Garner, J. T., & Peterson, B. L. (2017). Untangling the processes of leaving a member-abusive organization. *Management Communication Quarterly, 32*(2), 143–171. doi:10.1177/0893318917717816.

Glasser, I., Powers, J. T., & Zywiak, W. H. (2009). Military veterans at universities: A case of culture clash. *Anthropology News, 50*(5), 33–33. doi:10.1111/j.1556-3502.2009.50533.x.

Haidt, J., & Joseph, C. (2004). Intuitive ethics: How innately prepared intuitions generate culturally variable virtues. *Daedalus, 133*(4), 55–66. doi:10.1162/0011526042365555.

Hasian, Jr., M. (2007). Dangerous supplements, inventive dissent, and military critiques of the Bush administration's unitary executive authorities. *Presidential Studies Quarterly, 37*(4), 693–716. doi:10.1111/j.1741-5705.2007.02620.x.

Haslam, S. A., Ellemers, N., Reicher, S., Reynolds, K. J., & Schmitt, M. (2010). Social identity today: The impact of its defining ideas. In T. Postmes & N. R. Branscombe (Eds.), *Rediscovering social identity.* Hove: Psychology.

Hinderaker, A. (2015). Severing primary ties: Exit from totalistic organizations. *Western Journal of Communication, 79*(1), 92–115. doi:10.1080/10570314.2014.943422.

Hinderaker, A., & O'Connor, A. (2015). The long road out: Exit stories from the Church of Jesus Christ of Latter-day Saints. *Communication Studies, 66*(5), 509–527. doi:10.1080/10510974.2015.1072096.

Howe, W. T. (2020). *Veteran contempt for civilian communication (VCCC): The development and validation of a scale for use with military veterans.* Doctoral Dissertation, University of Oklahoma, SHAREOK. https://hdl.handle.net/11244/324933.

Howe, W. T., & Bisel, R. S. (2020). Third-party online organizational reviews: Explaining review-and-rating patterns of the United States military and large corporate organizations. *Computers in Human Behavior Reports, 1*(1), 1–6. doi:10.1016/j.chbr.2020.100006.

Howe, W. T., & Hinderaker, A. (2018). "The rule was the rule": New member socialization in rigidly structured totalistic organizations. *Atlantic Journal of Communication, 26*(3), 180–195. doi:10.1080/15456870.2018.1472093.

Howe, W. T., & Shpeer, M. (2019). From military member to student: An examination of the communicative challenges of veterans to perform communication accommodation in the university. *Journal of Intercultural Communication Research, 48*, 203–220. doi:10.1080/17475759.2019.1592770.

Igielnik, R., Parker, K., & Cilluffo, A. (2019, December 31). *Trump draws stronger support from veterans than from the public*. Pew Research Center. Retrieved January 3, 2020, from https://www.pewsocialtrends.org/2019/07/10/trump-draws-stronger-support-from-veterans-than-from-the-public-on-leadership-of-u-s-military/.

Jablin, F. M. (2001). Organizational entry, assimilation, and disengagement/exit. In F. Jablin & L. Putnam (Eds.), *The new handbook of organizational communication* (pp. 732–818). Sage.

Jetten, J., Branscombe, N. R., Schmitt, M. T., & Spears, R. (2001). Rebels with a cause: Group identification as a response to perceived discrimination from the mainstream. *Personality and Social Psychology Bulletin, 27*(9), 1204–1213. doi:10.1177/0146167201279012.

Jung, E., & Hecht, M. L. (2004). Elaborating the communication theory of identity: Identity gaps and communication outcomes. *Communication Quarterly, 52*(3), 265–283. doi:10.1080/01463370409370197.

Kees, M., Risk, B., Meadowbrooke, C., Nellett, T., & Spinner, J. (2017). Peer Advisors for Veteran Education (PAVE): Implementing a sustainable peer support program for student veterans on college campuses. *Metropolitan Universities, 28*(3). doi:10.18060/21541.

Klingler, J. D., & Chatagnier, J. T. (2014). Are you doing your part? *Armed Forces & Society, 40*(4), 673–695. doi:10.1177/0095327x12471932.

Knight, J. P. (1990). Literature as equipment for killing: Performance as rhetoric in military training camps. *Text & Performance Quarterly, 10*(2), 157–168. doi:10.1080/10462939009365965.

Koenig, C. J., Maguen, S., Monroy, J. D., Mayott, L., & Seal, K. H. (2014). Facilitating culture-centered communication between health care providers and veterans transitioning from military deployment to civilian life. *Patient Education and Counseling, 95*(3), 414–420. doi:10.1016/j.pec.2014.03.016.

Kramer, M. W. (2010). *Organizational socialization: Joining and leaving organizations*. Polity.

Kramer, M. W., Lee, S. K., & Guo, Y. (2018). Using communication technology to manage uncertainty during organizational assimilation: Information-seeking and information-giving. *Western Journal of Communication, 83*(3), 304–325. doi:10.1080/10570314.2018.1518538.

Leal, D. L., & Teigen, J. M. (2015, November 11). Recent veterans are more Republican than older ones. Why? *Washington Post*. Retrieved January 3, 2020, from https://www.washingtonpost.com/news/monkey-cage/wp/2015/11/11/recent-veterans-are-more-republican-than-older-ones-why/.

May, A., & McDermott, V. (2019). Cultural collision: A review of literature. In L. H. Hernández & J. N. Belding (Eds.), *Military spouses with graduate degrees: Interdisciplinary approaches to thriving amidst uncertainty*. Lexington Books.

McNamee, L. G., & Gould, K. K. (2019). Sensemaking after the mission: Totalistic yet terminal roles, identity, and exit. *Journal of Communication & Religion, 42*(4),

56–73. http://search.ebscohost.com.ezproxy.lib.ou.edu/login.aspx?direct=true&db=cms&AN=141856688&site=ehost-live.

Mehrabian, A. (1996). Relations among political attitudes, personality, and psychopathology assessed with new measures of libertarianism and conservatism. *Basic and Applied Social Psychology, 18*(4), 469–491. doi:10.1207/s15324834basp1804_7.

Milburn, A. R. (2010). Breaking ranks: Dissent and the military professional. *Joint Force Quarterly*, pp. 101–107. https://apps.dtic.mil/dtic/tr/fulltext/u2/a536591.pdf.

Mohr, S., Sørensen, B. R., Weisdorf, M., Ben-Ari, E., Hautzinger, S., McSorley, K., Scandlyn, J., & Wool, Z. (2019). Discussing empathy and critique in the ethnography of things military: A conversation. *Ethnos*, Advanced Online Publication. doi: 10.1080/00141844.2019.1687551.

Nail, P. R., & McGregor, I. (2009). Conservative shift among liberals and conservatives following 9/11/01. *Social Justice Research, 22*, 231–240. doi:10.1007/s11211-009-0098-z.

Newport, F. (2009, November 25). Military veterans of all ages tend to be more Republican. *Gallup*. Retrieved January 3, 2020, from https://news.gallup.com/poll/118684/military-veterans-ages-tend-republican.aspx.

Noza, V. S., Ananda, A., & Azmi, A. (2018). Correlation between political news consumption and civic education students' political awareness. *Proceedings of the Annual Civic Education Conference (ACEC 2018)*. doi:10.2991/acec-18.2018.83.

Ordway, D. M. (2018, August 27). Younger US veterans more likely to be affiliated with Republican Party. *Journalist's Resource*. Retrieved January 3, 2020, from https://journalistsresource.org/studies/government/security-military/veterans-republican-party-affiliation/.

Osborne, N. J. (2014). Veteran Ally: Practical strategies for closing the military-civilian gap on campus. *Innovation in Higher Education, 39*, 247–260. doi:10.1007/s10755-013-9274-z.

Parott, S., Albright, D. L., Dyche, C., & Steele, H. G. (2018). Hero, charity case, and victim: How U.S. news media frame military veterans on Twitter. *Armed Forces & Society, 45*(4), 702–722. doi:10.1177/0095327x18784238.

Phillips, J. (2013, April 19). *IG recommends getting 'back to basics,' values*. Retrieved December 8, 2019, from https://www.army.mil/article/101495/ig_recommends_getting_back_to_basics_values.

Philipsen, G. (1992). *Speaking culturally: Explorations in social communication*. SUNY Press.

Rasmussen Poll. (2018, June 5). *Most military voters see Trump as stronger Commander than recent presidents*. Retrieved March 4, 2020, from https://www.rasmussenreports.com/public_content/politics/general_politics/may_2018/most_military_voters_see_trump_as_stronger_commander_than_recent_presidents.

Rhidenour, K. B., Barrett, A. K., & Blackburn, K. G. (2017). Heroes or health victims?: Exploring how the elite media frames veterans on Veterans Day. *Health Communication, 34*(4), 371–382. doi:10.1080/10410236.2017.1405481.

Rozin, P., Lowery, L., Imada, S., & Haidt, J. (1999). The CAD triad hypothesis: A mapping between three moral emotions (contempt, anger, disgust) and three moral codes (community, autonomy, divinity). *Journal of Personality and Social Psychology, 76*(4), 574–586. doi:10.1037/0022-3514.76.4.574.

Schmidt, H. (2020). "Hero-worship" or "manipulative and oversimplifying": How America's current and former military service members perceive military-related news reporting. *Journal of Veterans Studies, 6*(1), 13–24. doi:10.21061/jvs.v6i1.156.

Shane, L. (2018, October 17). Troops see rising political tension in the ranks, poll shows. *Military Times.* Retrieved January 3, 2020, from https://www.militarytimes.com/news/pentagon-congress/2018/10/17/troops-see-rising-political-tension-in-the-ranks-poll-shows/.

Shane, L. (2019, December 19). *Half of active-duty service members are unhappy with Trump, new Military Times poll shows*. Retrieved March 4, 2020, from https://www.militarytimes.com/news/pentagon-congress/2019/12/17/half-of-active-duty-service-members-are-unhappy-with-trump-new-military-times-poll-shows/.

Shpeer, M., & Howe, W. T. (2020). Socialization, face negotiation, identity, and the United States Military. *International Journal of Communication, 14*, 726-744. https://ijoc.org/index.php/ijoc/article/view/11885.

Sørensen, H. (2011). Core values of Danish expeditionary soldiers. In F. Ã. Henrik & K. Ã. Gerhard (Eds.), *Core values and the expeditionary mindset: Armed Forces in metamorphosis* (pp. 179–191). Nomos Publishers.

Tajfel, H., Turner, J. C., Austin, W. G., & Worchel, S. (1979). An integrative theory of intergroup conflict. In M. J. Hatch & M. Schultz (Eds.), *Organizational identity: A reader* (pp. 56–65). Oxford University Press.

Thompson, M. (2012, November 5). Does the military vote really lean Republican? *TIME*. Accessed on March 4, 2020. https://swampland.time.com/2012/11/05/does-the-military-vote-really-lean-republican/.

Tilghman, A. (2015, August 20). DoD plans to send more officers to civilian grad schools. *Military Times*. Accessed on July 21, 2020. https://rebootcamp.militarytimes.com/education-transition/education/2015/08/20/dod-plans-to-send-more-officers-to-civilian-grad-schools/.

United States Congress. (2020). *Uniform Code of Military Justice*. Retrieved June 22, 2020 from https://uscode.house.gov/view.xhtml?path=/prelim@title10/subtitleA/part2/chapter47&edition=prelim.

U.S. Army Center of Military History. (2020). *Oaths of Enlistment and Oaths of Office*. Retrieved June 22, 2020 from https://history.army.mil/html/faq/oaths.html.

Wilson, S. R. (2020, January 16). *Four military-civilian gaps*. Military-Civilian Dialogue. https://militaryciviliandialogue.blogspot.com/2020/01/four-military-civilian-gaps.html.

Wilson, S. R., Hintz, E. A., MacDermid Wadsworth, S. M., Topp, D. B., Southwell, K. H., & Spoont, M. (2019). Female U.S. Military Veterans' (non)disclosure of mental health issues with family and friends: Privacy rules and boundary management. *Health Communication*, Advanced Online Publication. doi:10.1080/10410236.2019.1693128.

Chapter 5

Challenges for Military Partners Pursuing College and Resilience-Based Strategies to Support Their Success

Kelly R. Rossetto and Jennifer S. Owlett

Pursuing higher education can be very stressful for students because of academic pressure, financial burden, and increased autonomy from both teachers and families (Kadison & DiGeronimo, 2004). Recent findings from the American College Health Association Survey (ACHA, 2016) show that 50.4 percent of students report experiencing three or more life events that were considered traumatic or very difficult to handle (e.g., academics, family problems, intimate relationships, finances) within the past twelve months. Military-affiliated learners (MALs) may experience even more stressors than their civilian college peers due to the increased probability of working while attending classes, caring for dependents, and feeling alienated from social support networks (see Ford & Vignare, 2015). Veteran students with post-traumatic stress or depression might also have difficulty socializing outside of class and may struggle to fit in with peers (Cate, 2011). Current research on MALs tends to focus on active duty personnel and veterans, even though MAL definitions also include military partners and children (Ford et al., 2009; Stanford University Office for Military Communities, n.d.). Military partners and children share many of the same challenges as service members and veterans, underscoring the need for scholarly attention (Gleiman & Swearengen, 2012).

In this chapter, we will focus specifically on military partners pursuing college (MPPCs). "Eighty-nine percent of Active Duty spouses have at least some college education, with 28% having attained a 4-year degree and 14% having an advanced degree" (Department of Defense, 2018, p. 1). These numbers indicate that college campuses are educating a significant number of MPPCs, and various benefit programs are making educational pursuits more accessible. In 2009, under the Post 9/11 G.I. Bill, military spouses and family members became eligible for higher education benefits (U.S. Department of Defense, 2009). One example of these funds is the My Career Advancement Account Scholarship (MyCAA), a "workforce development program that provides up to $4,000 of financial assistance for licenses, certifications, or Associate's degrees to pursue an occupation or career field" (Military One Source, n.d., para. 1). In 2018, close to 30,000 spouses used these G.I. benefits and MyCAA funds (U.S. Department of Veteran Affairs, 2018; Military One Source, n.d.) to further their education.

With increased funding opportunities and more access to higher education (including on-campus and online programming), the number of MPPCs pursuing college education will likely continue to grow (American Council on Education, 2018). Despite the potentially increasing number of MPPCs, "little research has investigated the educational needs and barriers of military spouses as a separate student population" (Gleiman & Swearengen, 2012, p. 78). Using existing research along with data from interviews conducted with military spouses as part of a larger study, this chapter will first describe potential challenges military partners face along with ways these challenges may impact them during their college career. Then, based on these challenges and their impact, we will propose resilience-based strategies faculty can enact to better support MPPCs and improve the academic and personal resilience of this unique population of learners.

BACKGROUND AND METHOD

The first author conducted in-depth interviews ($N = 26$) with military partners (i.e., spouses, fiancés, long-term romantic partners) to better understand how they perceive and make meaning of their experiences with military life challenges, how they cope with their experiences, and what supportive resources are most helpful for them. Interview questions asked about the challenges military partners face, how they cope with the challenges, and types of support they receive. Interviews (range: 47–128 minutes; $M = 94.9$) were conducted face-to-face ($n = 20$) in a participant-chosen location or over the telephone ($n = 6$). Locations included the researchers' on-campus office, coffee shops, and the participants' (or their friends'/family's) homes in three different Southwest cities. Telephone interviews reached partners from four Southeast, Southwest, West, and Midwest cities.

Recruitment and Sample

Several techniques were used to recruit participants to reach partners across branches and demographic markers. First, the study was listed on Craigslist

.com, Meetup.com, and military spouse blogs. This approach required searching the "milblog" network (i.e., informal social networks created by members of the extended military community, dependents, active duty personnel, and/or veterans, to connect, share resources, or discuss topics) and Meetup groups, then posting announcements and requests for participants. Second, participants were recruited from different organizations and through people connected to military populations, including military groups (e.g., Family Readiness Groups), military website organizers, counselors, professors, and peers. Third, participants passed along information about the study to other military partners through email, phone, or blogs.

Although not a requirement, all participants who volunteered to participate in the study were female; ages ranged from twenty to forty years ($M = 27.42$). Participants reported racial/ethnic backgrounds: White/Caucasian (n = 21; 80.8 percent), Hispanic/Latina (n = 2; 7.7 percent), and mixed ethnicity (n = 3; 11.5 percent). Participants' partners' racial/ethnic backgrounds included White/Caucasian (n = 17; 65.4 percent), Black/African American (n = 1; 3.8 percent), Hispanic/Latino (n = 4; 15.4 percent), and mixed ethnicity (n = 3; 11.5 percent). One participant did not disclose her partner's racial/ethnic background (3.8 percent). According to 2018 demographics, these numbers align with active duty service members' reported ethnicities, with the exception of Black/African American which trends higher in national reports (U.S. Department of Defense, 2018). Most participants were married (n = 24) or engaged (n = 2). Relationships ranged in length for married participants from approximately one month to eleven years (M = 5.33). Engaged participants reported four month and two-year engagements. Sixteen women indicated having children with a range from zero to four (M = 1.23). In terms of joint family income, participants indicated the following ranges: less than $25,000 ($n$ = 2; 7.7 percent), $25,000 to $50,000 ($n$ = 9; 34.6 percent), $50,000 to $75,000 ($n$ = 8; 30.8 percent), $75,000 to $100,000 ($n$ = 5; 19.2 percent). Two participants did not report income (7.7 percent). Participants represented three branches of the military: Air Force, Marines, and Army. Proximity to a major Army base likely influenced the recruitment and sample, but online recruitment allowed for further reach to the additional branches.

Analysis

Once interviews were transcribed, the first author conducted a thematic analysis of the data, using constant comparison to locate, interpret, and define themes emerging from the data (Strauss & Corbin, 1990). The data were organized based on the guiding research question (i.e., military life/deployment challenges). Then, data were extracted from the transcriptions and organized into a spreadsheet to minimize, visualize, and analyze the data within and across participants. Using the organized and reduced data, the first author openly coded the interview data to discover, create, name, and explain multiple preliminary categories that emerged (Lindlof & Taylor, 2002). Next, data were analyzed and coded within appropriate categories, based on the properties of the initial category description. Finally, axial coding was conducted so that multiple categories could be integrated and/or collapsed into more inclusive and manageable themes (Strauss & Corbin, 1990). This step was done in an effort to present a more coherent and organized picture of the data, including all participants and based on broad relations between responses. Once the data were organized into clear and distinguishable themes, the results were written using participants' words to exemplify the results.

To ensure credibility, we triangulated our findings (see Baxter & Eyles, 1997) by analyzing four additional interviews conducted with military partners who had recently or were currently pursuing college. These data were collected as part of a separate study conducted by the first and second author; however, interview questions (e.g., challenges, coping, support) overlapped across the two studies. These participants were recruited online through military spousal groups on Facebook. Additional interviewees (i.e., MPPCs selected for the triangulation methods) were female and married to active duty military personnel (e.g., Army, Air Force, National Guard). To check consistency across data sets, the authors deductively analyzed the four additional interviews using the developed themes. This triangulation process helped ensure the challenges from the initial data set were relevant to MPPCs specifically. Overall, the additional interview data matched the initial themes. Exemplary quotes were integrated into the findings.

DISCUSSION OF DATA AND PREVIOUS LITERATURE

In this chapter, we highlight three issues (i.e., identity challenges, control concerns, emotional burdens) that emerged from the data analysis, along with the relevant literature on these topics. Our goal is to outline how these experiences might influence when and how MPPCs follow and complete higher education opportunities. Through approaching these challenges, we can offer resilience-based strategies to improve our capacity for empathy and support for this student population.

Challenges for Military Partners Pursuing College (MPPCs)

Issue 1: Control concerns. First, many military partners recognized changes in their *sense of control*. Most partners expressed a sense of uncontrollability when it came to military life. Although there was often an expressed micro sense of control (over tasks, household decisions, and so on) that could be both empowering and challenging, there was a macro sense of uncontrollability (over life structure, life decisions, and so on) that created uncertainty. Recent findings have highlighted several reasons why these concerns are salient. One major concern comes from frequent relocation. For military partners, permanent change of station (PCS) can occur frequently, as over 400,000 PCS orders are issued annually (Military OneSource, 2019). This concern becomes exacerbated when one considers current findings regarding employment for military spouses in comparison to civilian counterparts. In 2012, 52 percent of military spouses were employed full-time or part-time, 13 percent unemployed, and 35 percent out of the labor force (Defense Manpower Data Center, 2018). Moreover, frequent mobility can negatively influence educational and career advancement (Marshall, 2014). For example, Cooke and Speirs (2005) suggest "migration is associated with a 10% decline in employment among all civilian wives and a four-hour decline in hours worked per week among civilian wives who remain employed" (p. 343). In addition, the Defense Manpower Data Center (2018) notes that almost 25 percent of military spouses need ten months or more to secure a new job. Our data speaks to these challenges and their more personal implications.

For MPPCs, being a part of the military means giving up autonomy in terms of where they live, when they move, and when their partners are home or away. As noted by one research participant:

> You have to plan a lot more with the Army. It's not like if you work in corporate America, you ask, 'Hey, can I have March 5th through March 10th off?' You know? No. No, you don't do that with the Army. You plan, 'Ok, when's block leave? When can we do this?' You plan around the Army, so you know that's been a new thing as far as planning vacations and even like larger stuff, planning on having kids.

Furthermore, classes and graduation might not fit into the military's schedule for MPPCs. As one partner stated:

> I transferred to a school and most of my degrees in Education or, well, Spanish, and I have a minor in Education. But once I found out I'm going the military route, I didn't get certified, just because we knew we were moving to Texas with the job he had for the third part of his training.

This sentiment was echoed by a third interview subject, noting the following:

> If you want to, let's say be a nutritionist, the requirements, let's say in Tennessee, are different than the requirements anywhere else. So, once you kind

> of start going through the schooling process, I've noticed that by the time I get done with that schooling process, we've gone to another state. So, I've been really hesitant, actually, about getting another degree.

Control concerns were also discussed in terms of deployment. Military personnel and their families often view deployment as beyond their control (Huebner, Mancini, Wilcox, Grass, & Grass, 2007). In some cases, control came with independence (e.g., freedom to complete tasks; control over child-rearing). However, this control also became a burden because their main source of support was not present. For example, one partner described feeling more anxious during deployment:

> The stress, knowing that I'm responsible for everything. If anything breaks in the house, something happens with the dogs, there's a bill that got messed up or something, I have to take care of it. And that brings a lot of anxiety to me as far as knowing that if anything comes to the house, or something happens I have to take care of it. I am the sole provider. And knowing that if something happens to me, there's nobody here to take care of me.

Another echoed these feelings by noting the following:

> Having to manage everything . . . all the household chores, getting the lawn mowed, getting the floor mopped, and having to take care of the kids. Then dealing with other things, you know, you kind of would hope that life would stop a little bit during these things, but real life things just keep happening. So you have to deal with things on your own that would be easier to deal with as a couple.

Adding school to these lists of responsibilities can be a further challenge for MPPCs.

Issue 2: Identity and relational challenges. Second, and related to control concerns, many military partners felt *identity and relational challenges.* They became caught up in their partners' identities, jobs, and experiences rather than their own. One way in which to better understand these identity concerns is through exploring current research on military partners as "trailing spouses" (Bouchard, 2019; Cooke & Speirs, 2005; Friedman, Miller, & Evans, 2015). Trailing spouses are defined as "those who follow a partner due to relocation related to a partner's job" (Meadows, Griffin, Karney, & Pollak, 2016, p. 543). This labeling suggests that military partners are defined by their marital status and circumstances rather than personal characteristics. Following frequent relocation (as described above), trailing spouses can face additional adjustment stressors (e.g., loss of a job or career, finding employment, time for employment, new culture) (Cooke & Speirs, 2005; McNulty, 2012), which have the potential to affect well-being (Richmond et al., 2017). However, the literature regarding the impact of these moves on military families, especially military partners, and MPPCs, is limited (Agbisit, 2019; Blakeley, Hennessy, Chung, & Skirton, 2012).

Military spouses explain that military lifestyles can compound these relocation-related issues by leading to identity losses (Matchett-Taylor, 2018). For example, they may feel like they are "leading two lives," which could lead to burnout, alienation, and even marital conflict (May & McDermott, 2019, p. 76). In a study of expatriate trailing spouses, McNulty (2012) found that the link between dual-careers and identity can create feelings of isolation, loneliness, resentment, depression, and marital issues. This identity confusion can be even more salient within an academic setting and may be partially dependent on the culture of the academe in relation to the military community (see May & McDermott, 2019, for a review).

Research participants shared these perspectives and noted multiple ways in which identity concerns were apparent and had implications for other elements of military life. For example, one MPPC stated:

> School started the next day, and I went to two classes and it was like it reminded me that I had my own identity, I wasn't just [my husband's] wife, or that I had ideas and thoughts and when I introduce myself. I shouldn't introduce myself as, 'My husband is deployed'. I introduce myself as my name, and what I like to do, and all that stuff . . . It's like a constant struggle and you have to find that balance. Where I feel like I just dipped really far into what [my husband] was doing . . . I start defining myself as him and not me.

Concentrating on school may also be difficult, especially during deployment, due to split identities and feeling like part of the self is with the partner. One military partner characterized her experience by noting the following:

> I am more frazzled, a little more disorganized thinking and stuff and it takes more to get focused on things . . . As for the disorganized thinking you know it's constant distraction. No matter what I'm doing there's an underlying thought of what's going on over there. So half of my mind is here, and it is 12:30 in the morning there. So I'm constantly in two different time zones; I'm wondering thinking about what he might be doing. So it's you know I try to focus on the tasks that need to be done, but most of time they don't quite make it.

Here we can see the difficulties that might arise for students who are trying to concentrate on tasks and organize their time (two major challenges for college students generally), when a large part of their minds may be elsewhere.

Also related to identity struggles, some military partners felt it was harder to take time for their social lives. They became more withdrawn and socially isolated when their partners were away; in part because they had so many responsibilities, but also because they didn't want to miss out on opportunities to talk to their partners who might call. One partner stated,

> I really just I haven't made the time to go out and find friends just for sheer fact that that means I'll be gone. If I make new friends I'm going to be gone and out hanging out with them and I'm not going to be here when [my husband] gets on the computer.

Another said, "I'm very focused on work. That's all I do. I work, I sleep. It's not fun. I've become a little bit more withdrawn from my friends probably than I should have because I don't like people to see me upset." Being focused on (school) work may be helpful for MPPCs, and may help contribute to their personal identity building. However, being socially withdrawn from others on campus may negatively impact their sense of belonging, which can be deleterious for resilience and retention (see, for example, O'Keeffe, 2013).

Frequent transfers also influence the ability to make friends and feel a part of a community. Discussing a recent relocation, one partner said,

> I actually had a really hard time adjusting when we moved to Detroit. We didn't have a traditional military unit or community here . . . I suddenly became this person who might leave, and so people are less reluctant to invest time and friendship in you as well in certain circumstances.

Another participant reflected on her most recent relocation, "There's not that local community. So I think with social community, I think I would feel more—I think I would feel isolated because I wouldn't have those friends that know what it's like." Similar challenges likely exist for MPPCs because they may not be surrounded by other military partners on their campuses, and consequently can feel distant from their communities.

Issue 3: Emotional burdens. In addition to identity challenges, being a part of the military (and especially experiencing deployments) can add an *emotional burden* for military partners. Military life is likened to an emotional rollercoaster, filled with extreme highs and lows (e.g., Gleiman & Swearengen, 2012). Specifically, deployment can lead to feelings of helplessness, loss, disruption (Palmer, 2008), and high levels of fear and uncertainty (Huebner et al., 2007). The emotional ups and downs of military life likely compound with the emotional stress associated with college life. Mental health issues are on the rise for college students generally, with 54.7 percent of college students reporting feeling above average or tremendous levels of stress, 58.4 percent reporting feeling overwhelming anxiety, 65 percent reporting feeling very sad, and 36.7 percent reporting feeling so depressed it was difficult to function (ACHA, 2016). As such, it is imperative that we recognize the potential mental health challenges that MPPCs face as well. With this in mind, mental health professionals on college campuses may benefit from additional training to ensure they understand the unique needs of this population (Gorman, Blow, Ames, & Reed, 2011; Johnson et al., 2007).

Discussing the emotional experience of deployment, one participant noted, "[I feel] depression, neurotic, tendency to build up OCD, lonely, independent. It'll either make you or break you . . . You'll see what you're made of. Either you're strong or you're weak, but to be strong doesn't mean that you can never cry." Another said,

> With that loss you have all those emotions, the sadness, the depression, the things that go along with that, even if it is temporary. I think there are lots of people that go through that. And I go through periods of depression, and I go through periods of sadness and periods where I feel lost and unloved and lonely, lonely is huge.

Community and social support act as protective factors for military family members, which positively impacts their ability to maintain resilience (Wiens & Boss, 2006). However, it is well documented that military personnel and their families may not seek mental and behavioral help due to stigma, embarrassment, judgment, or lack of access (Johnson et al., 2007). It is also necessary to recognize that MPPCs may not be comfortable asking for support from non-military-affiliated people. One participant said,

> I wouldn't wanna have to justify or rationalize my emotions to somebody who didn't understand what it meant. It would sound weird to somebody who's not affiliated with the military or who's never been through a deployment to describe the fear that comes with your service member returning.

Even if MPPCs do seek formal support, colleges across the country are often unable to keep up with mental health care demands (Thielking, 2017).

SUMMARY AND RECOMMENDATIONS

Previous literature, and the findings from this current data, suggest that MPPCs are faced with multiple challenges such as balancing school, work, and identities in the context of an uncertain military lifestyle. Collectively, these issues might prevent MPPCs from experiencing a sense of belonging on campus. This sense of belonging, or feeling of respect, acceptance, inclusion, and support (Goodenow & Grady, 1993), can positively predict academic-related outcomes (Irvin, Meece, Byun, Farmer, & Hutchins, 2011) and resilience (Scarf et al., 2016). Studies have shown that students with a sense of belonging are more likely to develop positive academic attitudes and behaviors and have a sense of competence and responsibility (Battistisch et al., 1997; Cefai, 2007). As such, a focus on creating this sense of belonging and support is imperative when making changes to how we promote the most effective and encouraging learning environment for MPPCs.

One way to lessen the impact of control challenges, specifically, is to encourage programs to have licensures (and curricula) that can be applied to multiple states (institutions). Currently, six occupations (cosmetology, dental hygiene, massage therapy, mental health counseling, occupational therapy, and real estate commission) have licenses that can potentially be applied across state lines (see Richmond et al., 2017) but are contingent on state requirements. Moreover, transferring credits more seamlessly would also benefit students who need to transfer colleges. The inability to transfer qualifying credits can lead to extended degree completion periods and can possibly discourage partners from degree attainment (Matchett-Taylor, 2018). This transferability provides one way in which military partners can be supported in pursuing their own future career goals.

There are also opportunities for higher education institutions to offer academic support and mental health counseling for MPPCs and create a sense of belonging. But in order to actualize support, faculty and staff are challenged to acknowledge MPPCs on campus and the additional challenges they face. Veterans and spouses share many benefits and commonalities, so the services provided through the veterans' centers in colleges provide a great starting point to provide support for MPCCs (Matchett-Taylor, 2018); for example, providing professional development for faculty and students and curricular and co-curricular adjustments to address military-connected issues (Mercer, Thurston, & Hughey, 2014). With this training and co-curricular approach in mind, we offer a few evidence-based suggestions for faculty to bring resilience-based strategies into the classroom.

RESILIENCE-BASED STRATEGIES FOR SUPPORTING MPPCS

Although limited research suggests challenges for MALs considering college attendance, "the higher education community is without a sufficient understanding of their experiences and patterns related to access, persistence, and attainment" (Molina & Morse, 2015, p. 1). In order to retain MALs, additional support is needed. This is particularly true for military partners. Faculty and staff training, opening lines of communication, and engaging student veterans are core elements for supporting military-connected students academically and creating a MAL-friendly campus culture (American Council on Education, 2018). Our proposed strategies will expand the focus to include military partners using insight from resilience theory.

Focusing on student resilience is one way to approach better supporting MALs and MPPCs. Resilience is the process of adapting to stress (Masten, 2001) involving the interaction between risk and protective factors to facilitate adaptation, adjustment (Olsson, Bond, Burns, Vella-Brodrick, & Sawyer, 2003; Patterson, 2002), reintegration, and productive change through adversity, obstacles, and setbacks (Buzzanell, 2010). Put simply, resilience is "the process of reintegrating from disruptions in life" (Richardson, 2002, p. 309). College students in general are facing considerable challenges, as evidenced by increases in depression, anxiety, and stress (ACHA, 2016). Decreased student mental health can impact learning, performance, persistence, and success (Douce & Keeling, 2014; Hartley, 2010). Therefore, it is necessary for campus communities to address these issues and make efforts toward improving students' resilience and mental health.

MPPCs' challenges (i.e., risk factors) hold the potential to undermine their resilience, and less resilient students take fewer risks (i.e., intellectually and creatively) and do not work as well with others in learning environments (ACHA, 2016). Exercising protective factors (e.g., support, coping, self-efficacy), on the other hand, supports successful outcomes under high-risk circumstances (Masten & Reed, 2002). What can members of the higher education community do to promote resilience for MPPCs? We suggest three broad strategies that focus on the previously described challenges. Because prior literature has emphasized more structural strategies (e.g., accessibility through online offerings, credit transfer, licensure adaptation, extending veteran services), we will illuminate what can be done within the classroom to help build MPPCs' sense of connection and belonging in order to promote resilience and retention (Irvin et al., 2011; Scarf et al., 2016). These strategies (e.g., develop relationships, promote reflection, normalize challenge, and failure) integrate existing findings on student resilience with Buzzanell's (2010) communicative resilience processes. They are not mutually exclusive and should be integrated into the classroom as a collective unit.

Develop Relationships

Maintaining social networks involves the formation and utilization of relationships, which is a necessary component of resilience (Buzzanell, 2018). Developing a MAL network composed of students, faculty, and allies would help to unite the MAL community and create a greater sense of belonging and connection on a college campus and in online contexts. Feeling connected and affiliated with other students and teachers can create a more physically and emotionally safe environment characterized by trust rather than fear, as well as a sense of pride and a sense of ownership in students' classes (Cefai, 2007).

First, developing relationships between students and faculty mentors is important for MPPCs, and faculty should facilitate these connections. A few simple tips for building relationships with students include learning names, showing up for class early to talk informally, encouraging students to attend office hours, maintaining positivity, and offering advice when it is sought. In online environments, instructors can use many of these strategies, some with slight alterations. For example, uploading and replying to introduction videos (using Flipgrid or learning management systems) and hosting virtual office hours will enable more engagement with students online.

Building relationships can also extend to creating more meaningful mentoring opportunities. Studies (see Kezar & Maxey, 2014) have shown the importance of meaningful faculty-student interactions; the amount of time spent and the quality of these relationships decreases student dropout rates and increases their persistence toward degrees. Student-faculty interactions are especially valuable for first-generation students' realizing academic and social integration. The quality of these interactions is related to perceptions that they will accomplish short- and long-term academic goals (McKay & Estrella, 2008). High quality mentoring interactions could include engaging students in faculty research projects, attending conferences, and hosting mentoring sessions (online or face-to-face) where students can ask questions about careers, goals, or other important issues.

MPPCs may experience feelings of self-doubt and a lack of connection, so exposure to positive role models (especially those who can share insights of adversity), including MAL faculty, is an important step in providing support. These relationships can encourage students by offering an opportunity for dialogue, support, and resilience (McAllister & McKinnon, 2009). McNulty (2012) suggests that trailing spouses perceive professional support and social support to provide the greatest influence on reconstructing one's own identity, through finding a way to belong and create a life of one's own. Faculty (and staff) can provide professional and social support within the academic setting through engaging students in research projects and teaching assistantships, advising student clubs, and/or leading internships. Service-learning courses may also be particularly impactful in facilitating the positive outcomes of faculty-student interactions (McKay & Estrella, 2008).

Second, enabling MPPCs to develop relationships with their peers in (and out of) the classroom could positively impact their sense of connection. Many campuses can feel socially segregated. These feelings may be exacerbated for MPPCs who often represent a minoritized group on campus. Engaging with diverse students thus becomes a necessary component for creating a sense of belonging and connection (Shapiro, 2019). This is especially valuable given the links to resilience (Hartley, 2010), personal growth, and academic performance (Shapiro, 2019). As such, it is important to facilitate interactions among MPPCs and civilian students that can aid in building connections and potentially dispelling misconceptions or erroneous judgments about military spouses (Gleiman & Swearengen, 2012).

One way to encourage peer relationships is to make time for group work and collaborative learning in the classroom. Using learning management systems and document sharing (e.g., Google Docs) for discussion and group work would help facilitate these connections in virtual environments as well. Group work and team learning will help military spouses connect with other students and participate in meaningful dialogues where critical reflection can occur (Gleiman & Swearengen, 2012). When students interact with classmates and gain peer support they can enhance the shared experience of learning, which is associated with resilience (Hartley, 2010). It may also, however, feel uncomfortable at times (Shapiro, 2019), so faculty should be ready and willing to encourage problem-solving, perspective-taking, and dialogue when disagreements occur. Creating this space for inter-group dialogue—positive and negative—may help further the social connections among students that instructors are trying to promote (Shapiro, 2019), which ties to the next two strategies.

Promote Reflection

The use of narratives and self-reflection in writing assignments (e.g., journaling, reaction papers) may benefit military spouses because they will be able to bring their own identities into their work. Reflection offers the opportunity for MPPCs to develop and affirm identity anchors, which can aid resilience (Buzzanell, 2018). By affirming identity anchors with others, people enact the characteristics and roles most meaningful to them in times of difficulty (Buzzanell, 2018). Identity-building work in the classroom engages students in meaningful questions, including "Who am I?" and "What are my aspirations?" which can enrich educational experiences and promote positive identity and adaptation (McAllister & McKinnon, 2009). Completing personal reflections (e.g., minute exercises) on readings, for example, may allow MPPCs to discover how their own circumstances relate to the subject material (i.e., identification with course content). Furthermore, sharing personal narratives through storytelling in an accepting environment can offer therapeutic benefits and help individuals develop resilience (East, Jackson, O'Brien, &

Peters, 2010). When MPPCs disclose their emotional experiences, they may be better equipped to foreground positive action while backgrounding negative feelings or express the contradictions, frustrations, and hurt associated with feeling a lost sense of self amidst their struggles with identity, responsibilities, and school (Buzzanell, 2018).

These exercises can also illuminate specific needs and experiences of MPPCs, enabling faculty (and/or other students) to better understand this population. Through reading reflections and stories, faculty are also placed in a position to legitimize MPPCs' negative feelings and promote productive action (Buzzanell, 2018). In other words, upon reading personal reflections faculty can acknowledge adversity, and encourage students to reframe these challenges to focus on positives while maintaining a sense of hopefulness (Buzzanell, 2010). This could be particularly useful in circumstances when MPPCs may disclose feeling overwhelmed or the desire to quit school due to the emotional load they are bearing. MPPCs may also be more comfortable seeking support from faculty who know more about their personal circumstances. Developing MAL faculty networks might provide one way in which to better foster these feelings of trust and support. It is important to consider that relationships built between faculty and students will promote the trust needed to make reflection (and responses to it) more open and comfortable.

Normalize Challenges and Failure

Buzzanell (2018) suggests that crafting normalcy involves language and routines, interactions and rituals, and storytelling to construct a new normal that integrates one's challenging circumstances. MPPCs experience many challenges, so coping through predictable discourse, coming to terms with ongoing changes, and avoiding uncertainty may help them adjust to daily routines (Villagran, Canzona, & Ledford, 2013) and the additional strains of college. To help facilitate this process, faculty can create clear guidelines and hold students accountable for their responsibilities. This could help diminish ambiguity (Bondy, Ross, Gallingane, & Hambacher, 2007) and bring a sense of control to this aspect of MPPCs' lives. Faculty can also provide opportunities for *all* students to discuss challenges so that they can learn from each other's trials and strategies for overcoming them.

Because resilience involves adapting to stressful circumstances, which can sometimes involve failure, students must learn how to recognize and recover from these experiences. Providing low stakes opportunities for students to try and fail (or succeed) could help normalize the failure process; showcasing that most success does not come on the first attempt. Normalizing failure and offering support could help MPPCs avoid the "'sink or swim' attempts to cope with the stressors, strains, and daily hassles," which can occur when support for MPPCs experiences is inadequate (McNulty, 2012, p. 430). Furthermore, finding new ways to approach problems, challenge their current status, and fulfill responsibilities may enable MPPCs to put alternative logics to work—another process of resilience (Buzzanell, 2018). Scaffolding large projects, creating practice exams/quizzes, providing individual and group exercise opportunities, and promoting study sessions may help enable these practices of trying something new. Faculty can also share their own trials and errors to illustrate that failure and challenges are a normal part of the academic journey.

FUTURE RESEARCH/LIMITATIONS

It is important to note that the included recommendations are only a starting point for continuing to support the MAL (and specifically, MPPC) community. In order to strengthen these recommendations, additional research is needed that expands upon existing findings and suggestions with larger and more diverse samples. For example, our interview-based data included a sample that was composed of female participants. While our data mirrors larger trends in the military community, as 92 percent of active-duty military partners are female (Department of Defense, 2018), additional research is needed that examines male military partners' experiences (especially those who are pursuing college). Despite attempts to reach participants across the country, and military branches, our data only represent three military branches (Army, Marines, and Air Force). Experiences may differ across branches, particularly in terms of the community they reside within and challenges related to relocation, training schedules, or deployment. A larger, more diverse sample would enable comparisons and more nuanced recommendations to be made. Our study captured the experience of military partners rather than military personnel or family members (e.g., children), whose experiences and needs on campus may differ. Future studies would benefit from exploring and comparing a wider range of MAL experiences in higher education. It is especially important to collect data with more MALs and MPPCs who are currently enrolled in and experiencing the challenges and successes of pursuing a degree. As the data described in this chapter were collected as part of a larger study not specifically focused on experiences in higher education, information about employment and education status was not collected. This is a limitation that should be addressed with future research. Finally, future researchers should assess the effects (e.g., academic success, well-being, belonging, retention) of the suggested resilience-based strategies. Understanding how these strategies are influencing MALs and MPPCs will help further our understanding of best practices with this unique population.

CONCLUSION

The goal of higher education is to offer the opportunity for all students to succeed. While all students are faced with challenges, our data suggest that military partners experience complex (inter)personal struggles that may impede their ability to adapt to and succeed in college. We have suggested three broad strategies (e.g., develop relationships, promote reflection, normalize challenges, and failure) that faculty can integrate into their classroom routines to promote the support that students (MPPCs specifically) need to persist through college (Douce & Keeling, 2014). These strategies should be tested and validated prior to implementation to assess their impact on improving student resilience and retention, particularly for unique student populations.

REFERENCES

Agbisit, M. D. (2019). *Exploring the lived experiences of career oriented military spouses pursuing education for career advancement.* [Doctoral Dissertation, Old Dominion University]. ProQuest Dissertations and Theses.

American Council on Education. (2018). *Toolkit for veteran friendly institutions.* https://www.acenet.edu/Documents/Veterans-Toolkit-2018.pdf.

American College Health Association. (2016). *American college health association-national college health assessment II: Reference group executive summary.* American College Health Association.

Battistisch, V., Solomon, D., Watson, M., & Schaps, E. (1997). Caring school communities. *Educational Psychologist, 32*(3), 137–151.

Baxter, J., & Eyles, J. (1997). Evaluating qualitative research in social geography: Establishing 'rigour' in interview analysis. *Transactions of the Institute of British Geographers, 22*, 505–525.

Blakely, G., Hennessy, C., Chung, M. C., & Skirton, H. (2012). A systematic review of the impact of foreign postings on accompanying spouses of military personnel. *Nursing & Health Sciences, 14*(1), 121–132.

Bondy, E., Ross, D. D., Gallingane, C., & Hambacher, E. (2007). Creating environments of success and resilience: Culturally responsive classroom management and more. *Urban Education, 42*(4), 326–348.

Bouchard, M. (2019). Navigating academic and personal life: A personal journey. In L. H. Hernández & J. Belding (Eds.), *Military spouses with graduate degrees: Interdisciplinary approaches to thriving amidst uncertainty* (pp. 15–36). Lexington.

Buzzanell, P. M. (2010). Resilience: Talking, resisting, and imagining new normalcies into being. *Journal of Communication, 60*(1), 1–14.

Buzzanell, P. M. (2018). Communication theory of resilience: Enacting adaptive-transformative processes when families experience loss and disruption. In D. Braithwaite, E. Suter, & K. Floyd (Eds.), *Engaging theories in family communication* (2nd ed., pp. 98–109). Routledge.

Cate, C. A. (2011). *Student veterans' college experiences: Demographic comparisons, differences in academic experiences, and on-campus service utilization.* [Doctoral Dissertation, University of California]. ProQuest Dissertations and Theses.

Cefai, C. (2007). Resilience for all: A study of classrooms as protective contexts. *Emotional and Behavioural Difficulties, 12*(2), 119–134. doi:10.1080/13632750701315516.

Cooke, T. J., & Speirs, K. (2005). Migration and employment among the civilian spouses of military personnel. *Social Science Quarterly, 86*(2), 343–355.

Defense Manpower Data Center. (2018). *DoD personnel, workforce reports, and publications.* https://www.dmdc.osd.mil/apps/dwp/dwp_jsp.

Department of Defense. (2018). *Military spouse demographics and employment information.* https://www.whitehouse.gov/wp-content/uploads/2018/09/DoDMilitarySpouseDemographicsandEmployment_20180912.pdf.

Douce, L. A., & Keeling, R. P. (2014). *A strategic primer on college student mental health.* American Council on Education. https://www.naspa.org/images/uploads/main/Campus_Mental_Health_Primer_web_final.pdf.

East, L., Jackson, D., O'Brien, L., & Peters, K. (2010). Storytelling: An approach that can help to develop resilience. *Nurse Researcher, 17*(3), 17–25.

Friedman, E. M., Miller, L. L., & Evans, S. (2015). *Advancing the careers of military spouses: An assessment of education and employment goals and barriers facing military spouses eligible for MyCAA.* RAND Corporation.

Ford, D., Northrup, P., & Wiley, L. (2009). Connections, partnerships, opportunities, and programs to enhance success for military students. *New Directions for Student Services, 126*, 61–69.

Ford, K., & Vignare, K. (2015). The evolving military learner population: A review of the literature. *Online Learning, 19*(1), 7–30.

Gleiman, A., & Swearengen, S. (2012). Understanding the military spouse learner using theory and personal narratives. *New Directions for Adult and Continuing Education, 136*, 77–88.

Goodenow, C., & Grady, K. E. (1993). The relationship of school belonging and friends' values to academic motivation among urban adolescent students. *Journal of Experimental Education, 62*, 60–71.

Gorman, L. A., Blow, A. J., Ames, B. D., & Reed, P. L. (2011). National Guard families after combat: Mental health, health services, and perceived treatment barriers. *Psychiatric Services, 62*(1), 28–34.

Hartley, M. T. (2010). Increasing resilience: Strategies for reducing dropout rates for college students with psychiatric disabilities. *American Journal of Psychiatric Rehabilitation, 13*(4), 295–315.

Huebner, A. J., Mancini, J. A., Wilcox, R. M., Grass, S. R., & Grass, G. A. (2007). Parental deployment and youth in military families: Exploring uncertainty and ambiguous loss. *Family Relations, 56*, 112–122.

Irvin, M. J., Meece, J. L., Byun, S. Y., Farmer, T. W., & Hutchins, B. C. (2011). Relationship of school context to rural youth's educational achievement and aspirations. *Journal of Youth and Adolescence, 40*, 1225–1242.

Kadison, R., & DiGeronimo, T. (2004). *College of the overwhelmed: The campus mental health crisis and what to do about it*. Jossey-Bass.

Kezar, A., & Maxey, D. (2014). Faculty matter: So why doesn't everyone think so. *Thought & Action, 2014*, 29–44.

Johnson, S. J., Sherman, M. D., Hoffman, J. S., James, L. C., Johnson, P. L., Lochman, J. E., Magee, T. N., & Riggs, D. (2007, February). *The psychological needs of U.S. military service members and their families: A preliminary report.* American Psychological Association Presidential Task Force on Military Deployment Services for Youth, Families, and Service Members.

Lindlof, T. R., & Taylor, B. C. (2002). *Qualitative communication research methods.* Sage.

Marshall, S. L. (2014). Helping military spouses address employment challenges through career communications. *Career Planning and Adult Development Journal, 30*(2), 85–90.

Masten, A. S. (2001). Ordinary magic: Resilience processes in development. *American Psychologist, 56*(3), 227–238.

Masten, A. S., & Reed, M. J. (2002). Resilience in development. In C. R. Snyder & S. J. Lopez (Eds.), *Handbook of positive psychology* (pp. 74–88). Oxford Press.

Matchett-Taylor, S. T. (2018). *A phenomenological study on military spouses and their journeys to post-high school degree attainment.* [Doctoral Dissertation, Northcentral University]. ProQuest Dissertations and Theses.

May, A., & McDermott, V. (2019). Cultural collision: A review of literature. In L. H. Hernandez & J. Belding (Eds.), *Military spouses with graduate degrees: Interdisciplinary approaches to thriving amidst uncertainty* (pp. 59–85). Lexington Books.

McAllister, M., & McKinnon, J. (2009). The importance of teaching and learning resilience in the health disciplines: A critical review of the literature. *Nurse Education Today, 29*(4), 371–379.

McKay, V. C., & Estrella, J. (2008). First-generation student success: The role of faculty interaction in service learning courses. *Communication Education, 57*(3), 356–372.

McNulty, Y. (2012). 'Being dumped in to sink or swim': An empirical study of organizational support for the trailing spouse. *Human Resource Development International, 15*(4), 417–434.

McWhirter, K. (2019). *Empowering military spouse education and employment.* https://www.learnhowtobecome.org/career-resource-center/military-spouse-education-employment/.

Meadows, S. O., Griffin, B. A., Kearney, B. R., & Polla, K. J. (2016). Employment gaps between military spouses and matched civilians. *Armed Forces and Society, 42*(3), 542–561.

Mercer, D., Thurston, L., & Hughey, J. (2014). The college-wide military-connected learner initiative in K-State's College of Education. *The Advocate*, *22*(2), 24–30.

Military One Source. (2019). *PCS: The basics about permanent change of station.* https://www.militaryonesource.mil/moving-housing/moving/planning-your-move/pcs-the-basics-about-permanent-change-of-station.

Military One Source. (n.d.). *Spouse education and career opportunities.* https://mycaa.militaryonesource.mil/mycaa.

Molina, D., & Morse, A. (2015). *Military-connected undergraduates: The current state of research and future work. Convening summary*. American Council on Education. https://www.acenet.edu/Documents/Military-Connected-Undergraduates-Research-Convening-Summary.pdf.

O'Keeffe, P. (2013). A sense of belonging: Improving student retention. *College Student Journal, 47*(4), 605–613.

Olsson, C. A., Bond, L., Burns, J. M., Vella-Brodrick, D. A., & Sawyer, S. M. (2003). Adolescent resilience: a concept analysis. *Journal of Adolescence, 26*(1), 1–11.

Palmer, C. (2008). A theory of risk and resilience factors in military families. *Military Psychology, 20*, 205–217. doi:10.1080/08995600802118858.

Patterson, J. M. (2002). Understanding family resilience. *Journal of Clinical Psychology, 58*, 233–246.

Richardson, G. E. (2002). The metatheory of resilience and resiliency. *Journal of Clinical Psychology*, *58*, 307–321.

Richmond, A., Mischel, E., Smischney, T., Lu, Z., Root, H., Otto, M., … Borden, L. (2017). *Military spouse licensure portability examination.* Report submitted to the U.S. Department of Defense.

Scarf, D., Moradi, S., McGaw, K., Hewitt, J., Hayhurst, J. G., Boyes, M., Ruffman, T., & Hunter, J. A. (2016). Somewhere I belong: Long-term increases in adolescents' resilience are predicted by perceived belonging to the in-group. *British Journal of Social Psychology, 55*(3), 588–599.

Shapiro, S. (2019). Why building community – even through discomfort – could help stressed college students. *The Conversation.* http://theconversation.com/why-building-community-even-through-discomfort-could-help-stressed-college-students-121398.

Stanford University Office for Military-Affiliated Communities. (n.d.). *Are you military-affiliated? What does that mean?* https://military.stanford.edu/stafffaculty/are-you-military-affiliated-what-does-mean.

Strauss, A. L., & Corbin, J. (1990). *Basics of qualitative research.* Sage.

Thielking, M. (2017). *Surging demands for mental health care jams college services.* https://www.scientificamerican.com/article/surging-demand-for-mental-health-care-jams-college-services/.

U.S. Department of Defense. (2009). *Department of defense report of the 2nd quadrennial quality of life review*. Department of Defense Procurement and Personnel Statistics.

U.S. Department of Defense. (2018). *2018 demographics: Profile of the military community*. https://download.militaryonesource.mil/12038/MOS/Reports/2018-demographics-report.pdf.

U.S. Department of Veteran Affairs. (2018). *Annual benefits report: Education*. https://www.benefits.va.gov/REPORTS/abr/docs/2018-education.pdf.

Villagran, M., Canzona, M. R., & Ledford, C. J. (2013). The milspouse battle rhythm: Communicating resilience throughout the deployment cycle. *Health Communication, 28*(8), 778–788.

Wiens, T. W., & Boss, P. (2006). Maintaining family resiliency before, during, and after military separation. In C. Castro, A. Adler, & T. Britt (Eds.), *Military life: The psychology of serving in peace and combat (Vol. 3): The military family* (pp. 13–38). Praeger Security International.

Part II

EDUCATOR AND SUPPORT STAFF EXPERIENCES WITH MALs

Chapter 6

Speaking Chicana/o Secrets to Military Students

An Autoethnography of Teaching Senior Military Noncommissioned Officers

Frank G. Pérez

Mexican Americans have a distinguished history of service in the U.S. armed forces. Proportionately, they received the most U.S. Congressional Medals of Honor during World War II (Rivas-Rodríguez, 2005, p. xvii) and are among the most decorated military populations (Morin, 1963/2013). However, only a small number of academic studies (e.g., Mariscal, 1999; Rosales, 2017) and popular literature (e.g., Harrison, 2014; Ortega & Hernández, 2015) document their exploits. This chapter uses a cultural studies perspective and autoethnographic methodology to offer suggestions for incorporating critical/cultural studies, such as Chicana/o Studies, into the classroom curriculum to help students develop intercultural communication competence and challenge the dominant narrative by introducing different ways of knowing. My goal is to offer educators potentially useful strategies they may wish to employ to teach students about diverse subjects, such as African American Studies or Women's Studies. This chapter provides a brief overview of Mexican American military contributions and experiences to reveal a hidden or "secret" history missing from mainstream public memory and higher education curricula. I build on González's (1998) notion of speaking secrets (i.e., addressing sensitive or unknown topics in university and professional settings) to bring light to this important history. The chapter recounts my family's experiences in the U.S. military and its influence on me as a scholar. An autoethnographic retelling of my experiences provides examples that educators may find useful as they consider the speaking secrets pedagogy offered here. The chapter concludes with a summary of the lessons I have learned in a decade of teaching military students.

PUBLIC MEMORY AND MEXICAN AMERICANS IN THE U.S. MILITARY

U.S. public memory remains largely unaware of Mexican American sacrifices via military service. Worse yet, Mexican American combat veterans have repeatedly returned home to find U.S. racial structures and racism intact, denying them the freedoms they fought to protect. For example, after the Second World War, Mexican American veterans were denied the entitlements of the G.I. Bill because of racism against their ethnicity (Rosales, 2011). There is a small but growing amount of academic literature on Latinx peoples' military service (Lutz, 2008). Army veteran Raúl Morín (1963/2013) recounted Mexican American acts of military bravery in *Among the Valiant*. Scholar Rivas-Rodríguez (2005) edited an anthology exploring the Mexican American experience during World War II. Mariscal (1999) addressed sixty-six items related to the Mexican American experience during the Vietnam War. Ultimately, the U.S. mainstream culture lacks awareness of the sacrifices this demographic category has made in defense of the nation.

Moreover, negative stereotypes of Mexican Americans as "immoral, violent, given to fighting, dirty, unintelligent, improvident, irresponsible, and lazy" (Barnes, 1969, cited in Cross & Maldonado, 1971, p. 27) have been prevalent in U.S. popular culture (Acuña, 2000). Dating back to the 1820s, Mexican Americans have been stereotyped as greasy, swarthy, and inept (Acuña, 2000), and these stereotypes were incorporated into U.S. cinema in the 1890s (Woll, 1977). These negative media representations continue in the current era (Guo & Harlow, 2014; Hall, 2018; Mastro & Behm-Morawitz, 2005) and affect daily Mexican American interactions with other groups (Erba, 2018; Romero et al., 2015). Today, young Mexican American males remain stereotyped as a threat to the U.S. social order (Erba, 2018; Maldonado, 2011). Latina/o military personnel, regardless of their national origins, often struggle against a racially biased mainstream system. For example, in 2019, National Public Radio (NPR) reported on noncitizen veterans being deported by the U.S. government after they served in ongoing battles in the Middle East. Overall, the negative views many hold against Latinas/os, of which Mexicans/and Mexican Americans[1] are the numerical majority, hinders this social category's chances for upward social mobility (Carnevale et al., 2018; Chávez, 2011). Providing knowledge about this group's positive impact and struggles is a crucial step toward mainstream acceptance. Private Félix Z. Longoria's case is a key illustration. Longoria was killed in action during World War II. Yet, the only funeral home in his hometown, Three Rivers, Texas, denied his wife the use of its racially segregated chapel (Green, 1991). Owner Tom Kennedy explained, "The

Whites wouldn't like it" (Carrol, 2003, p. 46). This and other injustices triggered an era of Mexican American civil rights advocacy after WWII. Groups such as the American GI Forum and the League of United Latin American Citizens flourished in the 1950s. Later, the National Council of La Raza, the United Farmworkers, and others joined the struggle. By the 1980s, many "Hispanic"[2] organizations were forced to rely on corporate support and lost some independence (Griswold del Castillo, 1993). The contemporary era poses new challenges as anti-immigrant legislation (López, 2011) and anti-Latina/o hate crimes have spiked under the Trump administration (Koski & Bantley, 2020).

THEORETICAL FRAMEWORK

This chapter recounts my experiences teaching noncommissioned officers (NCOs) in a Master of Arts program at the University of Texas at El Paso (UTEP) from 2008 to 2018. It offers insights for including Chicana/o Studies or other non-mainstream academic areas in traditional classes. Insights from Chicana theorists and hegemonic theory guide the chapter. It employs a cultural studies approach that focuses on "speaking secrets," González's (1998) notion of bringing controversial, unpopular, or often overlooked discussions to light. In her work, González (1998) calls out Chicanos (politically aware Mexican American men) for their sexist attitudes at a panel discussion at the National Association for Chicana and Chicano Studies (NACCS). In doing so, she highlights the attitude change needed for the field of Chicana/o Studies to move past such open sexism. Her willingness to speak about "secrets" related to sexism on the panel was a key step in fostering social change at NACCS. Similar actions have taken place throughout academia. For example, Chicana scholar Córdova (1998) advocates for frank dialogue about difficult or taboo subjects to improve higher education spaces. She addresses issues faced by students and faculty of color by building on Foucault's (1972) *regimes of truth*—the idea that societies establish processes to determine who can speak "truth"—and combining it with colonialism. In Western societies, scholars have the unique responsibility of creating truth. However, the colonial legacy and the resultant Eurocentric bias in academic work distorts it.

To address this bias, Córdova (1998) categorizes academic roles via a colonial perspective. She identifies three core academic tenets that silence people of color. *Objectivism* operates via narrative and other structures that offer knowledge as truth, free from cultural bias. *Sanctity of the curriculum* refers to the notion that universities teach the best or important works in Western culture, preserving academia's Eurocentrism. Finally, *cries for standards* is code for minorities are undeserving benefactors of affirmative action on campus. This reality is further complicated because many comply with the system, willingly, knowingly, or not. Córdova (1998) calls for change by categorizing how academics and administrators execute their roles using colonial metaphors, among them rebels and missionaries. The Chicana/o experience remains a secret that is often forgotten or hidden in universities. I focus my pedagogy on being a rebel who challenges erroneous information about Mexican Americans. Also, as a missionary, I proselytize by spreading Chicana/o knowledge to the students. One useful strategy is to bring Chicana/o case studies, videos, and similar items into mainstream classes. Thirty years after González's (1998) and Córdova's (1998) landmark essays, the struggles in higher education continue (Carnevale et al., 2018; Ojeda et al., 2014); as such, I speak secrets. For example, students of color constitute a minority of the student population on many campuses, particularly at Tier I or elite universities. This underrepresentation is important because students who graduate from the nation's top schools earn disproportionately higher salaries and entrance to top tier universities. Often the access to elite colleges and universities is tied to social class and race (Carnevale et al., 2019; Carnevale & Strohl, 2013). A majority of the nation's people of color continue to attend less elite institutions widening the class/racial divide in U.S. society (Carnevale et al., 2019).

AUTOETHNOGRAPHY

My goal is to bring awareness to Mexican American military history and to share my experiences. This objective stemmed from my family's military tradition and my social justice orientation as a Chicano working-class academic (see Aguirre, 2017). If military senior leaders learn and share information about Mexican Americans' role in military history, they may act as opinion leaders, promoting this knowledge with their colleagues. This is an important task because "Chicanos have been silenced not only by the grave, but by political transformation, social dispossession, cultural rupture, and linguistic alienation" (Padilla, 1988, p. 286). Writing one's stories lends "textual permanence" (Padilla, 1998, p. 286) to these experiences. Literary testament is harder to ignore than oral accounts. Sharing one's experiences is a form of speaking secrets that validates one's experiences. It empowers others to learn from our successes and mistakes and is a central strategy in Chicana/o Studies (Noriega, 2004; Noriega et al., 2020).

This chapter uses autoethnography, relying on my notes from this time and my personal recollections and interpretations of the recounted events. The retelling of one's experiences to explain cultural norms in a systematically analytical manner is a core tenet of autoethnography. The method generates research that sheds light on unique experiences (Ellis et al., 2011; Ellis & Bochner, 2000). Chang (2013/2016) suggests that autoethnography needs to: (a) use the researcher's personal experiences and memories; (b) seek to increase the understanding of social life; and (c) can be written using various approaches. Specifically, I reviewed notes in a personal journal where I wrote about my experiences and focused on key events related to teaching at Fort Bliss. The notes triggered memories that I then revisited by writing down what I recalled and answering the following questions:

1. What happened?
2. Who was involved?
3. How did race/ethnicity, gender, rank, or other factors appear relevant to the event(s)?
4. How can these recollections illustrate my efforts to speak secrets as a pedagogy?

The answers were initially grouped by similarity of topic or context. They were then categorized by focusing on the context of what was said, leading to the development of three categories. *Professional duties* refers to comments about a person's current or imagined future job duties. *Educational links* refers to comments that suggested students linked their lived experiences with a class reading or topic. Finally, *political orientation* references when students shared comments that suggested their sociopolitical orientation, such as when students voiced support for a particular social or political idea. These findings are addressed throughout the chapter. Additionally, I was careful to avoid the two key concerns of autoethnographic research. First, unable to determine the readers' reactions to my recollections (Méndez, 2013), I chose examples that I believed would clearly relate key concepts or ideas. Second, I selected only items that I am comfortable sharing via an honest effort at self-disclosure (Méndez, 2013). My goal was to frame events as I remember them not in a way that benefits me or the people involved. Finally, the examples omit details and use alternate names to help preserve people's anonymity. I taught over 300 military personnel over a ten-year period. Because they are a tightly knit group of professionals, one primary goal was to share very few potentially identifying details.

Situating Myself and My Inquiry

In 2008, my college dean invited me to teach graduate classes for the previously mentioned leadership studies program UTEP had developed for NCOs completing command sergeant major training at nearby Fort Bliss. Teaching military members gave me the opportunity to speak openly and factually with them about Mexican American military and civilian successes. During this time, my classes sought to expand students' understanding of Mexican Americans and their achievements in the U.S. military. I often shared stories from my family's military experiences and my desire to "help in some small way" to show my support for our military. These interactions established my credibility with students.

My family has served in the U.S. military since at least the late 1800s. The Treaty of Guadalupe Hidalgo ended the U.S. War against Mexico on February 2, 1848, creating the first generation of Mexican Americans (Griswold del Castillo, 1992). Like many Mexican Americans, we have been in the United States for close to 175 years. My family's oral history traces my ancestors to the El Paso area since at least the 1850s. We have no knowledge of our forebears prior to that. An unknown ancestor served in the U.S. Army in the late 1800s. His story faded with my father's passing in 2015. Eulalio Pérez, my "*Tio Lalo*," served during World War I. A portrait of him at eighteen, in his Army uniform, hangs in our home. He drove a wagon during the war and had to place gas masks on the horses before he put on his own during chemical attacks. Men were cheaper and easier to replace than horses. Lalo returned to El Paso but died in his twenties from health complications from his exposure to mustard gas. During World War II (1939–1945), Ramón "Ray" Pérez, my great uncle, served in the Army in the Pacific theater. My father's cousins, María José "Chema" and Marcelino "Chalino" Pérez, served in the Army during this war. Chema was killed in action, and Chalino fought against the Nazis in Europe. My uncle Jessie Tony "Jeff" Jeffries, married to my aunt, Josephine "Josie" Pérez, also served in Europe and suffered life-altering injuries repairing a stalled tank during a firefight with Nazis. My father, Manuel Antonio Pérez, served in the Air Force during the Korean War, and my uncle and *padrino* [godfather] Guillermo "Memo" De La Rosa served in the Navy during the same conflict. My uncles Federico "Lico" and José De La Rosa and my cousin Jaime De La Rosa served in Vietnam. My cousin, Arturo "Art" De La Rosa, served in the Air Force immediately after Vietnam. Finally, my cousin, Xavier Aguilar, served in the Air Force during Desert Storm. It is to them, their service, and sacrifices that I dedicate this chapter.

Learning about the deplorable conditions that U.S. soldiers faced during World War I and relating these experiences to my Tío Lalo's sacrifices made me proud to be his distant nephew. As a child, my Aunt Josie would tell me about my Uncle Jeff's experiences fighting in the European theater. In the Pacific theater, my Uncle Ray woke to find the soldier next to him had been murdered during the night. According to Uncle Ray, the (presumably) Japanese soldier who committed this action spared my uncle to humiliate him. I do not know the extent of the event's psychological impact on him. Ray shared the story with my dad and believed the act was aimed at him as a U.S. soldier. Race was not a factor to Uncle Ray because he said other soldiers had faced similar acts. However, even during a world war, Ray did face racial discrimination. While in Australia, a U.S. Army captain called him a "wet back." The situation escalated and my uncle punched the captain, resulting in Ray's demotion from sergeant to corporal.

During the Korean Era, a sergeant told my father to clean an area outside the building where my father worked as an administration clerk. My father explained that his job was not to clean gardens. The sergeant became upset and told my father, "Look, Pancho [*sic*], the only thing you Mexicans are good for is gardening. So do as you're told!" My father walked away and reported the sergeant to the officer of the day (OD). A while later, the sergeant and my father were called to meet with him. The OD forced the sergeant to apologize and asked my father if he wanted to file a formal complaint. Such an action would likely result in the sergeant losing a rank or "stripe," according to the OD. My father replied, "No, sir, this man has a family and I would hate to negatively impact his wife and kids." As they left the meeting, my father approached the sergeant, pointed at the stripes on the man's sleeve, and said, "Don't ever forget a Mexican gave you that stripe." This and other stories where my family served in battle or fought against racism made me patriotic and social justice oriented. I am proud of my family's service and know of the racism they encountered while in uniform. Their contributions and those of many other Mexican Americans motivate me to help make their military service part of U.S. public memory. Such recognition would likely lessen the racism against Mexican Americans and other Latina/o peoples.

I skipped military service and opted for university. School made me a politically aware Mexican American or Chicano (see Garcia, 1997; Rinderle, 2006). Chicanas/os are Americans of Mexican descent who actively promote social justice. They are more politically active than Mexican Americans. This distinction is important because Chicanas/os are Mexican Americans, but not all Mexican Americans are Chicana/o (Urrea, 2007). As a Chicano, I use my academic position to correct misperceptions about Mexican Americans and to advocate for social justice. Autoethnographic accounts promote these goals by sharing personal experiences, particularly in areas with a paucity of academic information.

RESULTS AND DISCUSSION

My research led me to identify three key elements that I will address here and return to throughout the remainder of the chapter. First, my experiences teaching military students, combined with my interest in bringing cultural studies (i.e., Chicano Studies) into the classroom led me to develop a series of strategies to reach that goal. The result of these efforts is my speaking secrets pedagogy. Second, Chicano history remains hidden or a secret. After a decade of teaching military students, my observations suggest that few of my students are familiar with Mexican American contributions to the military or U.S. society. Finally, military culture provided a guiding framework for in-class discussions with military students. I relied on this culture's focus on mutual respect to address social justice issues after gaining their trust as an instructor. It is also important to allow the class to build a sense of community by allowing time for students to trust one another.

TEACHING CHICANA/O STUDIES TO MILITARY PERSONNEL

I taught classes in leadership communication, complexity theory, and small group communication in the MA program. Primary texts included academic books and bestsellers on leadership topics. I also implemented case studies, supplemental readings, and videos with a Chicana/o perspective whenever possible. For example, I used Charlie Trujillo's (2003) film *Soldados: Chicanos in Viet Nam* as a tool for talking about Post-Traumatic Stress (PTS). This first strategy for speaking secrets allowed me to speak Chicano secrets to students whose trust and respect I had earned. The Chicano veterans in the film spoke their own secrets as well. My students came to respect me and spoke positively to other students about my courses. My second strategy is to build trust with students. Not every instructor will have the benefit of students linked by space and cohort to speak positively about them to other students. Building trust with each student and as a class facilitates one's efforts to use this pedagogy. I repeatedly told my students that the classroom was a space to explore ideas, to follow "Army values," and to be respectful. Once I established trust, students learned about Chicanas/os in uniform and their service. This type of reading also helped students empathize with the lived experiences of people who may be different from them, using such readings as Rosales's (2011) essay on the denial of veterans' rights to Mexican Americans after WWII. I framed the discussion as something that many people know happened to Black soldiers then spoke about how Mexican American soldiers experienced the same injustices. I asked my students to consider the ethical and leadership aspects of this discrimination.

Today, Chicano history remains a secret, hidden or forgotten from public memory. On many campuses, one must actively seek out Chicano Studies courses to cultivate a comprehensive understanding of the Mexican American experience. Also, the term Chicano is taboo in many mainstream contexts and is relegated to a marginalized space. In the El Paso/Ciudad Juárez area, where UTEP is situated, for example, many students reject the label. The term is often linked to *cholos* (Mexican American gang members or other social undesirables), a low socioeconomic status, or to being a social agitator. After learning about the Chicano Movement, many students develop a more positive attitude toward the term, and some adopt it as a self-referent. As González (1998) argues, "If we are to change the institutions of higher education in this society, spaces need to exist for new dialogues" (p. 46). Building on this idea, I revealed Chicano secrets or knowledge about Mexican Americans to military students and created safe spaces for the discussion of brown bodies in green uniforms. Specifically, I exposed students to Mexican American military heroes, to borderlands culture, and to works that show how Chicano views are consistent with U.S. culture. For example, my classes included academic and non-academic readings about Mexican American military heroes (e.g., Pvt. Marcelino Serna, Cpl. Guy Gabaldon, MSG. Roy Benavides). We explored how colonial machinations laid a foundation for Spanish overlords in the late 1500s and Anglo domination of Mexicans in the Southwest in the 1850s. Like the Spanish, Anglos placed themselves structurally and culturally above Natives and mestizos on the social hierarchy (Gordon, 1999) and then blamed Mexican/Americans for their low socioeconomic status.

My emphasis on social justice leads me to challenge commonly held misperceptions about Mexican/Americans by educating people about structural racism, narrative paradigms, and intercultural communication issues. Resistance to some topics can be viewed via hegemonic theory, the idea that a dominant perspective, beneficial to those who hold power at any particular historical moment, permeates society (Hall, 1996). Hegemony is context-specific, tied to a series of economic, social, and historical events. Dominant thought permeates society, including the military. Hegemony is dialogic and includes a combination of premises about the world (Hall, 1996). These premises allow people to ignore or reconcile the contradictions inherent to their worldview. For example, a person may not see themselves as racist but may display prejudice against one or more ethnic groups. Hegemonic thought excuses such views, possibly at the societal level, to help maintain the sociocultural system.

Counterhegemonic voices stand in opposition to or critique this domination and are typically omitted from or downplayed in mainstream discourse. I used this framing to categorize and make sense of how my students commented on controversial topics such as whether all U.S. citizens should be entitled to a free college education. This framing was useful for creating the space to hear and explore counterhegemonic voices inside the classroom. First, the secrets spoken in class countered mainstream perceptions of Mexican Americans. They were counterhegemonic. Second, I noted that students often upheld a mainstream (hegemonic) framing of class topics and societal norms within the classroom. Outside of class, however, some students voiced more progressive or counter-hegemonic points of view in my interactions with them. They shared secrets with me in ways they could not in class. Hearing counterhegemonic accounts allowed me insights into what my students could not share in the classroom.

Military culture framed reality for many of my students along with a mainstream point of view that made some topics sacrosanct. For example, international conflicts or military errors were consistently blamed on actors other than the United States throughout the ten-year period addressed here. Scholarly and popular writers have commented on this U.S. worldview (e.g., Chomsky, 2017). Some of my students were political liberals but most seemed conservative, a finding supported by academic and mainstream research (Bachman et al., 2000; Dorman, 1976; Igielnik et al., 2019). This is not surprising considering the rigid structure of military life and how it differs from university cultural expectations that students question and push social boundaries (Howe & Shpeer, 2019). All of my military students were respectful, and I adapted a stance that encouraged but did not require them to act as traditional university students. My military students were predominantly White, male, middle-aged, senior military leaders. My key goal was to provide academically rigorous classes that provided them with the required content using Chicano examples where possible.

STUDENT CONCERNS AND EXPECTATIONS

Overall, students mostly accepted and embraced the speaking secrets pedagogy. Speaking secrets helped to build a connection between course materials and online discussions to students' common military experiences: combat, professional experiences/duties at different ranks, and the transition from military to civilian life. For example, one student stated he was pursuing a graduate degree because, in his words, "I drive a tank, Dr. Pérez, what I can I do with that?" I would often remind my students that as NCOs they had supervised personnel (perhaps in life-or-death situations), traveled, and experienced international settings. As such, these traits would make them very marketable in the civilian world. The tank driver's comments led me to talk about differences between individualistic and collectivistic cultures. It also helped me illustrate the value of a military background:

> Hey, Dr. Pérez, what do you think about this? I was in Pakistan and I think I see your point on culture. Part of the winning of hearts and minds was for us to visit village elders. We would offer assistance with local projects to show that the US

> are not the bad guys. One thing I never got is why every time we visited it took all day. The key elder always insisted we talk with him [through interpreters] for a long time. They would serve us tea. It would take another long while to drink the tea. Finally, he would let us talk about the day's topic.

Using the student's experience, I offered an explanation of the differences between individualistic and collectivist cultures. The student responded with a very positive reaction. Smiling and saying, "Yes, that's exactly it!" I then linked this concept to understanding communities of color in the United States, noting that many people of color come from collectivistic cultures. The tank driver ended this discussion by noting that he now had an additional way to sell himself to potential employers. The discussion helped my students gain insights into several topics, including some considerations on how to interact with Mexican/Americans and other people of color.

On rare occasions, students did not like my approach. For example, I once had students read two editorials about state-level funding declines for higher education. One from the *New York Times* and one from *FOX News*, both pieces lamented the decreases. As the class discussed the topic, John (not his real name) said, "You know, Dr. Pérez, I found this a very difficult reading." I asked why and he replied, "Well, I think people should work for what they get. If you want a college degree, you need to be ready to pay for it." I acknowledged his point and offered that no other developed nation bankrupts its citizens for pursuing a university education. John became visibly upset and said, "Fine, I'll stop here, sir, but you sound like a communist." Always a courteous and hardworking student, John never spoke to me outside of class after this incident. Interestingly, John had no apparent problem with race but was concerned about political ideology. Still, speaking the secrets of Mexican American military contributions to students worked out well. Students respected me as both a professor and as someone who empathized with their sacrifices in defense of the nation. They also accepted non-mainstream information and counterhegemonic narratives because I provided it fairly and as useful to them.

Documentaries as Vehicles for Speaking Secrets

Documentaries have the unique ability to bring light to important but often lesser-known social problems. Additionally, films are effective in getting students engaged in a topic. Linking the film to leadership as an initial discussion point often put students at ease. I repeatedly used *The Last Conquistador* (2007; PBS Video), a documentary about a public monument in El Paso, to show leadership agendas and their relations to economic privilege, as well as to municipal-level politics. It allowed the class to explore El Paso's ethnic composition, local leadership, and community politics. These factors were relevant to students as they would spend ten months in the city and many had not previously interacted with Mexican/Americans. Class discussions began with a focus on traditional leadership questions: Who won? Who lost? What do you think was the most effective use of communication by either side? As the discussions evolved, the questions became more complex: Why do you think a small group of White community leaders leads a city that is 83 percent Mexican American? What does this situation suggest about power dynamics in the city? I would then explain the city's history of racial segregation and the Southwest's two-tiered social system (Hampton & Ontiveros, 2019). These items led to other leadership topics, such as ethics (i.e., Are White people today obligated to change a system that benefits them? What can Mexican American leaders do to usher change?). This video showed students the neocolonial and intercultural challenges leaders encounter in many places haunted by a similar or darker past during an era of shifting demographics. Then, my background of research (Pérez, 2007; Pérez & Ortega, 2020, 2008; Pérez & Ruggiero, 2007) that critiques El Paso's neocolonial reality helped to inform and lead these discussions.

This strategy worked well in all but one case. I covered my wife's class in the same MA program and decided to screen *The Last Conquistador* (2007; PBS Video). When the film ended, I asked how it related to leadership. A nonmilitary student and supporter of the Oñate statue and its related romantic framing of the Spanish colonial experiment (see Pérez & Ortega, 2008, 2020), argued that the film amounted to reverse racism. I explained that I disagreed, but the class sided with the student. The student continued: "The sculptor [John Houser] wanted to make a statue and that's what he did." I suggested that perhaps the sculptor should have considered the impact of creating a four-story tall statue of a Spanish colonizer for a city that is 83 percent Mexican/American. One military student suggested that I was a communist who wanted to censor art. At that point, the class ended. Regrettably, my wife faced a very hostile class for the seven remaining weeks. In retrospect, I should have covered the regular class topics because it was not my class. Being in charge for one night made it impossible for me to develop the rapport and trust needed to prod students out of their comfort zones. This experience taught me to build trust first and to avoid implementing this pedagogy too quickly or in other professors' classes. I now use it only with my students after building a sense of trust with them and tolerance for challenging topics. Another film I regularly used is *Soldados* (2003) a twenty-eight-minute film that examines the lives of five Chicano Vietnam veterans, their families, and the long-term effects of combat, including PTS. The film worked well because the soldiers' and families' sacrifices were relatable to students, promoting a sense of empathy. This example relates to the elements noted in the results and discussion section earlier in the chapter.

Exploring questions on ethnicity and other social topics, as the discussion progressed, helped me gauge student attitudes as we navigated difficult discussions. Class discussions addressed students' experiences leading or being led in combat situations and their ways of coping with combat stress. As the conversation continued, I would ask them about Mexican American war heroes: Could they name any? We would then talk about Mexican American soldiers in different military conflicts, notably World Wars II and Vietnam. I would ask them about their experiences in the Middle East and whether they served with any Latina/o or Mexican American soldiers and what their experiences had been. I asked students why they thought Mexican Americans received such little recognition. Typical answers were, "I don't know" or "I guess that much of this was just so long ago that no one remembers." African American, the dominant label used by my students, were often more likely to openly consider racism a factor than were White students. One student soldier who self-identified as African American told the class that buffalo soldiers, early African American soldiers (his term), were used to fight "Indians and Mexicans." He framed his example as one of "the military" using "Blacks to kill other people of color (also his term)." Two students nodded to acknowledge his comments, but no other students commented. Furthermore, no one commented on the student's use of different labels—African American and Black—to reference the same population. African American is used to denote U.S.-born Black people; Black is used to describe immigrants, particularly those who are stereotypically economically affluent (Lindsay, 2015).

I noted the framing used by the African American student. He offered a counterhegemonic example but placed blame on "the military" as an organization. He did not refer to the "Whites" in charge of the military. In private conversations, students often told me about their desire to frame comments in a respectful manner so as not to upset their class peers. Speaking secrets takes courage, and this student did his best to share an African American secret but was mindful to be counterhegemonic and diplomatic.

Case Studies

Case studies serve as an effective teaching tool to link theory with lived experiences. Yin (2017) states that case studies are ideal when "your questions require an . . . 'in-depth' description of some social phenomenon" (p. 4). Because most of the students had twenty-plus years of service, we examined cases from the military or private sector. The first is an area they lived in and the latter was their future. Finding Chicano cases was not always easy. Still, I developed my own in relation to Private Marcelino Serna, about whom a paucity of academic texts exists. Students read a PDF copy of an article about him from the El Paso Community College student newspaper, *Tejano Tribune*. The article provided a short bio and recounted Serna's military heroism. When the class began, I asked which students had heard of Serna, Texas's most decorated WWI veteran. I can recall only one student ever replying in the affirmative. The class discussion framed Serna's legacy to leadership, integrity, and structural racism. An officer erroneously told Serna his limited English precluded him from receiving a Congressional Medal of Honor. He received the Distinguished Service Cross thanks to General John J. Pershing's intersession (see historycollection.com). Pershing's recognition of Serna served to foster discussions of organizational barriers and how good leaders transgressed them to champion their subordinates. I followed a similar strategy with other important Mexican Americans, for example, Master Sergeant Roy Benavides (Vietnam War hero), Raymond Telles (first Mexican American mayor of a large U.S. city and ambassador to Costa Rica), and Eva Longoria (actor and social activist).

Class Discussions

Chicana/o material works well in traditional classrooms particularly with strategic framing linking it to the class topic. One African American soldier appeared appreciative of my efforts to illustrate how Hollywood cinema perpetuates a view of national defense as the exclusive domain of White men. In his words, "Too often Hollywood makes it appear as though there were no Black or Mexican Soldiers. There's a lot more history here than many folks realize." A Mexican American Soldier added he was "excited to be learning about [his] ethnicity" in class. He earned a doctorate after retiring and is presently an assistant professor at a nationally ranked university.

The speaking secrets pedagogy, which I developed by building on González's (1998) and Córdova's (1998) insights, adds an intercultural communication component to any class. It brings knowledge about marginalized groups to mainstream classrooms. One must get to know their students and classroom groups to gauge their acceptance of or resistance to materials about people from marginalized social categories. Student comments and nonverbal cues are useful tools here. As a precursor, I often asked about topics related to social class, educational opportunities, and had students compare their experiences abroad with life in the United States. Once I found an "in," I provided my students with knowledge of Mexican American heroes new to many of them. By framing these individuals as Americans or Mexican American leaders, students learned about Chicano leadership and social justice struggles. For example, some classes read about organizer Emma Tenayuca's views on citizenship vis-a-vis American political thought (Holling, 2012) as well as leadership theories. I advise anyone using this approach to consider the importance of labels and word choice beyond being culturally sensitive.

In other cases, my primary goal was to provide students with a more complex understanding of border dynamics. This was especially important along the borderlands where soldiers cannot enter Mexico, yet they are surrounded by Mexican/Americans and exist in a binational space. The border is not the United States; it is not Mexico. Anzaldúa (1987/2012) compared the border to an open scar where the first and third worlds rub against one another. Furthermore, Martínez (1994) argued that the border is represented on and through people's bodies. Communication scholars have recently expanded this theme (DeChaine, 2012). Acuña (1996) reminds us that many Anglos believe they are more entitled to inhabit the South/West than Mexican Americans because they view us as apathetic and lazy, a threat to the Anglo social order (Mora, 2011). Anglos are largely oblivious to the structural racism that marginalizes Mexicans (Rodríguez et al., 2019). Regrettably, few Whites try to change the status quo (Bonilla-Silva, 2018; Diangelo, 2018; Eddo-Lodge, 2019). Thus, Whites continue to dominate top-level economic ventures (Acuña, 1996, 2000; Rodríguez et al., 2019).

Speaking Chicano or other secrets in the classroom is a first step to elucidate on these realities with, in this case, senior military leaders, few of which are Mexican/American. My classes explore marginalized people's experiences to lessen cultural gaps and inform students about important issues. They provide students with opportunities to talk about their experiences, as well. In addition to traditional academic sources, class discussions may address blogs, newspaper editorials or stories, podcasts, popular films, video clips, or documentaries. To learn about my students' lives and sociocultural backgrounds, I often assign a four-page essay where students answer one question: What is your ethnic identity and why? Student reactions to it have been overwhelmingly positive because the assignment allows them to reflect on their life experiences. It validates what they know from lived experiences. Another intercultural tool is to ask students to analyze or describe a favorite book, film, music video, or personal hero. In all cases, class discussions may include asking for volunteers to share their answers. The instructor can help students explore links between their essays and the class subject matter to help them understand and accept cultural differences, a primary aim of intercultural communication. Hall (1959/1990) suggested that intercultural gaps can be bridged by learning about a foreign culture and interacting with people from that culture. This tactic works in classrooms.

BRIDGING SECRETS WITH MILITARY STUDENTS

Instructors can bridge university and military cultures by speaking secrets, so long as they are sincere in their efforts and provide useful information. If one is authentic and informed on a topic, students will generally listen and be respectful, even in those instances where they disagree with the instructor. For example, a feminist scholar may share a case study with students about workplace discrimination, an African American scholar may share one about the Tuskegee Airmen, and so on. Learning about and being supportive of individuals in a profession characterized by service, sacrifice, and frequent travel is an effective primer for students to handle new information and to remain respectful. I self-identify as Chicano and make no apologies for talking about Mexican Americans in class. However, I also use strategic terms in class lectures (i.e., Mexican American, not Chicana/o) and show direct links between the cases and videos covered and the course topic. Students respect instructors who respect them, and they will usually explore topics new to them if one shows them its relevance. As such, I encourage all instructors interested in non-mainstream academic work to consider this pedagogy.

The examples that inspired this work reflect a social justice focus. González (1998) called out her professional peers for their sexist behaviors and attitudes. Córdova (1998) called out her own university! Similarly, my efforts are context-bound. Military bases are unique settings, and the opportunity to teach senior NCOs is indeed a privilege. Bringing to light the experiences and contributions of Chicana/o military heroes and veterans, particularly those brave enough to address PTS, helped me teach my students about Brown Americans' struggles against racism. Military students of color were usually receptive to this information and only a few White students resisted this method. I noted one example above and will now share another. A group of five veterans in their late 40s-early 50s in one class often challenged my comments about racism. One night, I entered a local pub and saw these students. One waved to say hello. I approached the table to say hi. As we chatted, one commented, "Dr. Pérez, we'd invite you to stay, but we know you don't like White people." The entire group laughed, as did I, but it was clear they perceived me negatively, and possibly even as racist. However, I am committed to this pedagogy because, after a decade of teaching about 300 military students, only six challenged this method. Most students openly engaged Chicana/o secrets.

Handling (Counter)Hegemonic Discussions

People typically speak differently in public than in private. Many of my military students of color are politically aware. Talking with them outside class showed me the adherence to respect was central. When students disagreed in class, they often remained quiet, changed the topic, or used humor to offer a different point of view. I recall a time a sergeant major said something a staff sergeant disagreed with. Rather than openly challenge the senior NCO, the staff sergeant interrupted with, "Sergeant major, you know I respect you but you were a staff sergeant so long ago that I think they still rode horses in the cavalry (the class laughed). I think things have changed." The lower-ranked NCO's framing, "I respect you" and "I think" avoided a direct confrontation. The sergeant major laughed and replied, "You may be right, staff sergeant. You may be right." Here, both NCOs mentioned rank. It was much more common for students to refer to one another by last name, but the diplomacy shows a concern for mutual respect.

The day Donald Trump announced his candidacy for president and referred to Mexicans as rapists, two students asked me if I would vote for him. I politely replied, "It's not likely I will vote for Mr. Trump." My wife is Mexican; our son has dual U.S.-Mexican citizenship. They knew this. Another student later told me he had approached his classmates later and asked, "Are you crazy? The man's wife is Mexican. His son is a U.S. and Mexican citizen. He speaks Spanish as well as he does English. What were you thinking?" I thanked the student and said, "Don't worry about it." The Trump question highlights the mainstream hegemonic thought common in many classrooms. The students who posed the question claimed they saw me as a successful professional. They assumed my middle-class status might lead me to overlook Trump's racist statement.

This situation highlights the hegemonic and counterhegemonic realities I observed. Professional decorum and army values led one student to talk to his peers outside the classroom. Another time, a female NCO said that she failed to understand why female army leaders needed their own social media outlets or spaces away from men. Most of the twelve-student class, two women and ten men, supported her assertion that all military members are the same. The other female said nothing during class or to me at any time. A few minutes before our next class, a male student approached me: "Dr. Pérez, I don't think she knows what she's saying. How can she not understand that as a woman she faces additional obstacles?" I replied that I agreed but that everyone has the right to their own views. He replied, "I know you agree with me but need to remain on good terms with everyone" and walked away. His demeanor suggested that he understood my role as instructor prevented me from openly agreeing with him.

THE SPEAKING SECRETS PEDAGOGY: A SUMMARY

I remind the reader that this pedagogy is adaptable to fit many diverse subjects and classroom contexts. It is based on my experiences and builds on the work of two important Chicana scholars, but is adaptable to other topics/contexts. Other instructors may want to address feminist thought (Anzaldúa, 2012; Collins, 2008), White privilege (Rothenberg, 2016), or any other topic the instructor desires. Here is a recap on how to speak secrets with students in the classroom. First, a secret can be any knowledge, fact, or series of facts that one wants to share with students about different ways of knowing. Here, one can use cultural studies research to share "secrets." Second, to share a secret I rely on and recommend that one use peer-reviewed literature, documentaries that aim for a balanced interpretation of a topic, and/or one's lived experience. Students respond well to this type of knowledge or secret if one is sincere and does so after building trust with them. Finally, listen to your students and allow them to share their own secrets, such as the student who commented on buffalo soldiers being used to attack Indians or the tank driver who was unsure how to transition to civilian life. Honest, well-informed discussion works well in military structure because of its emphasis on mutual respect and "selfless service," the idea of placing others before oneself.

In terms of application, first, early in meeting your students, share information that shows commonality between you and them. If you are there because you support our men and women in uniform let them know. Many of my students said they were relieved to hear me say I support our military on the first day of class. Second, ensure your supplemental topic, here Chicana/o Studies, is relevant to the core material in any class or seminar. The links between the case studies or other materials and the central topic must be obvious or explained. For example, I used borderlands case studies because students were spending almost a year in the region. Using materials where the link is not obvious can be detrimental. Third, be strategic in the terms you use and in the framing of controversial or difficult topics. For example, talking about "clashes" between pachucos and sailors during the Zoot Suit Riots works better than saying sailors "attacked." Your guidance will show these events were attacks without using potentially polarizing language. Fourth, ensure that class discussions begin with a focus on the central topic of the class and then advance to more complex and difficult questions. As my example above shows, too much, too fast, can lead to a class that resists the speaking secrets pedagogy. Finally, allow students to share their own thoughts on the materials and to speak about their own experiences. This strategy gives students ownership of the class and validates their experiences.

LIMITATIONS

The chapter has three limitations readers should consider. First, the quotes are purposely short to maintain my students' anonymity. Again, given the relatively small number of military personnel who completed my classes and their strong social bonds and networks, I opted for short examples. However, I believe they may still serve the reader well. Second, a co- or multi-authored chapter may have increased the number of examples, but none of my colleagues used the speaking secrets methodology during this time frame. I was developing it during this period and am only now sharing it with others. Finally, interviews about my approach with students may have yielded additional insights. I am currently working on such data collection.

Speaking secrets by sharing information about the Mexican American experience with military students educated them about the largest ethnic category in the United States, a group of people whose contributions are often omitted from mainstream academics and public memory. It is also a political act consistent with long-standing Chicano goals of "redefin[ing] ideas of nationhood and belonging" (Leen & Thornton, 2014, p. 2). Speaking academic secrets (i.e., forgotten or little-known knowledge) allows students to learn. It also opens the door for them to share information about their respective communities, fostering greater intercultural understanding. Ideally, this method will also motivate opinion leaders to share their newly acquired knowledge with others in their organization, allowing the experiences and concerns of marginalized peoples to move from secrets to mainstream acceptance.

CONCLUSION

My goal while teaching is to present useful and interesting information to students. The military students' course evaluations suggest that this objective was met successfully. The courses I taught addressed intercultural communication issues in leadership studies, using Mexican American examples where possible. Many other topics would have worked and I softened the language from my campus lectures, substituting Mexican American for Chicana/o. My focus on social justice challenges faced by women and people of color remained. Most students were completing a military program for promotion to command sergeant major (CSM), the highest NCO rank in the army. My students often commented on their need to retire within a few years of attaining this rank. They were also concerned with their upcoming civilian lives. Most wanted post-military careers in the private or nonprofit sectors, a few had political aspirations and at least one completed a PhD and entered academe.

The speaking secrets pedagogy worked well with them because I built trust and showed my students respect. Prior to addressing difficult topics/secrets, I allowed the class to also build trust and respect among themselves. Next, I spoke secrets or knowledge about Mexican Americans in the U.S. military, addressing the heroic actions of some and the institutional racism they encountered, while focusing on the leadership aspects of these individuals or institutions. Finally, I relied on military culture to speak secrets, relying on the atmosphere of respect commonly displayed by my students. I invite the reader to explore this pedagogy and to mold it to their needs. Many secrets that represent the lived experiences of countless people remain outside the scope of mainstream teaching and scholarship. Adding information that challenges the dominant narrative to this discourse can only strengthen our military with information that promotes intercultural understanding. The real power of the military is not the size of its weapons but its knowledge of its peoples and cultures. I would argue that learning about the secrets of its many members is an excellent first step toward this objective.

NOTES

1. Writing about both the Mexican and Mexican American populations forces authors to struggle with issues of redundancy, such as the frequent use of the "Mexican and Mexican populations." Elsewhere (Pérez & Ortega, 2020), inspired by Holling's (2012) use of Mexican (American) to denote both groups, I use Mexican to denote Mexican nationals. Mexican American is used in reference to Americans of Mexican descent and Mexican/American for both groups. Mexican/Americans is the plural form for both groups.

2. Many dispute the term Hispanic because it was adopted by the US Census Bureau under the Nixon administration. It blurs the distinctions between the different Spanish- and Portuguese-speaking (i.e., Brazilian) populations living in the United States. Many politically aware people from Spanish-speaking populations use terms that they believe better reflect their culture than the government-imposed Hispanic label (Tienda & Ortíz, 1986).

REFERENCES

Acuña, R. F. (2000). *Occupied America: A history of Chicanos* (4th ed.). Pearson.

Acuña, R. F. (1996). *Anything but Mexican: Chicanos in contemporary Los Angeles.* Verso.

Aguirre, N. (2017). A stranger in Charleston: Intersectional anxiety as a Latina in academe. *Journal of Working-Class Studies, 2*(2), 96–107.

Anzaldúa, G. (2012). *Borderlands/La frontera: The new mestiza* (4th ed.). Aunt Lute Books.

Bachman, J. G., Freedman-Doan, P., Segal, D. R., & O'Malley, P. M. (2000). Distinctive military attitudes Among U.S. enlistees, 1976–1997: Self-selection versus socialization. *Armed Forces & Society, 26*(4), 561–585.

Bonilla-Silva, E. (2018). *Racism without racists: Color-blind racism and the persistence of racial inequality in America* (5th ed.). Rowman & Littlefield.

Carnevale, A. P., Fasules, M. L., Quinn, M. C., & Campbell, K. P. (2019). *Born to win, schooled to lose: Why equally talented students don't get equal chances to be all they can be.* Georgetown University, Georgetown Center on Education and the Workforce.

Carnevale, A. P., & Strohl, J. (2013). *Separate and Unequal: How higher education reinforces the intergenerational reproduction of white racial privilege.* Georgetown University, Georgetown Public Policy Institute.

Carnevale, A. P., Van Der Werf, M., Quinn, M. C., Strohl, J., & Repnikov, D. (2018). *Our separate & unequal public colleges: How public colleges reinforce white racial privilege and marginalize Black and Latino students.* Georgetown University, Georgetown Center on Education and the Workforce.

Carrol, P. J. (2003). *Felix Longoria's wake: Bereavement, racism, and the rise of Mexican American activism.* University of Texas Press.

Chang, H. (2013/2016). Individual and collaborative autoethnography as method: A social scientist's perspective. In S. H. Jones, T. E. Adams, & C. Ellis (Eds.), *Handbook of autoethnography* (pp. 43–57). Routledge.

Chomsky, N. (2017). *Who rules the world?* Picador.

Collins, P. H. (2008). *Black feminist thought: Knowledge, consciousness, and the politics of empowerment.* Routledge.

Córdova, T. (1998). Power and knowledge: Colonialism in the academy. In C. Trujillo (Ed.), *Living Chicana theory* (pp. 1–24). Third Woman Press.

Cross, W. C., & Maldonado, B. (1971). The counselor, the Mexican American, and the stereotype. *Elementary School Guidance & Counseling, 6*(1), 27–31.

Diangelo, R. (2018). *White fragility: Why it's so hard for White people to talk about acism.* Beacon Press.

Dorman, J. E. (1976). ROTC cadet attitudes: A product of socialization or self-selection? *Journal of Political and Military Sociology, 4*, 203–216.

Eddo-Lodge, R. (2017). *Why I'm no longer talking to white people about race.* Bloomsbury Publishing.

Ellis, C., Adams, T. E., & Bochner, A. P. (2011). Autoethnography: An overview. *Historical Social Research, 36*(4), 273–290. doi:10.12759/hsr.36.2011.4.273-290.

Ellis, C., & Bochner, A. P. (2000). Autoethnography, personal narrative, reflexivity. In N. K. Denzin & Y. S. Lincoln (Eds.), *Handbook of qualitative research* (2nd ed., pp. 733–768). SAGE Publications.

Erba, J. (2018). Media representations of Latinas/os and Latino students' stereotype threat behavior. *Howard Journal of Communications, 29*(1), 83–102.

González, D. (1998). Speaking secrets: Living Chicana theory. In C. Trujillo (Ed.), *Living Chicana theory* (pp. 46–77). Third Woman Press.

Gordon, L. (1999). *The great Arizona orphan abduction.* Yale University Press.

Green, G. N. (1991). The Longoria affair. *Journal of Ethnic Studies, 19*(3), 23–49.

Guo, L., & Harlow, S. (2014). User-generated racism: An analysis of stereotypes of African Americans, Latinos, and Asians in YouTube videos. *Howard Journal of Communications, 25*(3), 281–302.

Hall, E. T. (1959/1990). *The silent language.* Anchor Books.

Hall, R. E. (2018). Media stereotypes and "coconut" colorism: Latino denigration vis-à-vis dark skin. *American Behavioral Scientist, 62*(4), 2007–2022.

Hall, S. (1996). The problem of ideology: Marxism without guarantees. In D. Morely & K. H. Chen (Eds.), *Stuart Hall: Critical dialogues in cultural studies* (pp. 25–46). Routledge.

Hampton, E., & Ontiveros, C. C. (2019). *Copper stain: ASARCO's legacy in El Paso.* University of Oklahoma Press.

Harrison, C. (2014). *The ghosts of hero street: How one small Mexican-American community gave so much in World War II and Korea.* Penguin.

historycollection.com. (n.d.). *Texas' most decorated World War I hero was an undocumented Mexican immigrant.* https://historycollection.co/texas-most-decorated-world-war-i-hero-was-an-undocumented-mexican-immigrant/3/.

Holling, M. A. (2012). A dispensational rhetoric in The Mexican question in the Southwest. In D. R. DeChaine (Ed.), *Border rhetorics: Charting enactments of citizenship and identity on the U.S. – Mexico Border* (pp. 65–85). University of Alabama Press.

Igielnik, R., Parker, K., & Cilluffo, A. (2019, December 31). *Trump draws stronger support from veterans than from the public.* Pew Research Center. https://www.pewsocialtrends.org/2019/07/10/trump-draws-stronger-support-from-veterans-than-from-the-public-on-leadership-of-u-s-military/.

Koski, S. V., & Bantley, K. (2019). "Dog whistle politics: The Trump administration's influence on hate crimes." *Seton Hall Legislative Journal, 44*(1), 39–60.

Leen, C., & Thornton, N. (2014). Introduction. In C. Leen & N. Thornton (Eds.), *International perspectives on Chicana/o studies: "This World is My Place"* (pp. 1–5). Routledge.

Lindsay, K. (2015). Beyond "model minority," "superwoman," and "endangered species:" Theorizing intersectional coalitions among Black immigrants, African American women, and African American men. *Journal of African American Studies, 19*(1), 18–35.

López, P. (2011). A response: Continuity and change in Hispanic identity. *Ethnicities, 11*(4), 570–577.

Lutz, A. (2008). Who joins the military? A look at race, class, and immigrant status. *Journal of Political and Military Sociology, 36*(2), 167–188.

Mariscal, G. (1999). *Aztlán and Viet Nam: Chicano and Chicana xperiences of the war.* University of California Press.

Martínez, O. J. (1994). *Border people: Life and society in the U.S. – Mexico Borderlands.* University of Arizona Press.

Mastro, D. E., & Behm-Morawitz, E. (2005). Latino representation on primetime television. *Journalism & Mass Communication Quarterly, 82*(1), 110–130.

Méndez, M. (2013). Autoethnography as a research method: Advantages, limitations, and criticisms. *Columbian Applied Linguistics Journal, 15*(2), 279–287.

Morín, R. (1963/2013). *Among the valiant: Mexican-Americans in WWII and Korea* (rev. ed.). Valiant Press.

National Public Radio. (21 June 2019). *Deported US veterans feel abandoned by the country they defended.* https://www.npr.org/local/309/2019/06/21/733371297/deported-u-s-veterans-feel-abandoned-by-the-country-they-defended.

Noriega, C., Belcher, W. L., & Villaseñor Black, C. (2020). *Autobiography without apology: The personal essay in Chicanx and Latinx Studies.* University of California Los Angeles, Chicano Studies Research Center Press.

Noriega, C. (Ed.) (2004). *I am Aztlán: The personal essay in Chicano Studies.* University of California Press.

Ojeda, L., Castillo, L. G., Meza, R. R., & Piña-Wilson, B. (2014). Mexican Americans in higher education: Cultural adaptation and marginalization as predictors of college persistence intentions and life satisfaction. *Journal of Hispanic Higher Education, 13*(1), 3–14.

Padilla, G. M. (1988). The recovery of Chicano Nineteenth-century autobiography. *American Quarterly, 40*(3), 286–306.

Pérez, F. G. (2007). Communicating for social justice: Juan de Oñate and the struggle for Chicano cultural representation in public art. *Noesis, 16*, 98–121.

Pérez, F. G., & Ruggiero, T. (2007). Juan de Oñate as public art: A case study of cultural representation and tourism. *Communication for Development and Social Change Journal, 1*(3), 233–250.

Pérez, F. G., & Ortega, C. (2008). Mediated debate, historical framing, and public art: The Juan de Oñate controversy in El Paso. *Aztlán, 33*(2), 121–140.

Pérez, F. G., & Ortega, C. (2019). *Decolonizing Eurocentric tourism and heritage narratives in Mexican American communities: Juan de Oñate as a West Texas icon.* Routledge.

Rinderle, S. (2005). The Mexican diaspora: A critical examination of signifiers. *Journal of Communication Inquiry, 29*(4), 294–316.

Rivas-Rodríguez, M. (2005). *Mexican Americans & World War II.* University of Texas Press.

Rodríguez, C. E., Staudt, K., Martínez, O. J., & Neill, R. (2019). *Who rules El Paso? Private gain, public policy, and the community interest.* Community First Coalition.

Romero, A., González, H., & Smith, B. (2015). Qualitative exploration of adolescent discrimination: Experiences and responses of Mexican American parents and teens. *Journal of Child & Family Studies, 24*(6), 1531–1543.

Rosales, S. (2011). Fighting the peace at home: Mexican American veterans and the 1944 GI Bill of Rights. *Pacific Historical Review, 80*(4), 597–627.

Rosales, S. (2017). *Soldados Razos at War: Chicano politics, identity, and masculinity in the U.S. Military from World War II to Vietnam.* University of Arizona Press.

Rothenberg, P. S. (Ed.) (2005/2016). *White privilege: Essential readings on the other side of racism* (5th ed.). Wadsworth.

Tienda, M., & Ortíz, V. (1986). "Hispanicity" and the 1980 census. *Social Science Quarterly, 67*(1), 3–20.

Trujillo, C. (2003). *Soldados: Chicanos in Viet Nam* [Documentary]. Chusma House.

Valadez, J. J. Dir. (2010). *The Longoria Affair.* PBS Video/Independent Lens.

Woll, A.J. (1977). *The Latin image in American Film.* Los Angeles, CA: University of California at Los Angeles, Latin American Center Publications.

Yin, R. K. (2017). *Case study research and applications: Designs and methods* (6th ed.). Sage.

Chapter 7

Nepantla, Liminality, and Teaching and Learning in the Space Between

A Borderlands Approach to Military-Affiliated Learner Pedagogy

Leandra Hinojosa Hernández, Sarah De Los Santos Upton, and Arthur A. Aguirre

Over the past several years, the military has become ever more present in the U.S. American[1] consciousness because of high-profile cases and public policy debates that highlight discrimination against the LGBTQ community, violence against women, and racial inequality. When President Obama took office in 2009, LGBTQ service members could not serve openly in the U.S. military because of Don't Ask, Don't Tell (DADT), a policy that originated under the Clinton administration and heavily influenced both the treatment of soldiers and public opinion as well. While this policy was repealed in 2010 under Obama, its legacy lives on, restricting LGBTQ soldier recruitment and, more importantly, hindering LGBTQ troops who serve in hiding (Lopez, 2017). Specifically, LGBTQ service members remain fearful of repercussions from their chain of command if they disclose their sexuality or gender identity (Myers, 2020). Fifty-nine percent of service members in a recent study did not feel comfortable being out at work (McNamara et al., 2020). Overall, LGBTQ service members:

> Seek a military in which disclosure will not subject them to negative career repercussions, burden them with feelings of differentness or expectations to teach others how to treat them, limit their ability to access needed resources for themselves or their family, and, ultimately, that their physical and personal integrity will not be endangered. (McNamara et al., 2020)

Recent developments with the Trump administration highlight how the fight for equality and recognition continues (Woods, 2019). From 2017 to 2019, the Trump administration attempted to restrict transgender individuals from serving in the military, with a tweet from Trump in 2017 stating, "Our military must be focused on decisive and overwhelming . . . victory and cannot be burdened with the tremendous medical costs and disruption that transgender in the military would entail" (Lopez, 2019, para. 5).

In addition to concerns of homophobia and transphobia surrounding LGBTQ service members, other events have highlighted the intertwined nature of racism and sexism surrounding service members like Vanessa Guillen, a soldier from Fort Hood who was harassed and then brutally murdered and whose body remained missing until media scrutiny brought the base under fire (Campion, 2020). Her case incited a national firestorm highlighting sexual assault in the U.S. military ensconced within larger racist and sexist policies that permitted her disappearance and murder to go unsolved for weeks. National coverage of her case has also brought to the forefront other cases of service member disappearance, assault, and murder that were not spotlighted or covered in larger media contexts. Moreover, Guillen's case inspired Domingo Garcia, the president of the League of United Latin American Citizens (LULAC), to urge Latinas not to join the U.S. military:

> We are asking all women, especially Latina women or their families: "Do not enlist in the Army until we have assurance they will be protected and taken care of when they serve our country," Garcia said in a statement. "And right now I just don't believe the military is capable of doing that because of what happened to Vanessa Guillen." (Moreno, 2020, para. 3)

Garcia's sentiments were amplified when a University of Wisconsin-Milwaukee lecturer, Betsy Schooeller, a former Wisconsin Air National Guard officer, posted inflammatory, misogynistic comments on Facebook about Spc. Guillen's murder saying, "sexual harassment is the price of admission for women in the military and 'if you're gonna cry like a snowflake about it, you're gonna pay the price'" (Associated Press, 2020, para. 2). In response to social media and journalistic critiques of her response, Schooeller stated:

> I was giving voice to the messaging that women hear in the culture of sexual harassment: The message we receive from the culture is not only will you suffer from sexual harassment, if you squawk about it, you will suffer even more. Because it isn't just the sexual harassment. That's just the beginning. Then comes the agonizing decision about reporting. Or not reporting. The pressure applied by friends who know about it and only want to help. Having to

> ultimately stand up to that culture of sexual harassment on your own. (Jones, 2020, para. 6-7)

Despite her public apology, her original comments highlight the very real threat of violence, lack of support, and retaliation women, especially women of color, face in service to their country.

These case studies highlight the violence service members face concerning their identities in racial, gendered, and sexuality contexts. Furthermore, these cases spotlight the lingering effects of the intertwined systems of cisheteropatriarchy, internalized misogyny, sexism, and racism that shape U.S. American service members' experiences. Long gone are the days when the U.S. American military consisted primarily of white men. Diversity in the U.S. American military also parallels diversity in higher education, particularly as military-affiliated learners (MALs) of diverse backgrounds and identities are increasingly pursuing college degrees. As such, frameworks for diversity, inclusion, and equality in both military contexts *and* when reentering the civilian realm (i.e., higher education) are needed to support today's military. A borderlands approach to MAL pedagogy can be a particularly effective framework for supporting MALs, given its disruption of traditional, white, western, colonial, masculine approaches to higher education curriculum development and instruction. In this chapter, as three borderlands communication scholars of color, we first discuss the changing racial, ethnic, and gender/sexuality identity demographics of U.S. American service members. Next, we propose a borderlands approach to MAL pedagogy that can cultivate a space of awareness, inclusion, and equity. A borderlands pedagogical perspective, rooted in Chicana feminisms and academic equity and justice, honors students' vivencias, or lived realities and experiences, and encourages students to draw from their vivencias to make sense of their academic experiences. Finally, we conclude with three case studies that illustrate a borderlands approach to MAL pedagogy framework in practice and strategies that educators can utilize when working with MALs to both acknowledge MAL experiences and create supportive, inclusive spaces for enhancing MALs' educational success.

THE CHANGING PROFILE OF THE U.S. MILITARY

The past several decades have seen an increase in multiple types of diversity in the U.S. military, although the increase in diverse identities, backgrounds, and experiences hasn't always been met with comparable inclusion policies and efforts to dismantle racist, sexist, misogynistic, and exclusionary practices. As Rohall et al. (2017) describe, diversity can sometimes be perceived as a "managerial challenge" because organizational decision-making processes, procedural decisions, and shifts impact individuals from different backgrounds and lived experiences in different ways (p. 4). Consider, for example, recent developments and efforts in the U.S. military to become more accepting and inclusive of transgender service members, including but not limited to efforts to use more inclusive language and pronouns by shifting traditionally gendered "military speak" to recognize preferred pronouns. Indeed, within the early years of our military's creation such developments in the name of diversity and inclusion perhaps would not have been foreseen as being present or necessary, although military developments in diversity and inclusion at the time this book went to press are being combated by the Trump presidential administration (Human Rights Campaign, 2020). As Han (2017) asserted, "having an adequate ratio of races within the military does not necessarily mean that forces are equitable," a powerfully accurate statement when considering the extent to which representation matters for diversity in all forms (p. 25).

Race, Ethnicity, Gender, and U.S. Military Service

Several significant and historical trends pertaining to race/ethnicity and gender situate our analysis and illustrate the racial and gendered dynamics that have shaped the evolution of diversity in the U.S. military. From a historical perspective, racial/ethnic minoritized groups, women, and LGBTQ populations were barred from service to varying degrees and have unique service timelines based upon discriminatory policies and cultural relationships to citizenship. Racist and segregational policies originally barred African-Americans from joining the military during colonial times, and the Selective Service Act of 1917 significantly expanded African-American participation in the U.S. Armed Forces (Han, 2017), with African-American men's participation in the Armed Forces increasing through World War II and beyond. Similarly, Hispanic/Latino/a service members have a long history of military participation in every major American conflict beginning with the Revolutionary War (De Angelis, 2017). However, as Pérez describes in chapter 6, they have faced systemic barriers at every turn, such as language barriers, less formal recognition of their achievements and valor, and racial/ethnic segregation both while serving and also when home from war (De Angelis, 2017). Both of Leandra's grandmothers, for example, were excluded from restaurants and other local institutions because of skin color and racial/ethnic segregationist policies in Texas, but her grandfathers—two Army staff sergeants and WWII veterans—were allowed to enter restaurants because, due to their light skin and light-colored eyes, they were perceived by others to be white racially, although they were also Mexican-American ethnically.

Sarah's grandfather, a Mexican-American WWII veteran, on the other hand, was denied entry into a Texas restaurant while in uniform and was instead asked to eat in the kitchen where he would not be seen by the other patrons.

Similar to African-American and Hispanic/Latino/a service members, Asian-American and Native American service members were barred from service, as well. This statement by Sohoni (2017) provides an important link between the experiences of all minoritized groups who have served in the U.S. military, as the duality between stereotypes of "true American" and "foreigner" persist:

> As with other minority groups, military service has at times provided Asian-Americans the opportunity to prove themselves "true" Americans and deserving of all the legal rights of American citizenship. Yet frequently, pre-existing racialized stereotypes of Asian-Americans as "permanent foreigners," unable or unwilling to assimilate into American society, have led to discriminatory politics that constrained their participation in the military, as well as limited the benefits they received when they served. (p. 57)

Sohoni (2017) further describes how individuals with Asian ancestry have served the U.S. military since the War of 1812, yet it was not until the early 1900s that the first widespread incorporation of Asians in the U.S. military commenced.[2] Finally, Native Americans have been involved in every war in American history and, despite lacking U.S. citizenship, over 20,000 American Indians fought in the Civil War (Meadows, 2017). Ultimately, racial inequality in the military has manifested in many contexts, primarily through "representation in different ranks and grades that determine power, authority, prestige, and compensation" (Han, 2017, p. 25), as well as through the rate of promotion service members experience within their respective groups, retention disparities (Harris, 2009), administration of military justice, risk of death in combat, and health care that wounded soldiers receive (Burk & Espinoza, 2012). As this micro-history illustrates, racially and ethnically minoritized groups in the United States have a complicated relationship with the U.S. military characterized by a long, rich history of service alongside the fight for visibility, recognition, and belonging.

Currently, today's active-duty military is more ethnically and racially diverse than it has ever been, a development and expansion that brings with it more diversity in family structure, cultural beliefs, lived experience, and professional experience (Barroso, 2019). In addition to more ethnic and racial diversity, gender dynamics are changing as well. In 2017, nearly one-in-five commissioned officers in the U.S. military were women, a trend that has been increasing since the mid-1970s (Barroso, 2019). Women also have a complicated history with the U.S. military. Although men have long served the U.S.

military for the past few centuries, the first woman to enlist, Loretta Walsh, only did so in 1917. Before Loretta Walsh, other women, such as Deborah Sampson, disguised themselves as men so that they could serve in the Armed Forces (Katz, 2019). It wasn't until the Women's Armed Services Integration Act, which passed in 1948, that women were legally permitted to serve as permanent, regular members of the U.S. Armed Forces (Ziobro, n.d.). From this act, in 1976, 119 women joined the Corps of Cadets, together comprising the first class of women cadets at West Point. Of these 119 cadets, 62 graduated in 1980, becoming Second Lieutenants while making important historical strides for women in the process (Schloesser, 2010). While cisgender women have slowly gained access to military service, the LGBTQ experience has been drastically different.

While LGBTQ service members have served in the U.S. military since its inception, scholars note that the history of their service is relatively unknown because of centuries of both unofficial and official anti-LGBTQ policies that "defined LGBT service members (LGBT SMs) as criminals, mentally ill, and unfit for duty" (Ramirez & Sterzing, 2017, p. 68). As a result of military anti-LGBTQ policies, historically, LGBTQ service members were sent to psychiatric hospitals, military incarceration facilities, or were discharged altogether (Ramirez & Sterzing, 2017). Between 1940 and the late 1980s, "approximately 100,000 male and female service members received discharges related to their non-heterosexual identity and/or behavior, averaging 1,500 to 2,000 per year" (Berube, 1990, as is cited in Ramirez & Sterzing, 2017, p. 68). As a result of DADT, over 12,500 service members have been discharged because of their sexuality (Shane, 2009). Additionally, according to the 2015 National Center for Transgender Equality's U.S. Transgender Survey (USTS), "USTS respondents served in the military at nearly twice the rate as the people in the U.S. population overall" (para. 2). Of the survey's respondents nationwide (27,715 respondents total), 18 percent have served in the U.S. military, and 15 percent were veterans (nearly 2 times the rate in the U.S. population [8 percent]) (National Center for Transgender Equality, n. d.).

Ultimately, from a gender identity and sexuality perspective, as we discussed earlier in the chapter, recent policy and system strides in creating more inclusive spaces for LGBTQ service members have been paramount from an equality perspective, yet such policies are still under attack politically and administratively as this book goes to press. Forty-seven percent of the 2015 National Center for Transgender Equality's U.S. Transgender Survey stated that their leadership supported their name change, and 36 percent said their leadership supported their transition-related medical treatments. However, 19 percent reported they believed their separation from the military was partially or solely due to being transgender, and an additional 19 percent of respondents stated they separated from the military so they could avoid mistreatment and harassment (National Center for Transgender Equality, n.d.). Such findings illustrate the continued fight for transgender recognition and acceptance in the U.S. military. Taken together, such historical trends illustrate how diversity has been a cornerstone of the U.S. military's service members for centuries. Although less information is known about MALs who are spouses or children, our fundamental argument is that there is a rich diversity of lived experiences for MALs, which, as we discuss further throughout the chapter, necessitates a framework that makes space for such lived experiences in the higher education classroom.

U.S. Military Demographics and The Culture Divide

According to the Department of Defense (DoD) *2018 Demographics: Profile of the Military Community* report, the total number of military personnel is almost at 3.5 million, including active-duty personnel (1,304,418), reserve and retired reserve members, and DoD Appropriated and Non-Appropriated Funds civilian personnel. From a race and gender perspective, "Women, who number 215,834, comprise 16.5 percent of the DoD Active Duty force, while 1,088,584 men comprise 83.5 percent of the DoD Active Duty force" (*2018 Demographics*, p. iii). Racially and ethnically, "nearly one-third (31.0%) or 404,851 of Active Duty members identify themselves as a racial minority (i.e., Black or African American, Asian, American Indian or Alaska Native, Native Hawaiian or Other Pacific Islander, Multi-racial, or Other/Unknown)" (*2018 Demographics*, p. iii). Compared to 2010, the percentage of Active Duty members who identify themselves as a racial minority is higher in 2018: "31.4% of enlisted members and 22.7% of officers in 2010, to 32.7% of enlisted members and 23.5% of officers in 2018" (*2018 Demographics*, p. iii), with Hispanic/Latino service members as the fastest-growing ethnic minority in the U.S. military (Barroso, 2019). From an educational perspective, most enlisted members at the time of data reporting (80.5 percent) had a high school diploma and/or some college experience, while fewer had an associate's degree (10.5 percent) or a bachelor's degree or higher (8.4 percent) (*2018 Demographics*, p. iv). Enrollment trends for active-duty enlisted personnel, noncommissioned personnel, and commissioned personnel shows an increase in percentages of bachelor's degrees or higher from 2010 to present (*2018 Demographics*, p. 44). In terms of current enrollment, 2,917,825 active-duty service members and dependents were enrolled in higher education in 2017 (*2017 Demographics*), which highlights the vast number of MALs in higher education and the necessity of continuing to develop frameworks to ensure MAL resilience, retention, and success.

As other chapters in this volume have discussed (in this edited volume see Chiara, Chiara, & Ashlock, chapter 3; Hernández & Belding, chapter

11; May & McDermott, chapter 2), a cultural divide exists among military, civilian, and academic cultures that impact retention and success for MALs in higher education. As May and McDermott (2019) illustrate, the military-academic cultural divide exists with regard to structure and hierarchy, values and belief systems, and practices that guide action and behavior. Historically, "the military has been defined by conformity and uniformity, while the academy has pushed for freedom of thought and social change" (May & McDermott, 2019, p. 59). Military culture can be defined by "a strict hierarchy, order-based decision-making, and the emphasis on acting in favor of the greater good" (May & McDermott, 2019, p. 60), whereas academic culture can be characterized by liberal thinking and academic freedom to practice as one pleases without interference or mandates from one's chain of command. This divide impacts MAL identity formation (in this edited volume see Howe, chapter 4; May & McDermott, chapter 2) and MAL experiences in higher education tremendously, ranging from the way MALs are socialized to write and study to the policies and procedures in place that mandate frequent relocations and other disruptions that impact the academic experience (in this edited volume see Cobos, chapter 12; Hernández & Belding, chapter 11; Ludewig, chapter 14).

Military service affects all MALs from a variety of vantage points and stressors. Such stressors and barriers include frequent relocations, family separations, spousal unemployment and underemployment, post-traumatic stress concerns and associated mental health concerns, reintegration issues in both society writ large and also the academic classroom, struggles with uncertainty management, lack of knowledge about available institutions and programs, lack of knowledge about how to navigate the educational process (e.g., from selecting an institution to applying and then graduating successfully; in this edited volume, see Moullet & Belding, chapter 8), and more (Drummet et al., 2003; Hernández, 2019; Hernández & Belding, 2019; Hill et al., 2019; May & McDermott, 2018; Sollitto & Cole, 2019). Specifically for military spouses, Sollitto and Cole (2019) found that in higher education contexts, military spouses and significant others experienced uncertainty in three specific contexts: program logistics, academic/professional success, and family consideration. Program logistics entailed length and timeline of academic programs, tuition and academic costs, required assignments and program expectations, stressors about academic program completion, how to develop relationships with faculty members and other mentors, and fears associated with school-life balance. Trained professionals, however, such as Veterans Affairs (VA) coordinators, behavioral therapists, and counselors help military spouses flourish amidst uncertainty.

Interviewing eighty-five military spouses, Hernández (2019) found similar anxieties and struggles in her study of how military spouses with graduate degrees navigated academic and professional life within a military context. Spouses and significant others expressed concerns about professional licensing and credentialing, navigating the transition between face-to-face and online programs, and completing professional and academic requirements while "holding down the home fort," so to speak, during deployments, missions, and family separation periods. Ultimately, after conversing with the military spouse authors and researchers who contributed to their edited volume, Belding and Hernández (2019) found that military spouses yearn to be taken seriously and need flexibility and understanding from academic faculty members and staff. Academic policies that would facilitate the inclusion of military spouses in higher education include processes that streamline and facilitate the transfer of undergraduate and graduate coursework, permit courses to be taken out of sequence (considering family relocation and permanent change of station [PCS] moves), and develop and support military-friendly networks.

Inclusion, Exclusion, and Higher Education

Thus far, we have chronicled a brief history of the evolution of racial/ethnic and gender diversity in the U.S. military, a complicated history with concerns about racism and sexism that has evolved parallel to the evolution of inclusion and exclusion in higher education. The development of higher education in the U.S., much like the U.S. military at a micro-level and U.S. American history at a macro-level, has long struggled with racism, sexism, neoliberal bureaucracy, and problematic practices that shape perceptions of who belongs in academia. For example, such perceptions include what bodies are even deemed worthy to be considered a student (see McDermott, Stewart, Walker, May, & Hernández, chapter 1 in this edited volume), what bodies are admitted for study, what bodies are hired (and retained) as educators, what bodies progress through the ranks to become high-level administrators, and associated concerns with publishing and teaching practices (i.e., racist and sexist gatekeeping practices in journal article publishing and professors who either explicitly or implicitly perpetuate whiteness and hegemonic masculinity in higher education through the construction of their syllabi, the texts selected for inclusion, and the topics taught). As Harris (2015) reminds us, "colleges and universities historically have supported hierarchies of race and other forms of difference from their founding in the colonial era through the civil rights struggles of the late-twentieth century" (para. 5). Racism, sexism, homophobia, and transphobia manifest both insidiously and explicitly in academia, and such manifestations have resulted in resistance from students and educators alike. For example, although the U.S. Military Academy at West Point, the U.S. Air Force Academy in Colorado Springs, and the U.S.

Naval Academy in Annapolis have increased percentages of women and nonwhites enrolling and graduating (a finding that has increased since 1992) (Jurby et al., 2010), efforts to racially diversify the U.S. military academies are a recent phenomenon (Hansen, 1999), and racial struggles and strife persist. West Point made news headlines recently because West Point graduates wrote a policy proposal calling for a more "anti-racist West Point" (Glenn, 2020). In this proposal, alumni called for "removing names, monuments, and art honoring the Confederacy; investigating the disciplinary system for racially discriminatory punishments; and improving anti-racism training" to provide more equitable, inclusive, and anti-racist spaces for cadets of color (Glen, 2020, para. 4). Other Black cadets have come forward, describing how "they were subjected to racism and weren't taken seriously when they reported it to authorities" (Choi, 2020, para. 1).

Similar modes of resistance are occurring in the communication discipline, with transformative impacts on discourses and practices pertaining to research, pedagogy, and mentorship. In recent years, the hashtag and journal article #CommunicationSoWhite took the communication discipline by storm as it called to attention racial inequalities and disparities in academic publishing in the communication discipline (Chakravartty et al., 2018). The study found that non-White scholars are underrepresented in publication rates, citation rates, and editorial positions, highlighting the insidious, intertwined nature of racism, white hegemony, and power in the communication discipline (Chakravartty et al., 2018). Similarly, Hernández and Martinez (2019) found that Latina/o/x populations are highly underrepresented in research in the communication discipline's top health communication journals. In line with #CommunicationSoWhite, other discourses were created that highlight white hegemony in sub-disciplines, such as #HealthCommSoWhite, #RhetoricSoWhite, and #BlackInTheIvory (Asante, 2019; Na'puti, 2019; Pham, 2019; Sowards, 2019; Subbaraman, 2020; Tarin et al., 2019; Wanzer-Serrano, 2019). Both studies highlight not only the problematic, hegemonic nature of academic publishing but also the problematic implications of such white hegemony on our pedagogical practices. Taken together, such studies (in a sea of others) argue that representation matters—how can our students, particularly those who are MALs who identify as anything other than white, cis-gender, heterosexual males, understand the communication discipline and their experiences as being important when their communities and lived experiences are not represented in the research being published or the readings and assignments selected for inclusion in their classrooms? Manning (2020) found, for example, that several foundational interpersonal communication textbooks were dominated by whiteness in the images presented, the theories presented, and the ways in which diverse groups were othered. Ultimately, we contend with Smith (2019), who asserts that communication pedagogy is a "double-edged sword that simultaneously liberates and subjugates" (p. 35). Smith (2019) urges communication faculty members to examine our pedagogical strategies and decenter whiteness; we heed this call and extend it to argue that our pedagogies must decenter hegemonic white masculinity and heteronormativity, as well, particularly for MALs who find themselves situated within the military-civilian-academic divide and lie at the intersection of so many subject positions, identities, and experiences. In the next section, we propose a borderlands approach to MAL pedagogy that seeks to achieve such calls, in addition to providing a more inclusive pedagogical space for the MAL experience.

A BORDERLANDS APPROACH TO MAL PEDAGOGY

Given the issues in higher education pertaining to diversity, inclusivity, and the MAL experience, we propose a borderlands approach to MAL pedagogy, a framework that creates a more inclusive space conscious of MALs' multiple, intersecting identities and experiences. Leandra, for example, is a third-/fifth-generation pansexual Mexican-American (Tejana) woman who was born and raised in Texas.[3] She has several family members who proudly served in the U.S. military and is the spouse of a veteran who served ten years in the U.S. Navy. Leandra was first a MAL when she completed her graduate education and later transitioned to educating, mentoring, and collaborating with MALs in research contexts while teaching at various higher education institutions with large MAL student populations. Sarah is a third-generation Mexican-American who was born, raised, and continues to live on the Mexico-U.S. border. Her Mexican-American grandfather was able to obtain a college degree because of his service in World War II. Sarah's father, who is white, is a veteran of the United States Air Force; however, she has never considered herself a MAL because she lived with her mother on the border growing up, and this is where she ultimately attended college. While she does not consider herself a MAL, her father's affiliation with the military, as well as their differences in cultural and ethnic identities, necessitated her learning to navigate intercultural communication from a young age and thus informs her approach to research on identity construction and borderlands pedagogy. Art is also a third-generation Mexican-American born and raised on the border and continues to live and work there. However, when he applied for and interviewed for a visiting assistant professor position in the Leadership and Community Engagement program at his institution, he had no idea that this department had a contract with the Department of Defense. He was also unaware that a majority of their students were in the military, that many were dual-enrolled in the United States Army Sergeant

Majors Academy (USASMA), and that he would be teaching on base at Fort Bliss. This was his first experience with MALs. Once hired, he took on these challenges and also went from teaching undergraduate students that were often Latinx and between eighteen and twenty years old to graduate students who were typically older than him and mostly white. Here, we acknowledge that our own process of writing ourselves into existence by sharing our narratives highlights the diverse, proud, and sometimes tenuous relationships that we have as MALs and MAL educators with the U.S. military in personal, relational, and educational spaces.

As we have encountered physical and psychological borders in our own lived experiences, our perspectives ultimately privilege fluidity and hybridity. In other words, we exist among two or more cultures, languages, and countries; in order to adapt, we find ways to move in and out of language and culture, back and forth between territories, and create ways of moving through the world that combine these experiences to create identities grounded in the mixture of these elements. We are, therefore, equipped with the knowledge and appreciation that MALs have diverse, rich lived experiences and, as such, need pedagogical frameworks that provide a space to honor the diversity of their lived experiences and their academic needs. Based on our experiences both inside and outside of classrooms, as students and as faculty members, our approaches to teaching and learning are informed by a pedagogy of the borderlands (Elenes, 2011; Hernández & De Los Santos Upton, 2020; De Los Santos Upton & Hernández, 2020). Because we perpetually exist in the in-between state of nepantla[4] we are accustomed to ambiguities and learn to not only embrace but also flourish in chaos, fluidity, and hybridity (Anzaldúa, 2007). As Anzaldúa (2007) explains, "nepantla is a liminal, in-between stage characterized by chaos and disorientation, where individuals experience disassociations, breakdowns, and buildups of their identities" (De Los Santos Upton, 2019, p. 124). A borderlands pedagogical perspective honors the potential and possibility inherent to nepantla (De Los Santos Upton, 2019) and allows us to draw from the vivencias (Trinidad Galván, 2001), or lived realities, of our students in order to "do justice to their lived experiences and equip them with tools to help them move through academia and through the world more effectively" (Hernández & De Los Santos Upton, 2020, p. 92). As Galván (2001) explains, our students enter into our classrooms equipped with the knowledge created with their families at their kitchen tables, yet know all too well that these diverse forms of knowledge are not always welcome or taken seriously in college classrooms.

In addition to the potential for valuing vivencias, borderland pedagogy also creates possibilities for transformation both on and off campus. Elenes (2011) offers the concept of border/transformative pedagogies as an approach to knowledge creation that honors multiple subject positions while simultaneously working to interrogate and transform social inequalities (Elenes, 2011).

She argues that much can be learned from rooting pedagogical approaches in the experiences of marginalized groups: “Certainly, understanding Chicanas’ subjectivity as active agents in their lives who are capable of offering a critique and struggle against multiple forms of oppression is necessary for progressive educational practices” (Elenes, 2011, p. 3). While we recognize that a majority of MALs do not identify as Chicana specifically, we argue that a borderlands approach to pedagogy can decentralize the hyper-focus on the military component of MALs (i.e., treating MALs as *only* active-duty service members or veterans or spouses) and can instead provide a more intersectional approach to identity, community, and educational experiences. As communication scholars and professors, we are especially attuned to the ways in which conflicts over meaning are embedded within classroom discussions. Elenes (2011) explains that these conflicts over meaning “result from distinctive ideological positions within different social actors and the power to decide which of these ideologies has more currency. Indeed, these conflicts represent different views on values, morality, spirituality, and social justice” (Elenes, 2011, p. 3). A borderlands pedagogical perspective creates space for these conflicts over meaning and the differing perspectives that students bring with them into the classroom, while at the same time privileging first and foremost social justice.

While these principles broadly guide our approaches to teaching and learning in the classroom, there are specific strategies we utilize to actively put these concepts into practice in our classrooms. As we work to prepare our syllabi, we are conscious about choosing readings that will deconstruct the white, colonial canon, and we instead privilege including decolonial works by authors that have experienced marginalization in the field of communication (Clark, 2020).[5] For example, in Leandra’s Health Communication & Culture course, she purposefully includes readings by black, Chicana, Latinx, indigenous, Asian-American, queer, and critical health communication researchers that highlight the intertwined nature of power, hegemony, and racism in shaping health experiences and histories. We adopt assignments and modes of teaching and assessment that are anti-racist and anti-sexist (Hernández & De Los Santos Upton, 2020; De Los Santos Upton & Hernández, 2020). For example, we have used photo essays, poetry, and testimonios in the form of journals and short reflections, finding that these assignments disrupt the rigid academic standards that far too often privilege white ways of knowing. These particular assignments make space for variations in language use, grammar, and most importantly, questions about what counts as knowledge. In the creation of assignments and facilitation of classroom discussions, we provide space for acknowledging and celebrating multiple identities. By recognizing that chaos and ambiguity are a part of nepantla life, we help students navigate difficult conversations in our classrooms and honor multiple perspectives while creating safe and brave spaces that validate the identities and vivencias of our students. Ultimately, as hooks (1994) asserts, "To teach in a manner that respects and cares for the souls of our students is essential if we are to provide the necessary conditions where learning can most deeply and intimately begin" (p. 13). We contend with hooks (1994) that such an approach lays the foundation for a borderlands pedagogy, and in the remainder of this chapter we will discuss specific examples of how we have used borderlands pedagogy with MALs in our own courses.

A BORDERLANDS MAL PEDAGOGY IN PRACTICE

This section provides a brief glimpse into how we as three borderlands communication scholars with diverse links to the MAL experience utilize a borderlands MAL pedagogical approach in the communication classroom. We acknowledge that our approaches are not generalizable from a borderlands pedagogy perspective, but rather serve as micro-case studies. Our hope is that these case studies provide a few examples for future pedagogical inquiry and practice.

Sarah

While I have never considered myself a MAL, or considered my classrooms specifically targeted toward MALs, I recognize that the very nature of teaching from a borderlands pedagogical perspective has created space for my students to critically interrogate their MAL identities as they intersect with other facets of their lived experiences. For example, vivencias play an important role in my course on interpersonal communication. Students come to class prepared to talk about the development of their identities, reflect on their perceptions of others, and make sense of the important relationships in their lives. As they work through the course readings, they write journal entries focusing on what parts of the chapters and articles were most meaningful for them and centering their own lived experiences. The journal assignment intentionally lacks a prompt so that students are instead encouraged to see the value of their vivencias in creating knowledge about the world around them. One former student, whose husband was deployed during the semester she took my class, immediately recognized the value of interpersonal communication in navigating the struggles of her (temporary) long-distance marriage. She often reflected on how concepts from our course were helping her and her spouse have more effective and satisfying communication that helped them feel connected across a great distance; at the same time, the insights she was gaining from this experience helped to complicate traditional concepts of interpersonal communication in important ways. In this course, I am also able to draw from my own MAL experience with my father in ways I did not expect. For example, when we discuss nonverbal communication and the concept of time, I often share that my father and I have completely different understandings of time as his military service has instilled a monochronic approach (another component of the military-culture-academia divide), while my lived experiences have led me to a polychronic orientation. In other words, his approach is punctual, strict, and rigid, while mine is more flexible and fluid. This time distinction exemplifies the divide I sometimes feel in our relationship resulting from our radically different approaches to culture, identity, and communication, with his largely being informed by his military service and mine largely being informed by my upbringing in a Mexican-American household on the Mexico-U.S. border. Through the lens of borderland pedagogy, I share these examples of a kind of border that I must cross within my family, which provides an example to help students creatively explore the borders of their own experiences: borders surrounding their military experiences and affiliations; their race, gender, ethnicity, and sexual identities, among others; and their own culturally informed approaches to communication and sensemaking. This example also leads students who may not have previously identified themselves as MALs to connect to my experience and share their own.

Art

As Anzaldúa (2015) explains, my identity was broken down and built up by the inherent push and pull of the border. More importantly, my search for el árbol de la vida[6] allowed me to create my identity while combining my ancestors, my culture, the people around me, and the environment in which I was raised and in which I live (De Los Santos Upton, 2019). My experience code-switching between English and Spanish, crossing between El Paso and Ciudad Juárez, and explaining my work academically with professors and colleagues and informally with family and friends provided a suitable borderlands framework to explain concepts like racial inequality, privilege, hegemony, and whiteness to students who often show reticence and sometimes outright indignation toward the "belief" in such concepts.

This approach I take utilizing my nepantla experience is best explained through my lecture on privilege. In my first instance of covering this topic with MALs, I realized even mentioning the word "privilege" prevented some students from even engaging in the conversation and discussion. Many of my MALs would explain that they are fortunate that the military "does a good job of making us a team that does not focus on race or gender," a perception that differs from Leandra's experiences with certain MALs in the classroom. It was then I realized that I needed to bring them back to the border in which they live their day-to-day lives. While the majority of their time is spent in the military, they cross the border between their military lives and their civilian lives nearly every day. Given that I was teaching on base, it was fairly easy for them to position this discussion on privilege within their military framework. I decided I had to frame the conversation differently and verbally walk the border between the civilian and military worlds. My first decision was to not use the word "privilege." Second, I had to start with privileges that were easier for cis-white males to see and understand, as that was primarily the demographic of my MALs.

First, I explained my own personal experience when a mentor of mine during my doctoral program showed me the privilege I have with regard to my sexuality, in that it is not something I have to think about or give much consideration. I then explain my experience during an employee onboarding seminar where some female colleagues explained their experience with sexual harassment and assault. More specifically, my female colleagues explained that at a young age, they had family members and friends give them advice on what to do to avoid being sexually assaulted when walking to their car at night (i.e., park under a light, put their keys between their fingers, carry mace, and so on). I explain that I have never had these types of conversations because as a male, I do not have the same worries or concerns about my safety as a female. This teaching moment also has important parallels to the military, as sexual assault reform in the military is currently underway. I segue this second point into the discussion on white privilege and say that similarly, people of color have been and are being treated differently because of the color of their skin.

In previous seminars, this lecture, and more specifically the mention of privilege, would yield a litany of rebuttals. A number of students have responded with something along the lines of, "Well just because I'm white/straight/a male doesn't mean everything was easy. We grew up very poor." In one course, I had a student respond to this lecture by vehemently denying that people of color get treated unfairly. In fact, this student claimed, "the Blacks actually have it better than us now; they have the Black Caucus and affirmative action." Anticipating these rebuttals, I added a section to my lecture where I explain that these privileges do not mean their lives won't be difficult, nor does it mean they won't have to deal with hardships. It just means that their sexuality, gender, or skin color will not make their lives that much harder.

This winding conversation with points of reference and familiarity in order for them to understand the critical tenets of my lecture is similar to the winding conversations I had with my grandma where my tongue crossed the border between English and Spanish in order for sensemaking to take place. This conversation echoes the exchanges with family members and friends where I had to use cultural or personal references to get them to understand the academic work I was doing. This in-between I lived and continue to live in prepared me to engage in these conversations where sensemaking and understanding are predicated on difficult, and often painful, realizations that lead to transformation. More importantly, though MALs often boast about the military's ability to create a team/family mentality, this mentality often conflates patriotism and whiteness. In turn, this conflation leads to the erasure of other identities. Thus, this borderlands pedagogical approach requires MALs to mentally and verbally traverse the border between civilian and military life and between the oppressed and the oppressor.

Leandra

Similarly to Sarah and Art, my identity ebbs and flows between and among several subject positions and identity categories at any given moment: Mexican-American, Chicana feminist, regionally identified Tejana, pansexual, MAL, and now MAL educator. My spouse, who was a Naval officer for ten years, recently transitioned to civilian life; as a result of my marriage, I was first a MAL and am now a MAL educator. From 2014 to 2019, I taught student populations that were almost exclusively MALs (mostly service members and veterans, although there were several spouses and children); through my own personal experiences and my educational/mentoring experiences, I have fiercely advocated for MALs at every institution with which I have been affiliated. As a result of my husband's military service, we spent five of his ten years of service physically separated while I completed my graduate work and when he underwent trainings, missions, and deployments. Combined, Sarah, Art, and I, whether on the side of the MAL or the side of the MAL educator, understand how the military lifestyle can impact MALs for better or for worse in educational spaces, and thus we are passionate about this borderlands pedagogy framework and its transformative possibilities.

Similar to Art, I, too, have had several MALs boast about the military family mentality, otherwise known as the "brotherhood," unknowingly conflating whiteness with patriotism and also unknowingly erasing their non-male and non-masculine counterparts from this larger framework. As I have discussed elsewhere (Hernández, 2020), the "military classroom battlefield" necessitates the use of critical love (Calafell & Gutierrez-Perez, 2018; Cummins & Griffin, 2012; Griffin, 2012) and extra care, precisely because of the military-civilian-academic divide discussed here and in other chapters in this volume. A borderlands pedagogical approach rooted in critical love builds upon Lorde's (1984) work and positions "love as a means to nourish critical intercultural selfhood and build coalitions among identity differences [to incite] a sense of hope" and connection (Griffin, 2012, p. 206). Critical love acknowledges the emotional labor, toil, and trauma associated with having difficult conversations in the college classroom, acknowledges and supports these emotions for all involved, and supports humanization through dialogue and shared efforts to build mutual understanding and respect (Griffin, 2012; Hernández, 2020).

While on the one hand, the deployment of the "military brotherhood" rhetorical phrase (at least as I have experienced in different classrooms) engenders service member pride, solidarity, and an intense sense of social support and connection; however, several women, non-binary, and LGBTQ MALs in the classroom have expressed both privately and publicly their dis-identification with the term because it is masculine-specific and, some worry, excludes women and non-binary service members in problematic ways. This discussion once manifested in an online discussion forum for a media representation class, a discussion that I did not see until the next day. A male student in the class was known to use the "brotherhood" phrase consistently, and he later made a homophobic, sexist remark, to which a woman classmate resisted by stating that his comment directly conflicted with his consistent use of the "brotherhood" phrase. She then questioned, quite directly, if that phrase was meant only for his male comrades or if she and her fellow women and non-binary service members were included in the phrase as well. That was a community learning moment for the entire class, myself included; as a result, in future sections of media representation, health communication, and gender and communication, I have devised assignments, both reflective and discussion-based, where I encourage MALs to deconstruct the broader identity implications of that phrase in relation to the growing diversity in the U.S. military. In my Media Representation course, for example, in the unit on identity politics and representations, students are asked to analyze popular media representations of themselves (whether that be active duty, veteran, spouse, or military child) in relation to their identity experiences from a race/ethnicity, gender, class, sexuality, and disability point of view. Using the rhetoric of "the military brotherhood" as the guiding concept, students are provided the space to safely and constructively deconstruct both their own identities and media representations of the military more broadly. Upon assignment completion, I dedicate at least one class period to assignment debriefing where, from an intersectional perspective, we map MAL identities in relation to media representations. MALs have shared that this assignment has helped them envision identity through a broader, more dynamic lens, and it also helps them communicate more effectively with their fellow MALs. The assignments also provide space for MALs who are not service members to think about what identity phrases and "rallying cries," so to speak, would work most effectively for them as they work to communicate the multiple, intertwined categories of their identities.

CONCLUDING THOUGHTS

In this chapter, we have discussed the development of the parallel histories of the U.S. military and higher education with a focus on racism, sexism, homophobia, transphobia, and larger concerns associated with inclusion, diversity, and exclusion. As MALs and MAL educators, we have described what a borderlands pedagogy is and how it can be applied to MALs in higher education to facilitate and foster more inclusive, productive spaces for MALs in communication contexts to de/construct their identities, build solidarity with one another, and ultimately find that (their) representation matters. Put simply, as three borderlands communication scholars, we care about the success of our MALs and propose that a borderlands approach can be one tool in a kit of many to provide a space where MALs see that they both belong and matter. Ultimately, through honoring our MALs' vivencias, we hope that a borderlands pedagogy approach in the MAL classroom can contribute to larger conversations about transformative, equitable approaches to pedagogy and more transformative relationships and experiences for educators and MALs alike.

NOTES

1. We recognize that the term "American" refers to North, Central, and South America, rather than one country. We, therefore, purposefully make the distinction "U.S. American" to avoid centering the United States as a representation of what counts as "America."
2. Although it is outside the scope of this chapter to discuss the relationship between citizenship and military service for all minoritized groups mentioned here, chapters by Han (2017), De Angelis (2017), Sohoni (2017), and Meadows (2017) further describe the complicated relationships among racism, racialized constructions, stereotypes, and systematic exclusion that all groups have faced since the 1800s that characterize the tenuous relationship between military service and citizenship.
3. Leandra uses the phrase "third-/fifth-generation" to signify that she is third-generation Mexican-America on her father's side and fifth-generation on her mother's side, which illustrates the regional identity angle of her Tejana identity in relation to Sarah and Art's identities. While Sarah, Art, and Leandra were all born in Texas, the Mexican-American and Tejana/o /x identities manifest and are experienced differently based upon city/state location and familial historias.
4. In an effort to avoid "taming wild tongues" (Anzaldua, 2007) and privileging the English language, we have purposefully chosen not to italicize or translate words from Spanish in this chapter. For a discussion of the politics of language and the labor of translation, please see Hernández, De Los Santos Upton, Bowen and Martinez (2019) and Sowards (2019).
5. At the time of publication, Leandra and Sarah were the Immediate Past Chair and Chair (respectively) of the National Communication Association (NCA) Latina/o

Communication Studies Division and the La Raza Caucus. In efforts to promote pedagogical and citational activism, here is the link to the Latina/o Communication Studies Division Working Bibliography, a resource with readings and citations written by Latina/o/x Communication Studies scholars: https://docs.google.com/document/d/1dfwXSeiHZEufMFFwXIBqH8mqspljEwu7rZUD-yVw-T4/edit?fbclid=IwAR3vgrifOcLQUZvS2MRnxBxheLd3-RZ7owhgQSPPTT0b0hUuwlr47rJN1Fk. Additionally, here is the link to the NCA African American Communication and Culture Division 2017–2018 Bibliography: https://drive.google.com/file/d/1pAVhDXqGgSd5Vov6Ea8SjUn1CTUSyAo4/view?fbclid=IwAR2n4qw7eVcZ_Q5hz3Ek9grkGCeypfJg7004MmvJWjZcz-Txo31rRZFkInY.

6. Anzaldúa (2002) describes el árbol de la vida as another option for mestizajes outside of assimilation and separation. It is the ability to merge the past, the present, people, culture(s), as well as all living beings in your environment.

REFERENCES

2018 demographics: Profile of the military community. (2018). Military One Source. https://download.militaryonesource.mil/12038/MOS/Reports/2018-demographics-report.pdf.

Anzaldúa, G. (2002). Now let us shift … the path of conocimiento … inner work, public acts. In G. Anzaldúa & A. L. Keating (Eds.), *This bridge we call home: Radical visions for transformation* (pp. 540–578). Routledge.

Anzaldúa, G. (2015). *Light in the dark/Luz en lo oscuro: Rewriting identity, spirituality, reality*. A. Keating (Ed.). Duke University Press.

Asante, G. A. (2019). #RhetoricSoWhite and US centered: Reflections on challenges and opportunities. *Quarterly Journal of Speech, 105*(4), 484–488.

Associated Press. (2020, July 4). Former Wisconsin air national guard officer under fire for comments on Vanessa Guillen. *Stars and Stripes*. https://www.stripes.com/news/us/former-wisconsin-air-national-guard-officer-under-fire-for-comments-on-vanessa-guillen-1.636376.

Barroso, A. (2019, September 10). *The Changing Profile of the U.S. Military: Smaller in Size, More Diverse, More Women in Leadership*. Pew Research Center. https://www.pewresearch.org/fact-tank/2019/09/10/the-changing-profile-of-the-u-s-military/#:~:text=The%20changing%20profile%20of%20the%20U.S.%20military%3A%20Smaller%20in%20size,diverse%2C%20more%20women%20in%20leadership&text=Today's%20active%20duty%20military%20is,diverse%20than%20in%20previous%20generations.&text=In%202017%2C%20women%20represented%2016,and%20just%201%25%20in%201970.

Belding, J. N., & Hernández, L. H. (2019). Conclusion: Advice from the trenches. In L. H. Hernández & J. N. Belding (Eds.), *Military spouses with graduate degrees: Interdisciplinary approaches to thriving amidst uncertainty* (pp. 259–267). Lexington Books.

Berube, A. (1990). *Coming out under fire: The history of gay men and women in World War II*. Free Press.

Burk, J., & Espinoza, E. (2012). Race relations within the US military. *Annual Review of Sociology, 38*, 401–422.

Calafell, B. M., & Gutierrez-Perez, R. (2018). (Critical) love is a battlefield: Implications for a critical intercultural pedagogical approach. In A. Atay & S. Toyosaki (Eds.), *Critical intercultural communication pedagogy* (pp. 49–63). Lexington Books.

Campion, S. (2020, July 29). Vanessa Guillen's family pushes for sexual harassment reform in military. *ABC 30 News*. https://abc30.com/vanessa-guillens-family-pushes-for-sexual-harassment-reform/6336955/.

Chiara, N. A., Chiara, M., & Ashlock, M. Z. (2020). What is impacting the military-affiliated learner experience in higher education. In V. McDermott, L. H. Hernandez, & A. May (Eds.), *Supporting the military-affiliated learner: Communication approaches to military pedagogy & education* (pp. 47–65). Lexington Books.

Choi, D. (2020, July 6). Black US Army cadets say they were called the n-word and 'shunned' for reporting discrimination at West Point. *Business Insider*. https://www.businessinsider.com/black-us-army-soldiers-n-word-at-west-point-2020-7.

Clark, M. D. (2020). Remaking the #Syllabus: Crowdsourcing resistance praxis as critical public pedagogy. *Communication, Culture & Critique, 13*(2), 222–241.

Cobos, A. (2020). Two-way communication as a constructivist pedagogy tool for enhancing meaning making in digital learning classrooms with military affiliated learners. In V. McDermott, L. H. Hernandez, & A. May (Eds.), *Supporting the military-affiliated learner: Communication approaches to military pedagogy & education* (pp. 257-278). Lexington Books.

Cummins, M. W., & Griffin, R. A. (2012). Critical race theory and critical communication pedagogy: Articulating pedagogy as an act of love from black male perspectives. *Liminalities: A Journal of Performance Studies, 8*(5), 85–106.

De Angelis, K. (2017). Rising minority: Hispanics in the U.S. military. In D. E. Rohall, M. G. Ender, & M. D. Matthews (Eds.), *Inclusion in the American Military: A force for diversity* (pp. 37–56). Lexington Books.

De Los Santos Upton, S. (2019). Communicating nepantla: An Anzaldúan theory of identity. In L. H. Hernández & R. Gutierrez-Perez (Eds.), *This bridge we call communication: Anzaldúan approaches to theory, method, and praxis* (pp. 123–142). Lexington Books.

De Los Santos Upton, S., & Hernández, L. H. (2020). Teaching Anzaldúa: Strategies in the physical and psychological borderlands in civilian and military classrooms. In M. Cantú-Sanchez, C. de León-Zepeda, & N. E. Cantú (Eds.), *Teaching Gloria E. Anzaldúa: Pedagogy and practice for our classrooms and communities* (pp. 152–164). The University of Arizona Press.

Drummet, A. R., Coleman, M., & Cable, S. (2003). Military families under stress: Implications for family life education. *Family Relations, 52*(3), 279–287. doi:10.1111/j.1741-3729.2003.00279.x.

Elenes, C. A. (2010). *Transforming borders: Chicana/o popular culture and pedagogy*. Lexington Books.

Glenn, H. (2020, July 6). West Point graduates' letter calls for Academy to address racism. *NPR*. https://www.npr.org/sections/live-updates-protests-for-racial-justice/

2020/07/06/887540591/west-point-graduates-letter-calls-for-academy-to-address-racism.

Griffin, R. A. (2012). Navigating the politics of identity/identities and exploring the promise of critical love. In N. Bardham & M. Orbe (Eds.), *Identity research and intercultural communication: Reflections and future direction* (pp. 207–221). Lexington Books.

Han, J. (2017). African-Americans in the U.S. military. In D. E. Rohall, M. G. Ender, & M. D. Matthews (Eds.), *Inclusion in the American Military: A force for diversity* (pp. 19–36). Lexington Books.

Hansen, S. (1999). The racial history of the U.S. military academies. *The Journal of Blacks in Higher Education, 26*, 111–116.

Harris, G. L. A. (2009). Recruiting, retention, and race in the military. *International Journal of Public Administration, 32*, 803–828.

Harris, L. M. (2015, March 26). The long, ugly history of racism at American universities. *The New Republic*. https://newrepublic.com/article/121382/forgotten-racist-past-american-universities.

Hernández, L. H. (2019). "We get it done because we have to": Military spouses with advanced degrees, career and educational experiences, and grit amongst uncertainty. In L. H. Hernández & J. N. Belding (Eds.), *Military spouses with graduate degrees: Interdisciplinary approaches to thriving amidst uncertainty* (pp. 239–258). Lexington Books.

Hernández, L. H. (2020). Identity politics in the classroom: An application of critical love to teaching difficult topics in the military classroom battlefield. In C. R. Webber (Ed.), *Working in the margins: Domestic and international minority women in higher education* (pp. 15–32). Peter Lang.

Hernández, L. H., & Belding, J. N. (2020). Best practices in supporting military-affiliated learners: Suggestions from communication and psychological sciences. In V. McDermott, L. H. Hernández, & A. May (Eds.), *Supporting the military-affiliated learner: Communication approaches to military pedagogy & education* (pp. 231–256). Lexington Books.

Hernández, L. H., & Belding, J. N. (2019). Introduction: Military spouses with advanced degrees: Current findings and pathways moving forward. In L. H. Hernández & J. N. Belding (Eds.), *Military spouses with graduate degrees: Interdisciplinary approaches to thriving amidst uncertainty* (pp. 1–12). Lexington Books.

Hernández, L. H., & De Los Santos Upton, S. (2020). Honoring vivencias: A borderlands approach to higher education pedagogy justice. In K. C. Blinne (Ed.), *Grading justice: Teacher-activist approaches to assessment* (pp. 91–107). Lexington Books.

Hernández, L. H., De Los Santos Upton, S., Bowen, D. I., & Martinez, A. R. (2019). Latina/o/x communication studies: Current considerations and future directions. In L. H. Hernández, D. Bowen, A. R. Martinez, & S. De Los Santos Upton (Eds.), *Latina/o/x communication studies: Theories, methods, and practice* (pp. 1–43). Lexington Books.

Hernández, L. H., & Martinez, A. R. (2019). Latinidad, heterogeneity, and health communication scholarship: A literature review analysis of Latina/o/x representation.

In L. H. Hernández, D. Bowen, A. R. Martinez, & S. De Los Santos Upton (Eds.), *Latina/o/x communication studies: Theories, methods, and practice* (pp. 97–115). Lexington Books.

Hill, C. B., Kurzweil, M., Pisacreta, E. D., & Schwarz, E. (2019). *Enrolling more veterans at high-graduation-rate colleges and universities.* https://sr.ithaka.org/publications/enrolling-more-veterans-at-high-graduation-rate-colleges-and-universities/.

hooks, b. (1994). *Teaching to transgress: Education as the practice of freedom.* Routledge.

Howe, W. T. (2020). Supporting military veterans in the graduate classroom: A case study of political beliefs and discussion. In V. McDermott, L. H. Hernández, & A. May (Eds.), *Supporting the military-affiliated learner: Communication approaches to military pedagogy & education* (pp. 67–86). Lexington Books.

Human Rights Campaign. (2020, March 2). *Transgender military service.* https://www.hrc.org/resources/transgender-military-service.

Jones, M. (2020, July 5). UWM instructor and retired Wisconsin air national guard colonel apologizes for her comments about slain soldier. *Milwaukee Journal Sentinel.* https://www.jsonline.com/story/news/local/wisconsin/2020/07/05/uwm-lecturer-retired-national-guard-colonel-says-remarks-sexual-harassment-military-misinterpreted/5380797002/.

Katz, B. (2019, July 2). Diary sheds light on Deborah Sampson, who fought in the Revolutionary War. *Smithsonian Magazine.* https://www.smithsonianmag.com/smart-news/diary-sheds-light-deborah-sampson-who-fought-revolutionary-war-180972547/.

Kirby, S. N., Thie, H. J., Naftel, S., & Adelson, M. (2010). *Diversity and the success of entering classes at the U.S. Service Academies.* RAND Corporation. https://www.rand.org/pubs/research_briefs/RB9496.html.

Lopez, G. (2017, January 17). Letting LGBTQ soldiers serve isn't just for equality. It also makes the military stronger. *Vox.* https://www.vox.com/2016/7/1/12063304/lgbtq-transgender-military.

Lopez, G. (2019, January 22). Trump's ban on transgender troops, explained. *Vox.* fhttps://www.vox.com/identities/2017/7/26/16034366/trump-transgender-military-ban.

Lorde, A. (1984). *Sister outsider: Essays and speeches.* Crossing Press.

Ludewig, A. (2020). Collaborative apprenticeship approaches to writing instruction in the military. In V. McDermott, L. H. Hernández, & A. May (Eds.), *Supporting the military-affiliated learner: Communication approaches to military pedagogy & education* (pp. 311–336). Lexington Books.

Manning, J. (2020). Examining whiteness in interpersonal communication textbooks. *Communication, Culture, & Critique, 13*, 1–5, First Published Online.

May, A., & McDermott, V. (2020). Understanding dual identity formation: Identifying commonalities and resolving "differentness" for military affiliated learners. In V. McDermott, L. H. Hernández, & A. May (Eds.), *Supporting the military-affiliated learner: Communication approaches to military pedagogy & education* (pp. 11–45). Lexington Books.

May, A., & McDermott, V. (2019). Cultural collision: A review of literature. In L. H. Hernández & J. N. Belding (Eds.), *Military spouses with graduate degrees: Interdisciplinary approaches to thriving amidst uncertainty* (pp. 59–85). Lexington Books.

McDermott, V., Stewart, M. C., Walker, C. M., May, A., & Hernández, L. H. (2020). Setting the stage: MALs in higher education. In V. McDermott, L. H. Hernández, & A. May (Eds.), *Supporting the military-affiliated learner: Communication approaches to military pedagogy & education* (pp. 3–10). Lexington Books.

McNamara, K. A., Lucas, C. L., Goldbach, J. T., Holloway, I. W., & Castro, C. A. (2020). You don't want to be a candidate for punishment": A qualitative analysis of LGBT service member "outness." *Sexuality Research and Social Policy*, First Published Online.

Moreno, J. E. (2020, July 3). LULAC president urges Latinas not to join the military after disappearance of Fort Hood soldier. *The Hill.* https://thehill.com/homenews/state-watch/505819-lulac-president-urges-latinas-not-to-join-the-military-after.

Moullet, Z., & Belding, J. N. (2020). Supporting the military-affiliated learner's medical school aspirations. In V. McDermott, L. H. Hernández, & A. May (Eds.), *Supporting the military-affiliated learner: Communication approaches to military pedagogy & education* (pp. 159–180). Lexington Books.

Myers, J. (2020, June 15). LGBT service members are allowed to be out and proud, but a fear of repercussions persists. *Military Times.* https://www.militarytimes.com/news/your-military/2020/06/15/lgbt-service-members-are-allowed-to-be-out-and-proud-but-a-fear-of-repercussions-persists/.

Na'puti, T. R. (2019). Speaking of indigeneity: Navigating genealogies against erasure and #RhetoricSoWhite. *Quarterly Journal of Speech, 105*(4), 495–501.

National Center for Transgender Equality. (n.d.). *Military service by transgender people: Data from the 2015 U.S. Transgender Survey.* https://www.transequality.org/sites/default/files/docs/usts/USTS-VeteransDayReport.pdf.

Pérez, F. G. (2020). Speaking Chicana/o secrets to military students: An autoethnography of teaching senior military noncommissioned officers. In V. McDermott, L. H. Hernández, & A. May (Eds.), *Supporting the military-affiliated learner: Communication approaches to military pedagogy & education* (pp. 109–131). Lexington Books.

Pham, V. N. (2019). The threat of #RhetoricNotSoWhite. *Quarterly Journal of Speech, 105*(4), 489–494.

Ramirez, M. H., & Sterzing, P. R. (2017). Coming out in camouflage: A queer theory perspective on the strength, resilience, and resistance of lesbian, gay, bisexual, and transgender service members and veterans. *Journal of Gay & Lesbian Social Services, 29*(1), 68–86.

Rohall, D. E., Ender, M. G., & Matthews, M. D. (2017). Diversity in the military. In D. E. Rohall, M. G. Ender, & M. D. Matthews (Eds.), *Inclusion in the American Military: A force for diversity* (pp. 1–15). Lexington Books.

Schloesser, K. (2010, October 27). The first women of West Point. *Army.mil.* https://www.army.mil/article/47238/the_first_women_of_west_point.

Shane, L. (2009, January 16). Obama wants to end "Don't Ask, Don't Tell" policy. *Stars and Stripes*. http://www.stripes.com/news/obama-wants-to-end-don-t-ask-dont-tell-policy-1.87294.

Smith, L. (2019). Can we share the light? De-centering communication whiteness with communication pedagogy. *Departures in Critical Qualitative Research, 8*(4), 35–40.

Snyder, R. C. (2003). The citizen-soldier tradition and gender integration of the US military. *Armed Forces & Society, 29*(2), 185–204.

Sohoni, D. (2017). Fighting to belong: Asian-American military service and American citizenship. In D. E. Rohall, M. G. Ender, & M. D. Matthews (Eds.), *Inclusion in the American Military: A force for diversity* (pp. 57–81). Lexington Books.

Sollitto, M., & Cole, C. (2019). Military spouses' uncertainty management: Navigating academic goals and military needs. In L. H. Hernández & J. N. Belding (Eds.), *Military spouses with graduate degrees: Interdisciplinary approaches to thriving amidst uncertainty* (pp. 125–142). Lexington Books.

Sowards, S. K. (2019). #RhetoricSoEnglishOnly: Decolonizing rhetorical studies through multilingualism. *Quarterly Journal of Speech, 105*(4), 477–483.

Sowards, S. K. (2019). Linguistic capital: Latinx and English/Spanish language fluency. In L. H. Hernández, D. Bowen, A. R. Martinez, & S. De Los Santos Upton (Eds.), *Latina/o/x communication studies: Theories, methods, and practice* (pp. 73–95). Lexington Books.

Subbaraman, N. (2020, June 11). How #BlackInTheIvory put a spotlight on racism in academia. *Nature*. https://www.nature.com/articles/d41586-020-01741-7.

Tarin, C. A., Hernández, L. H., De Los Santos Upton, S., & Sowards, S. K. (2019). Across bridges and islands: Borderlands perspectives on merit, whiteness, and privilege. *Departures in Critical Qualitative Research, 8*(4), 69–75.

Vaccaro, A. (2015). "It's not one size fits all": Diversity among student veterans. *Journal of Student Affairs Research and Practice, 52*(4), 347–358.

Wanzer-Serrano, D. (2019). Rhetoric's rac(e/ist) problems. *Quarterly Journal of Speech, 105*(4), 465–476.

Woods, T. (2019, June 30). I was kicked out of the military for being gay. *The Nation*. https://www.thenation.com/article/archive/trans-military-dadt/.

Ziobro, M. (n.d.). *"Skirted soldiers": The Women's Army Corps and gender integration of the U.S. Army during World War II*. The Army Historical Foundation. https://armyhistory.org/skirted-soldiers-the-womens-army-corps-and-gender-integration-of-the-u-s-army-during-world-war-ii/.

Chapter 8

Supporting the Military-Affiliated Learner's Medical School Aspirations

Zea Moullet and Jennifer N. Belding

INTRODUCTION

Although the frequency of medical school applicants has been steadily rising over the past decade (Association of American Medical Colleges [AAMC], 2019), estimates suggest that there will be a sizeable gap of more than 46,000 physicians available to provide healthcare by 2032 (Dall, Reynolds, Jones, Chakrabarti, & Iacobucci, 2019). The medical school application process is competitive, with only 21,622 matriculants into M.D. programs out of a total of 52,777 applicants in 2018 alone (40.97 percent; AAMC, 2019). Success in the application process is based on a combination of cognitive factors (e.g., those assessed by the MCAT as well as GPAs) and non-cognitive factors (e.g., perceptions of motivation; Gilliland et al., 2012). One domain in which the lack of available healthcare providers is particularly problematic is in military settings, especially during times of war (Holmes, Lee, Charny, Guthrie, & Knight, 2009). Indeed, recent estimates suggest that only approximately 5 percent of medical school matriculants reported intending to participate in military service during their career, the majority of whom already had some military affiliation (AAMC, 2018). Thus, it is necessary to provide support to military-affiliated learners (MALs), including active duty service members, veterans, and their dependents (e.g., spouses and children), who are interested in becoming medical providers. Not only will supporting these students assist them in pursuing their goals, but it may also be one avenue to increase the number of qualified military medical providers.[1]

The path to becoming a doctor is arduous and is an exercise in delayed gratification in that it takes more than a decade to complete. Consider a student who has decided they want to be an orthopedic surgeon in the Navy.

In order to be successful, they must earn their bachelor's degree, apply to medical school, and be admitted. They spend four years in the highly stressful and challenging medical school environment. Regardless of their desired specialty, they will be required to learn about nearly all specialties of medicine. In their fourth year of medical school, they will also have to select and apply for residency programs in orthopedic surgery specifically. They will then go through a match process, in which they may or may not be selected for an internship and/or residency program, as this is an especially competitive specialty. If not selected for an orthopedic surgery residency, they may be forced to choose a different medical specialty or take time off and reapply for residency the following year. If they are selected for residency, they will then complete a five-year program. Taken together, they will spend at least thirteen years in higher education prior to achieving their goal. This path can take even longer if they opt to take time off between their undergraduate studies and medical school, if they need to serve as a General Medical Officer in the Navy prior to residency, or if they want to specialize further (e.g., in foot and ankle surgery) after residency.

While the path to becoming a doctor is hard for all aspiring physicians, MALs may encounter unique challenges at the undergraduate level and while applying for admission to medical school. MALs are often different from traditional medical school applicants in that they are older, are more likely to have financial constraints that require them to work while attending school, and may have a spouse and/or children (Ford & Vignare, 2015). These learners may also be subject to uncontrollable and ever-changing demands on their time (e.g., unexpected deployments), which may increase the likelihood of withdrawing from coursework or negatively impact their graduation timeline (Bond et al., 2016; Neelands, 2014). In particular, the geographic mobility associated with one's military service, remote nature of some military bases, and challenging operational tempo (Bond, Lewis, Leonard, Pollak, Guo, & Rostker, 2016; National Security Advisory Group, 2006) may make it harder for military-affiliated students to complete and document necessary prerequisites, study for the MCAT, obtain strong letters of recommendation, conduct research, and more, all of which are necessary for obtaining admission to medical school. Nonetheless, these students may be particularly resilient compared to their civilian counterparts and add a great deal of value to medical programs despite the unique challenges they may face.

In this chapter, we identify how to best support MALs with medical school aspirations by identifying common gaps in their knowledge and suggesting ways to fill those gaps. Because relatively little research to date has examined the experiences of hopeful (civilian) premedical students and none to date has focused on MALs (Lin, Parnami, Fuhrel-Forbis, Anspach, Crawford, & De Vries, 2013), we sought to use an autoethnographic approach in which we systematically examined our own experiences serving as a student support specialist and Director of Admissions for a medical school (Zea) and an undergraduate psychology and post-baccalaureate premedical program educator (Jennifer), as well as our shared experiences as spouses of United States Navy physicians.

METHOD

Autoethnography is a particularly useful method in this context because it is a method in which researchers can leverage personal experience to understand cultural or structural experiences; it is especially valuable when relatively rare experiences are of interest (Ellis, Adams, & Bochner, 2011). Furthermore, it has been used effectively for the purpose of medical education in the past (e.g., Foster, 2013). Although we set out to use narrative autoethnography, which is a systematic means of examining one's personal experiences to understand common cultural experiences (Bochner, 2012; Ellis et al., 2011), our research quickly transformed into an analytic autoethnography. Although the two types of autoethnography share many characteristics, analytic autoethnography is focused primarily on objective writing and analysis rather than demonstrating empathy and resonance with the subject (i.e., medically aspiring MALs in our case) or the reader (e.g., undergraduate instructors; McIlveen, 2008).

Anderson (2006) proposed that there are five key features of analytic autoethnography, including: "(1) complete member researcher (CMR) status, (2) analytic reflexivity, (3) narrative visibility of the researcher's self, (4) dialogue with informants beyond the self, and (5) commitment to theoretical analysis" (p. 378). Unlike various forms of narrative or evocative autoethnographies, analytic autoethnography need not focus on a single personal story or experience told in an emotionally laden way, but can rather build on a series of similar experiences over time (Anderson, 2006). In part because we are members of the research group or setting (i.e., those who support medically aspiring MALs), we maintained our analytic reflexivity (through our self-awareness and reflection of how we have directly interacted with medically aspiring MALs), and we remained committed to theoretical analysis. As such, analytic autoethnography was the ideal autoethnographic method of choice (Anderson, 2006).

At the outset of our analytic autoethnography, we generated three prompts to systematically guide our reflections in order to identify common barriers and facilitators to medically aspiring MALs' educational pursuits. Specifically, we sought to identify: (1) what questions do self-identified medically aspiring MALs ask regarding their educational pursuits, (2) what are the knowledge gaps experienced by these MALs, and (3) what advice is perceived by these MALs to be particularly beneficial. We independently reflected on our personal experiences and transcribed salient responses to each question over the course of approximately two weeks in preparation for an in-person meeting to review our individual reflections. During this review process, we discussed a variety of examples of personal recollections exemplifying our experiences and organized these reflections around common themes based on a mostly chronological order reflecting the career trajectory of these learners.

One of the purposes of autoethnography is to use understanding gained from personal experience to reach a broad audience (Ellis, Adams, & Bochner, 2011). As a result, we as researchers should understand the audience that we are trying to reach (Bochner, 2012). While writing this autoethnography, we focused on tailoring our understanding to those who are in positions to support medically aspiring MALs, such as instructors and administrators at institutions of higher education, rather than medically aspiring MALs themselves. While some have argued that autoethnography ought to be emotional, evocative, and engaging (e.g., Ellis, Adams, & Bochner, 2011), others have suggested that not all autoethnography must have this form (Anderson, 2006). While an analytic autoethnography results from engaged dialogue rather than detached discovery (Anderson, 2006), our experiences are inherently more detached as supporters of medically aspiring MALs (rather than being one of these MALs ourselves) because we do not directly experience their struggles, yet we understand and sympathize with them as members of a close, affiliated community. Our autoethnography is less emotionally laden than traditional autoethnographies primarily because the support role we provide to students is inherently more distanced than if we were the medically aspiring MALs ourselves.

RESULTS AND DISCUSSION

As we discussed our responses to the research questions, we realized that our responses were far more integrated than we initially anticipated. We expected the responses to our three questions to be distinct and correspond to different themes, such that the questions MALs asked would differ from their gaps in knowledge, but instead found that we were almost constantly reiterating the same motifs for each of the three questions. For example, we found that MALs often ask questions about the things that are also their gaps in knowledge, yet the way they ask their questions is often insufficient for the purpose of obtaining the requisite knowledge to alleviate the gap. Although we reviewed findings for each research question individually, we present them here in an order representing the chronological progression MALs may face on their journey to medical school. We begin with a discussion of the questions and gaps students have with regard to the preapplication phase of their careers, move on to the medical school application phase, and conclude with a discussion of themes relating to what we termed "superstar advice" (i.e., advice our medically aspiring MALs have told us was the most beneficial to them).

Preapplication Phase

Making Appropriate Choices

The most striking theme that emerged with regard to the preapplication phase was that medically aspiring MALs, like traditional students, may struggle to make choices. When deciding to pursue an undergraduate degree, these students face many questions, some of which they asked us and some of which they didn't. Which undergraduate institution should they choose? Should they go to school part time or full time? What major should they select? Should they work toward a minor? How will they finance their undergraduate education? How many classes should they take simultaneously?

A student's choice of undergraduate institution is an early and tremendously important decision in that not all undergraduate institutions are created equal in the eyes of an admissions committee. For instance, students who obtain a degree from a purely online institution are typically evaluated with more caution than those from a brick-and-mortar institution (Adams, 2009). While online courses can provide quality learning, they can be difficult to standardize, and some medical schools may not accept online courses. Although it is not impossible for students from purely online schools to gain acceptance into medical school, it is rare and typically requires much more effort from the student (Adams, 2009; Lieberman, 2019). Similarly, students at community and junior colleges can face some of the same challenges, as admissions committees may view them as less rigorous than courses taken at a brick-and-mortar four-year institution (Adams, 2009). At the same time, these community and junior colleges may provide a more attainable entrance into college for some MALs, including more affordable tuition and accessible class schedules, which should not be automatically discounted.

While these institutions may be options for the beginning of one's college career, it is necessary to eventually transfer to a four-year institution, which can be stressful (Rhine, Milligan, & Nelson, 2010). Although the transfer from one institution to the next is not necessarily a harmful element of one's medical school application, it is possible that repeated transitions may appear as if one is "school hopping" and can raise questions for admissions committee members. In any case, it is imperative that the student show strong academic performance to demonstrate evidence of resilience and efficacy to an admissions committee. Even so, many MALs appear to us to be relatively unaware of these considerations.

When MALs make these decisions, they may seek out educational advice from relevant DoD and/or contract staff working at military support offices (e.g., the Navy's Fleet and Family Support Center), but the way such learners phrase their questions may result in getting information of limited utility with regard to selecting appropriate institutions for their desired path. The responses they hear are likely generalized to any MALs and may neglect considerations for medical school applications and preparation specifically. Rather, someone directly familiar with medical education, such as a premedical adviser at an undergraduate institution the MAL is considering, would be a more useful source of information (Elam, Taylor, & Nelson, 1996; Wang, Lin, Lewis, Fetterman, & Gesundheit, 2015). Another helpful option may be to contact an admissions adviser for a medical school the MAL would someday hope to attend and ask them what they require from an undergraduate degree. In this case, the prospective student should ask about whether certain majors are preferred, online courses are accepted, and if there is an "expiration" of past coursework and tests. This information may not be easily accessible online and can serve as great conversation starters as opposed to annoyances for busy admissions departments. The specific requirements will vary between schools and may change slightly before the student applies, but this information will be helpful in making decisions. Obtaining answers to these questions as early as possible in one's education can provide useful motivation and guidance on important decisions (e.g., course selection and enrollment) along the way when roadblocks arise, as we discuss subsequently.

Once students enroll in their undergraduate institution, they must declare their major. In the United States as a whole, successful medical school applicants tend to major in biological sciences, physical sciences, social sciences, humanities, specialized health sciences, and math and statistics (Murphy, 2019). However, medical schools welcome students from all majors and may even evaluate less common majors more positively as a mechanism to increase diversity in their student body. What is more important and influential to an admissions committee is the student's performance (i.e., grades). Students should not be compelled to major in a discipline in which they struggle; a GPA of 3.78 in Literature will usually be more impressive than a GPA of 3.12 in Biology, provided that the student adequately completed all prerequisites. Although medical school applicants (including MALs) frequently stress over the decision of whether to pursue a minor in a particular discipline, such decisions rarely make a difference in the eyes of an admissions committee (Murphy, 2019).

Another question these students face is which courses to take and when. Many students fail to realize that events in their personal lives influence their academic pursuits; students should deviate from norms or "standards" when their personal situations require it. For example, imagine the spouse of a Marine who has two children and must decide how many classes to take in the fall when their spouse will be deployed. They intend to take Organic Chemistry, as well as two challenging courses for their major. This would be a heavy course load for any student, but the added consideration of effectively being a single parent during the deployment may limit their ability to study consistently. In this case, the student may benefit greatly from spreading out the courses over a few academic terms to maximize the likelihood of earning a strong grade in each. Of course, this matter may be further complicated if such courses are only offered during specific academic terms that require taking them concurrently or face delays in degree progress and ultimately graduation.

On a related note, one critically important component of success in the preapplication phase is the successful completion of medical school prerequisites. Applicants simply must do well in terms of their grade *and* mastery of the material (Sklar, 2013). In our experience, students may be woefully misguided about the importance of the knowledge gained when taking a "tough" professor and instead opt to take the class with an "easy" professor to get an A. This is a mistake; students should be encouraged to take the hard professor and put in the work to get the A, as this will prepare them for the admissions process (e.g., taking the MCAT), as well as the study of medicine. Medical school courses are very challenging, and a hard-earned A will better prepare the student for the rigors of the advanced curriculum (Sklar, 2013).

Understanding the Path to Medical School

Although many medically aspiring MALs believe they know what it takes to be a doctor, the concrete steps to obtaining their medical degrees may be unclear. While they may understand the general process of applying (i.e., taking the MCAT, filling out primary applications, filling out secondary applications, being interviewed), they have several knowledge gaps. Applicants often underestimate the importance of certain objective criteria in their application, such as MCAT scores and GPA. While these metrics are subject to limitations (e.g., Evans & Wen, 2007), they are still widely used in admissions decisions as a mechanism to reduce the number of applications to evaluate. If these scores are too low, the applicant is automatically out of consideration, which means the admissions committee may not realize the strength of the rest of the application.

Additionally, MALs may struggle to identify or be inaccurate in their perception of their strengths and weaknesses on paper. We could each recount specific examples in which a potential medical school applicant had overly positive views of their aptitude and chances of being admitted to medical school, as well as an example where a truly stellar applicant believed their chances of success to be minimal, though the nature of these misperceptions were incredibly varied and differed by student. Such misperceptions may be due to limited access to accurate and reliable information about the medical school applications and the training trajectory. For example, MALs may not form the same rapport with college instructors who can help guide them to useful and important information (e.g., due to transfers from one institution to the next, limited one-on-one interaction with faculty). On the other hand, MALs who joined the military after high school and have not yet had success in higher education may not be able to envision themselves embarking on such a difficult journey and may consequently miss out on an exciting and fulfilling career. Fostering self-efficacy in these students will have profound implications for student performance and ultimate long-term success.

Furthermore, MALs may suffer from other knowledge gaps that relate to the specific details of the study of medicine. Prospective medical students, particularly MALs, may lack comprehensive awareness of the variety of options for a career in medicine. For example, it is important for applicants to understand the differences and similarities between a Doctor of Medicine (M.D.) and Doctor of Osteopathic Medicine (D.O.). When most people think of a doctor, they think of an allopathic physician who has obtained their doctorate of medicine. However, osteopathic physicians are growing in number, and 25 percent of U.S. medical students are enrolled in D.O. programs (American Osteopathic Association, 2020). Both types of providers can practice in any specialty, have much of the same content knowledge, and have similar residency requirements. Additionally, one program is not generally viewed as more prestigious than the other. The primary differences are in their approach; osteopathic physicians tend to use a whole-person perspective focused on prevention, rather than focusing solely on the patient's specific medical condition or diseases, and receive training in osteopathic manipulative medicine. Ultimately, each student should choose the school that they believe will best prepare them to be a successful physician and the path that is the best fit for the way they would like to practice medicine.

We have also noticed multiple cases in which students were unaware that medical school curricula is much more structured and prescriptive than an undergraduate degree. Nearly all programs begin in the fall, and it is rarely feasible to drop a class and retake it later. It is expected that a cohort of medical students will take the same courses at the same time. Even for the strongest students, medical school is very challenging—both in material and volume of workload—and many students struggle until they find the study methods that work for them in this new environment. If a student finds that they are having a hard time keeping up, it is essential to ask for help as quickly as possible to avoid falling even further behind.

Due to the fact that the workload in medical school is so great, it is nearly impossible and highly discouraged for medical students to be employed simultaneously. The majority of medical students will finance their education through student loans (Greysen, Chen, & Mullan, 2011; Marcu, Kellerman, Hunter, Curtis, Rice, & Wilensky, 2017). The idea of borrowing such a large sum of money can be difficult, especially for MALs, who may have envisioned the G.I. Bill paying the full cost of their education. While the G.I. Bill is an excellent benefit for the MALs who are entitled to it, it is not intended to pay for this considerable level of education. Specifically, the G.I. Bill covers only approximately about 45 percent of the cost of medical school tuition and related costs (Graves, Seagle, Kocherginsky, Yang, Grace, & Shehabi, 2018). Additionally, many MALs will exhaust their G.I. Bill benefits while earning their baccalaureate degree. Some scholarship options are available, although significantly fewer scholarships are available for medical school compared to undergraduate degrees. There are options for service in exchange for tuition or loan payments, such as the Health Professions Scholarship Program (HPSP), which pays medical school tuition, fees, equipment, and potentially a housing allowance in exchange for service as a physician in the military after graduation (e.g., Ragel et al., 2011). This is a fairly popular option for medically aspiring MALs who are already familiar with the military culture and may be attracted to relatively less debt in exchange for service. Other service-related options may also be available, such as the Public Service Loan Forgiveness program and the National Health Service Corps (Friedman, Grischkan, Dorsey, & George, 2016; Saxton & Johns, 2009). It has been suggested that one's desired specialty often influences decisions about how to finance one's medical education because some specialties, such as surgery or radiology, earn substantially higher salaries and are able to repay their debt more quickly than others (Marcu et al., 2017). However, these specialties tend to be competitive, and there is no guarantee that students will ultimately achieve such high salaries. In our experience, MALs may not be aware of all of these considerations.

MCAT

One crucial element of the preapplication phase is preparing for the Medical College Admission Test (MCAT), which is a comprehensive and challenging exam designed to demonstrate an applicant's preparedness for medical school. Most applicants take the MCAT near the routine application cycle (often late in their junior year or early in their senior year). Admissions committees generally view it as one of the most important predictors of student success in medical school. Because of this, the stakes are high, and there is a lot of pressure to perform well on the exam (e.g., Bolger, 1990). Additionally, the test can be costly in terms of financial cost, time, and stress. For example, the cost to register to take the MCAT in 2020 was $320. Although the average costs for American students have not been determined recently, estimates from Canadian medical school matriculants averaged over $6,000 in total costs (including test registration, preparation materials, test day expenses such as travel to the testing facility, and opportunity costs such as lost wages for time spent studying; Fortin, Kulasegaram, Kancir, Moineau, Carpentier, & Hanson, 2018).

It is essential to study for this exam. Students often presume that their other strategies and study skills will translate effectively to studying for the MCAT when they might not. In particular, students are plagued by the planning fallacy, or the tendency to underestimate the amount of time it takes to achieve one's goals (Buehler, Griffin, & Ross, 1994). For example, students may assume they can cover an entire topic of MCAT-related material in a week when it actually takes them two weeks. As a result, they may quickly fall behind on study plans. However, many MALs, particularly those who have recently returned to academia after serving in the military, may not realize that they might lack an understanding of fundamental principles of how to study and learn effectively. They may opt to use massed practice (i.e., studying in large blocks of time immediately before an exam) rather than distributed practice studying (i.e., studying in smaller time blocks distributed over a longer period of time), despite the latter being far more effective (e.g., Seabrook, Brown, & Solity, 2004). Additionally, they may not use study groups or quiz themselves, despite the correlation between such behaviors and performance (e.g., Roediger, Agarwal, McDaniel, & McDermott, 2011). We concluded that these students may benefit greatly from having open and honest dialogues with their instructors, advisers, and mentors about study strategies and skills. We found ourselves conflicted about whether MALs may be both more and less likely to initiate such conversations than their traditional college student counterparts under different circumstances; we concluded that there were too many additional variables that can influence whether or not MALs may be more or less likely to initiate such conversations.

Letters of Recommendation

MALs also ask a great deal of questions about how to obtain letters of recommendation for medical school. The average medical school requires three letters of recommendation, which are weighed heavily (Monroe, Quinn,

Samuelson, Dunleavy, & Dowd, 2013). Some schools have requirements on who is appropriate to write the letters (e.g., a science professor, a physician). Although schools will state that they accept "three or more" letters of recommendation, submitting only three is the norm and is often preferred for extremely busy members of an admissions committee who must read all submitted documentation. The typical letter of recommendation follows a common format in which it provides context about how the author knows the applicant, objective measures of performance (e.g., grades), subjective measures of performance and capability, and typically ends with a clear statement of the strength of their recommendation for medical school matriculation (Range, Menyhert, Walsh, Hardin, Ellis, & Craddick, 1991). There is one universal truth: these letters must be as strong as possible.

While we found that military-affiliated and civilian applicants share similar gaps in knowledge regarding the letter of recommendation element of their application, MALs are more likely to have someone outside of academia write their letters. For example, corpsmen and medics often request a letter of recommendation from the medical officer to whom they reported. While this choice is appropriate as the letter writer can be a credible source of information and a good judge of effective medical school performance, most of these letter writers are woefully uneducated in how to actually write the letter. If left to their own devices, military medical providers and/or military personnel more generally prepare letters of recommendation according to military conventions and doctrine (e.g., Department of Defense Manual 5110.04). For example, these letters may be short with numbered pseudo-paragraphs and often refer to the applicant by name and rank, rather than first name or colloquialisms. In the end, they are often devoid of any true connection with the applicant. This is a poor template for a letter of recommendation for medical school, yet many MALs and their letter writers are oblivious to this. In fact, they may experience a great deal of stress to have the letter serialized despite such lengths being entirely meaningless to medical school admissions committees. In short, these applicants and their letter writers simply don't know what a letter of recommendation for academic purposes is supposed to look like or how to obtain it. Clarifying these expectations is one clear way to support medically aspiring MALs.

Even when MALs opt to have someone within academia write their letter of recommendation, the same elements of their military service that affect their coursework and college career trajectory may also hinder their ability to get strong letters of recommendation. In particular, they may have fewer options for letter writers with whom they have excellent rapport, and they may have not interacted with those letter writers recently when they are applying to medical school. As a result, by the time they request a letter, the writer may have limited memory of the student and their potential, resulting in a subpar letter if they agree to write one at all. One solution to this problem is to ask letter writers to write a letter upon the completion of some interaction (e.g., a college course, a tour of duty at a specific unit) and keep it on file until it is required. Letters written while specific details about the student are clear in the writer's mind are often substantially stronger than letters written a year or more later.

Extracurricular Activities

Another area in which MALs ask questions and have many knowledge gaps is the benefit of extracurricular activities for medical school applications. For most students, extracurricular activities are beneficial, but they are not critical. A variety of extracurricular activities, for instance, does not make up for subpar GPAs or MCAT scores (Monroe et al., 2013). The most helpful extracurricular activities are those that demonstrate knowledge, skills, and abilities relevant to medical school or the pursuit of a medical career. For example, shadowing medical providers helps applicants demonstrate commitment to the field of medicine and helps convince an admissions committee that the applicant truly understands all that is involved in being a physician. For this reason, extensive shadowing across many settings is crucial. Paid positions in medical fields (e.g., being an Army medic, a nurse's aide, or medical scribe) can serve as a reasonable (and perhaps even preferred) substitute for shadowing while allowing the applicant to earn a salary. Additionally, volunteering at a free clinic or taking a medical mission trip can also help demonstrate commitment to the field of medicine. Moreover, obtaining research experience and developing associated skills is often an implicitly required extracurricular activity because it helps build necessary skills to read and evaluate peer-reviewed scientific studies necessary to succeed as a medical provider. In particular, applicants who are able to demonstrate an ability to conduct research and present their findings orally, as well as in published peer-reviewed venues, are particularly competitive in the eyes of admissions committees (Monroe et al., 2013). Even so, many MALs are not aware of the relative importance of these activities.

APPLICATION PHASE: UNDERSTANDING THE APPLICATION PROCESS

Once students choose to begin the application process, they must decide where to apply. The application process is extensive, and 47 percent of matriculants to M.D. schools were accepted into only one school in 2019 (AAMC, 2019). Because of this, most applicants apply to many schools to increase their chances of being accepted (AAMC, 2019). As with undergraduate institutions, not all medical schools provide a good fit for their students; applicants should consider a wide range of criteria. Does the school have a strong reputation of successful graduates? Does the school have high average board scores for its students, which are heavily weighted during the residency match process? Does the school offer innovative and state-of-the-art teaching and equipment? Is the location of the school a place the applicant would like to live? Does the mission of the school align with the applicant's values? Are there support services available to students? Is the student's application sufficiently competitive for this school? Once the applicant has identified possible institutions, they need to submit their initial application via the American Medical College Application Service (AMCAS) for M.D. programs and the American Association of Colleges of Osteopathic Medicine Application Service (AACOMAS) for D.O. programs.

As part of their primary application, students must write a single strong personal statement. This is an opportunity for the applicant to sell themselves, and it is the admissions committee's first chance to evaluate the applicant as a person, not just a combination of test scores and GPA. Additionally, applicants should use it to highlight the characteristics or experiences (e.g., military service, volunteer experience) that make them unique, strong applicants with concision. Prospective applicants often fail to appreciate how much time they should spend writing their personal statement (i.e., months, not weeks or days). They should have several qualified people, ideally medical providers and professors, read it and provide feedback. The same is true for questions included in supplementary applications. Supplementary applications will be unique to each school and should be treated as such. Applicants should write each response anew and not simply borrow relevant sections from their personal statement or submit generic responses, which can be viewed as lazy and not committed to the application process.

Many MALs wonder if they should directly discuss their military affiliation when writing their personal statements. Such decisions are highly personal and the search for a one-size-fits-all prescription is futile. It is our belief that to the extent that one's military affiliation highlights their aptitude for resilience in medical school and the applicant's subsequent career, it could be a welcome addition to a personal statement. However, care must be taken to present one's experiences in a way that is compelling and inspiring to the reader; they should not engender wariness or apprehension. Such adverse outcomes can easily be avoided when applicants ask several trusted others to read their statements and provide honest and direct feedback, which students should then consider carefully as they revise their statement.

If the student's objective scores and personal statement culminate in additional interest in the applicant, they may be invited to interview. The interview is a chance to show that the in-person applicant is just as good a fit as the on-paper applicant. What surprised us both is that MALs are unaware of what the interview process entails and how formal to be. In many ways, a medical school interview should be treated like any job interview with a business professional dress code. Applicants should walk a line between being appropriately formal, yet friendly and conversational. This may be hard for MALs for whom formality may result in being laconic. Importantly, MALs should be encouraged to refer to their military experience in interviews provided that it relates directly to the questions they are asked and serves to highlight their maturity, diversity, and calm under pressure.

Unfortunately, this application process is expensive and MALs may be particularly uninformed about the extent of the expense. There are application fees for each school, as well as supplemental fees and additional costs. The median cost of secondary applications alone was $1,200 for M.D. matriculants in 2019 (AAMC, 2019). If an applicant is fortunate enough to be invited for interviews, they will also need to pay to travel to those schools, including hotel, flights, etc., the median cost of which was $700 (AAMC, 2019). On top of this, once an applicant is offered admission, a seat deposit will be due to hold the spot regardless of whether the applicant ultimately decides to attend a different school. These costs cannot be included in medical school loans, since they will all be paid beforehand, so it is necessary to be financially prepared. If an applicant is unsuccessful in applying one year and decides to reapply, they will have to pay all of these costs again.

Superstar Advice

The third question we asked ourselves was what is the most helpful advice that we have provided to medically aspiring MALs, either because we think it is particularly important or because they have told us it has been especially impactful to them. In general, the content we try to communicate focuses on how to transition to studying effectively, clarifying that applications to medical school are acts of persuasive communication, providing honest feedback about the strength of students' applications, reassuring students that they have sufficient grit to succeed, providing information on the financial cost of medical school, and reminding them of the general importance of work-life balance. In our experience, this feedback is mostly delivered in one-on-one, face-to-face meetings with students.

Transitioning Back to Studying

We often have conversations with MALs prior to or just upon their entry or re-entry into academic settings. Going from a military lifestyle to a student lifestyle is quite a culture shock that can present a challenge for even the most gifted students (May & McDermott, 2019). Transitioning to reading and studying carefully can be vexing. Students frequently say that one of the most helpful pieces of advice we have discussed with them is how to take advantage of study resources and strategies. For example, we have provided information on study skills workshops and seminars offered by our institutions to teach students the basics of studying or directed students toward fantastic YouTube videos for effective studying (e.g., those by Samford University's Dr. Stephen Chew). By helping students connect with these resources, we empower them to help themselves as they pursue their education. Again, we make a point in these discussions of highlighting that students are capable of successful studying if they use the right strategies. This pays dividends when students apply for medical school because it is incumbent on the student to keep up with the material in order to earn good grades and be competitive for the admissions process.

Medical School Applications Are Exercises in Communication

Admissions committees review thousands of applications every year, and if an applicant wants to be seriously considered they must submit a stellar application. Admissions committees have a very large and difficult task and must find ways to narrow down the applications into a smaller pool of the most qualified and desirable candidates. Applications are evaluated quickly and must effectively persuade committee members of the applicant's propensity for success. How applicants present themselves in terms of the quality of their writing, motivation for pursuing a career in medicine, and reasons for applying to a specific school is paramount. Small details matter, such as punctuation, grammar, and editing, as does reading instructions carefully. Additionally, intelligent and capable people are passed over on a regular basis, simply because they did not apply correctly (e.g., failing to complete all sections of an application), and, therefore, missed their chance to persuade the admissions committee that they should be chosen. It is incumbent upon the applicant to demonstrate that they have both the intellectual capability and interpersonal/communication skills to succeed. Medical schools want physicians who have both the required intelligence and a positive bedside manner. Thus, an applicant who can demonstrate both will be the most likely to obtain at least one acceptance.

It is our belief that by conveying to MALs how this exercise in communication works from the perspective of a busy admissions committee, we provide a critical support role. Because students know themselves and their background, they may find it hard to take a truly objective look at their application materials. MALs in particular may assume that admissions committees will understand more about their military service than they often do. By reminding these students to assume no background knowledge on their personal experiences (including their military service), we can help students submit the most persuasive applications possible.

Honest Application Feedback

In addition to coming to us with struggles about their studies, students often come to us to discuss the strength of their application for medical school, and some of the most helpful information we provide to students is honest feedback. It is important to note that not everyone will successfully earn admission to medical school, and not all those that do will ultimately graduate. Indeed, the most useful feedback people can receive is often focused on negative events or characteristics as such feedback is often more constructive and helps highlight how to improve over time, though hearing such feedback is hard (Belding, Naufel, & Fujita, 2015).

MALs often have issues that may weaken the strength of their application if not addressed. For example, consider a military spouse who has two terms in their undergraduate career in which they withdrew entirely from classes. Typically, this would be a red flag to any admissions committee. However, if that same student found a way to describe how the withdrawal from one term was due to a permanent change of station (PCS; i.e., moving on military orders) and the other was due to providing care for their spouse who had been severely injured in combat, the admissions committee may disregard the withdrawals or potentially even view the candidate more positively. Even more helpful to that student would be a plan that prepares for these contingencies, and lessens the need for these withdrawals in advance. For instance, if that spouse is able to plan classes to avoid the PCS and transfer credits to a school in the new location, then their transcript can be saved. The American Council on Education provides great support for MALs in these types of situations on their public website. By sitting down and providing each individual student with clear, concrete suggestions for how to address the weakest elements of their application, we as educators can truly support them.

Grit

In addition to providing honest feedback about students' weaknesses, we must also provide them with reassurance that they have the potential to succeed (if they do). As someone who worked in several roles related to student support, Zea regularly helped students identify resources to directly combat problems at hand. As a psychologist, Jennifer regularly discusses psychological concepts in the classroom, including issues related to stress, self-esteem, and motivation. As a result, students have frequently come to us to have an open dialogue about their struggles. One of the most important things that we feel that we share with students on a regular basis is the concept of grit, or a personality characteristic in which people are capable of overcoming barriers to achieving their goals (e.g., Duckworth, Kirby, Tsukayama, Berstein, & Ericsson, 2011). We frequently remind students that everyone encounters roadblocks and that they have what it takes to get through it even when they doubt themselves. We believe that this reassurance, particularly from someone capable of accurately evaluating students' potential for success, serves a valuable support role, and students have told us this time and time again. Indeed, MALs in particular can write strong personal statements demonstrating their grit, which is highly valued by admissions committees because it shows the resilience that is required to succeed in a medical career.

Putting the Financial Cost into Perspective

Another piece of information that has been remarkably well-received by students is careful consideration of the financial cost of medical school applications, tuition, and other expenses. In recent years, college students have relied more and more on loans rather than other forms of income to pay for their undergraduate education (Denhart, 2013). Additionally, many feel the need to work one or more jobs while taking classes. Unfortunately, medical school applications are pricey, and the cost of medical school tuition is astronomical with average costs of $38,000 for in-state tuition and $52,000 for out-of-state tuition per year (Gil, Park, & Daniels, 2015). As a result, students are often compelled to put additional stress on themselves by working extensive hours.

When we discuss these issues with students, we try to highlight a few key pieces of information. First, the cost of pursuing a medical degree is high, yet doctors often earn a substantial salary that allows them to pay back their loans (Marcu et al., 2017). We encourage students to consider carefully what sorts of trade-offs they may have to make in order to afford their pursuits. For example, a student who works thirty hours per week while pursuing their undergraduate degree may get worse grades, thereby decreasing their chances of admission to medical schools. As a result, they may have to apply more times and take additional classes to be successful. In some cases, choosing to take out additional loans so that students can take an extra year to truly focus on their studies may end up being less costly in the long run. When we communicate this, we try to explain to students that some of these costs are "a drop in the bucket" and that it is extremely rare for medical doctors to graduate with no debt (Marcu et al., 2017). Indeed, the population that is most likely to graduate with minimal debt are those that obtain a scholarship from the military for school (Greysen et al., 2011; Marcu et al., 2017; Ragel et al., 2011).

Recognizing Personal Life Stressors

Finally, another piece of essential feedback we have provided to students is that their personal lives, such as taking care of themselves and their families, are important. Obligations outside of the classroom can both help and hurt their academic performance. In our experience, many aspiring medical students often put so much stress upon themselves and their desire for As in their classes that they often let the other important elements of their life slide. We have found it important to remind them regularly that it is okay to take a break from school in order to handle and even enjoy their personal obligations. In fact, we frequently communicate that it may actually be more efficient to take an afternoon off from studying in order to spend time with their spouse and/or kids because this can reenergize them for their studies.

CONCLUSION

In this chapter, we sought to leverage our experiences as a medical school director of admissions, a former student services and career support specialist, an undergraduate and post-baccalaureate instructor, and spouses of Navy physicians to identify and articulate how we can effectively support medically aspiring MALs using an analytic autoethnographic approach. In order to speak directly to others who are in positions to support medically aspiring MALs (e.g., university instructors and administrators, military policy makers), our autoethnography deviates slightly from traditional autoethnography in that it is less emotionally laden and focuses on themes identified reflexively from our past experiences rather than individual or specific events. We have argued that medical schools want to select the strongest applicants for their student body and that MALs are intelligent, hardworking, mature, and resilient, and may be wonderful additions to any medical program. Their military experience often provides them with barriers to pursuing a medical career, but it also affords them very valuable tools to make them stronger students and better physicians.

Nonetheless, the path to becoming a doctor is complex and challenging. While all aspiring physicians have questions about how to pursue this career path, MALs may be unique in their limited access to information and/or support. Drawing on our personal experiences, we suggest that those who desire to support medically aspiring MALs can do so by ensuring that these students have adequate and accurate information about their career path and have identified common gaps in their knowledge. Additionally, we have articulated specific, concrete suggestions that we as supporters can convey to medically aspiring MALs along their educational journey.

LIMITATIONS AND FUTURE RESEARCH

Even so, our research is subject to several limitations. For example, although we have both supported medically aspiring MALs in both personal and professional capacities, we ourselves have not pursued careers as physicians and thus may not be able to adequately reflect on all of the unique barriers and facilitators such learners experience in their pursuits. Second, analytical autoethnography typically requires dialogue with informants beyond the self (Anderson, 2006). In conducting our autoethnography, we did not consult with others but rather drew on our past experiences working directly with medically aspiring MALs. Unfortunately, we know of no published research to date on medically aspiring MALs specifically beyond this chapter. Additional research is needed in order to determine how to best support this important population. By taking the time to both encourage and challenge MALs to be the best applicant they can be, we can support them and help them find their way to being welcomed as valuable members into this exciting and meaningful profession in both civilian and military settings.

NOTE

1. For the purpose of this chapter, references to medical school imply both allopathic (i.e., M.D.) and osteopathic (i.e., D.O.) programs unless otherwise noted.

REFERENCES

Adams, J (2009). The acceptability of online courses as criteria for admission to medical school. *Ochsner Journal, 9*(1), 4–10. Retrieved from http://www.ochsnerjournal.org/content/9/1/4.abstract.

American Osteopathic Association. (2020). *OMP 2019: Osteopathic Medical Profession Report.* Retrieved from https://osteopathic.org/wp-content/uploads/OMP2019-Report_Web_FINAL.pdf.

Anderson, L. (2006). Analytic autoethnography. *Journal of Contemporary Ethnography, 35*(4), 373–395. doi:10.1177/0891241605280449.

Association of American Medical Colleges. (2019). *Matriculating Student Questionnaire: 2019 All Schools Summary Report.* Retrieved from https://www.aamc.org/system/files/2019-12/2019%20MSQ%20All%20Schools%20Summary%20Report.pdf.

Belding, J. N., Naufel, K. Z., & Fujita, K. (2015). Using high-level construal and perceptions of changeability to promote self-change over self-protection motives in response to negative feedback. *Personality and Social Psychology Bulletin, 41*(6), 822–838. doi:10.1177/0146167215580776.

Bochner, A. P. (2012). On first-person narrative scholarship: Autoethnography as acts of meaning. *Narrative Inquiry, 22*(1), 155–164. doi:10.1075/ni.22.1.10b0c.

Bolger, N. (1990). Coping as a personality process: A prospective study. *Journal of Personality and Social Psychology, 59*(3), 525–537. doi:10.1037/0022-3514.59.3.525.

Bond, C. A., Lewis, J. L., Leonard, H. A., Pollak, J., Guo, C., & Rostker, B. D. (2016). *Tour lengths, permanent changes of station, and alternatives for savings and improved stability*. RAND Corporation.

Buehler, R., Griffin, D., & Ross, M. (1994). Exploring the "planning fallacy": Why people underestimate their task completion times. *Journal of Personality and Social Psychology, 67*(3), 366–381. doi:10.1037/0022-3514.67.3.366.

Denhart, C. (2013). How the $1.2 Trillion college debt crisis is crippling students, parents, and the economy. *Forbes Magazine*.

Department of Defense. (2020). *DoD Manual 5110.04, Volume 1: Manual for Written Material: Correspondence Management*.

Duckworth, A. L., Kirby, T. A., Tsukayama, E., Berstein, H., & Ericsson, K. A. (2011). Deliberate practice spells success: Why grittier competitors triumph at the National Spelling Bee. *Social Psychological and Personality Science, 2*(2), 174–181. doi:10.1177/1948550610385872.

Elam, C. L., Taylor, E. D., & Strother, E. N. (1996). Preparing for medical school and the medical profession: Advice to advisors. *NACADA Journal, 16*(2), 34–38. doi:10.12930/0271-9517-16.2.34.

Ellis, C., Adams, T. E., & Bochner, A. P. (2011). Autoethnography: An overview. *Historical Social Research/Historische Sozialforschung, 36*(4 (138)), 273–290. Retrieved from www.jstor.org/stable/23032294.

Evans, P., & Wen, F. K. (2007). Does the medical college admission test predict global academic performance in osteopathic medical school? *Journal of the American Osteopathic Association, 107*(4), 157–162.

Ford, K., & Vignare, K. (2015). The evolving military learner population: A review of the literature. *Online Learning, 19*(1), 7–30.

Fortin, Y., Kulasegaram, K., Kancir, J. N., Moineau, G., Carpentier, A., & Hanson, M. D. (2018). The direct economic and opportunity costs of the Medical College Admissions Test (MCAT) for Canadian medical students. *MedEdPublish*. doi:10.15694/mep.2018.0000243.1.

Foster, E. (2013). Values and the transformation of medical education: The promise of autoethnographic research. *Journal of Medicine and the Person, 11*(1), 19–23. doi:10.1007/s12682-013-0142-5.

Friedman, A. B., Grischkan, J. A., Dorsey, E. R., & George, B. P. (2016). Forgiven but not relieved: US physician workforce consequences of changes to public service loan forgiveness. *Journal of General Internal Medicine, 31*, 1237–1241. doi:10.1007/s11606-016-3767-2.

Gil, J. A., Park, S. H., & Daniels, A. H. (2015). Variability in United States allopathic medical school tuition. *The American Journal of Medicine, 128*(11), 1257–1262. doi:10.1016/j.amjmed.2015.07.022.

Graves, S., Seagle, B.-L. L., Kocherginsky, M., Yang, J., Grace, A., & Shehabi, S. (2018). Scholarship support for veterans enrolling in MD, JD, and MBA programs.

Journal of the American Medical Association, 320(11), 1197–1198. doi:10.1001/jama.2018.10260.

Greysen, S. R., Chen, C., & Mullan, F. (2011). A history of medical student debt: Observations and implications for the future of medical education. *Academic Medicine, 86*(7), 840–845. doi:10.1097/ACM.0b013e31821daf03.

Lieberman, M. (2019, May 1). Many medical colleges still frown at online. *Inside Higher Education*. https://www.insidehighered.com/digital-learning/article/2019/05/01/many-medical-colleges-reject-students-who-took-prerequisite.

Lin, K. Y., Parnami, S., Fuhrel-Forbis, A., Anspach, R. R., Crawford, B., & De Vries, R. G. (2013). The undergraduate premedical experience in the United States: A critical review. *International Journal of Medical Education, 4*, 26–37. doi:10.5116/ijme.5103.a8d3.

Marcu, M. I., Kellermann, A. L., Hunter, C., Curtis, J., Rice, C., & Wilensky, G. R. (2017). Borrow or serve? An economic analysis of options for financing a medical school education. *Academic Medicine: Journal of the Association of American Medical Colleges, 92*(7), 966–975. doi:10.1097/ACM.0000000000001572.

May, A., & McDermott, V. (2019). Cultural collision: A review of literature. In L. Hinojosa Hernández & J. Belding (Eds.), *Military spouses with graduate degrees: Interdisciplinary approaches to thriving amidst uncertainty* (pp. 59–85). Lexington Books.

McIlveen, P. (2008). Autoethnography as a method for reflexive research and practice in vocational psychology. *Australian Journal of Career Development, 17*(2), 13–20.

Monroe, A., Quinn, E., Samuelson, W., Dunleavy, D. M., & Dowd, K. W. (2013). An overview of the medical school admission process and use of applicant data in decision making: What has changed since the 1980s? *Academic Medicine, 88*(5), 672–681. doi:10.1097/ACM.0b013e31828bf252.

Murphy, B. (2019). *Which undergrad majors are best for med school?* Retrieved from https://www.ama-assn.org/residents-students/preparing-medical-school/which-undergrad-majors-are-best-med-school.

National Security Advisory Group. (2006). *The U.S. Military: Under strain and at risk*. Retrieved from https://www.democrats.senate.gov/pdfs/NationalSecurityReport_01252006.pdf.

Neelands, M. L. (2014). *Active-duty military online student retention and factors leading to attrition: A qualitative study* (Publication No. 3648286), PhD Doctoral Dissertation, University of Phoenix.

Ragel, B. T., Klimo, P., Grant, G. A., Taggard, D. A., Nute, D., McCafferty, R. R., & Ellenbogen, R. G. (2011). Economic analysis of the military health professions scholarship program for neurosurgeons. *Neurosurgery, 69*(3), 525–532. doi:10.1227/NEU.0b013e3182181ec2.

Range, L. M., Menyhert, A., Walsh, M. L., Hardin, K. N., Ellis, J. B., & Craddick, R. (1991). Letters of recommendation: Perspectives, recommendations, and ethics. *Professional Psychology: Research and Practice, 22*(5), 389–392. doi:10.1037/0735-7028.22.5.389.

Rhine, T. J., Milligan, D. M., & Nelson, L. R. (2000). Alleviating transfer shock: Creating an environment for more successful transfer students. *Community College Journal of Research and Practice, 24*(6), 443–453. doi:10.1080/10668920050137228.

Roediger, H. L., Agarwal, P. K., McDaniel, M. A., & McDermott, K. B. (2011). Test-enhanced learning in the classroom: Long-term improvements from quizzing. *Journal of Experimental Psychology: Applied, 17*(4), 382–395. doi:10.1037/a0026252.

Saxton, J. F., & Johns, M. M. E. (2019). Grow the US National Health Service Corps. *Journal of the American Medical Association, 301*(18), 1925–1926. doi:10.1001/jama.2009.660.

Seabrook, R., Brown, G. D. A., & Solity, J. E. (2005). Distributed and massed practice: From laboratory to classroom. *Applied Cognitive Psychology, 19*(1), 107–122. doi:10.1002/acp.1066.

Sklar, D. P. (2013). Preparation for medical school: Reflections on the MCAT exam, premedical education, and the medical school application process. *Academic Medicine, 88*(5), 553–554. doi:10.1097/ACM.0b013e31828b7c8f.

Wang, J. Y., Lin, H., Lewis, P. Y., Fetterman, D. M., & Gesundheit, N. (2015). Is a career in medicine the right choice? The impact of a physician shadowing program on undergraduate premedical students. *Academic Medicine, 90*(5), 629–633. doi:10.1097/ACM.0000000000000615.

Chapter 9

Nuclear War!

Theorizing the Negative Impacts of Policy Debate's Tropological Mistreatment of War for Military-Affiliated Learners

Nick J. Sciullo

Debaters are often faced with discussing complex international relation scenarios that usually end in some type of violent confrontation; quite often it is nuclear war. These discussions occur frequently in policy debate where matters of federal government policy are proposed and argued against; however, a major problem with these debates is the way debaters discuss war with little regard for the potential impact that the discussion may have on those who have experience with war. Often, these discussions have the potential to exclude those affected by war, especially military-affiliated learners (MALs) who debate.

In academic policy debate there are often discussions of the costs and benefits of a team's proposed plan of action. These discussions often revolve around a team proposing some sort of federal government action, and the other team arguing against it. While the team proposing an action claims advantages to that action, the team arguing against the action proposes disadvantages. This disadvantage is the argumentative form in which nuclear war most commonly enters a debate round. Although other wars are also discussed in policy debate and are worthy of criticism as well, debaters widely understand "nuclear war" as a metonymic symbol for the focus on high-magnitude and low-probability scenarios that are often discussed with poor evidence and a lack of reasoning. This chapter will discuss the hypothetical nature of debating war in policy debate, and the ways these discussions may affect participants. Then, the chapter will discuss how the abstract discussion of nuclear war may alienate and harm MALs. Next, the chapter will provide alternatives to the popular policy debate discourses of (nuclear) war. Finally, the chapter will discuss further reforms to policy debate that can make the activity more respectful to and of MALs.

Proponents of debate reference better test scores, attendance, self-esteem, research skills, public speaking skills, and a range of other benefits to participation (Zwarensteyn, 2012). Both the United States Military Academy (West Point) and the United States Naval Academy (USNA) field successful teams in policy debate regularly, and these teams include both cadets and midshipmen along with prior enlisted students. The United States Air Force Academy occasionally has a debater that participates in policy debate with another service academy member (in what is called a swing team where two debaters from different schools debate together), but more regularly competes in parliamentary debate. The author has had the privilege of working as an Assistant Coach to the USNA Debate Team for several years and attests to the complexity of navigating the policy debate environment given those student-debaters' unique relationship to war. Many of these service academy debaters were amazed, if not shocked, at how little other debaters knew about war, military policy, and military practice. Specifically, debaters who had served in various capacities prior to enrollment at USNA and those with parents and grandparents who served in war zones were appalled by the ways in which people involved in war were ignored throughout war discussions occurring in policy debate. Quite a few midshipmen have called into question, in debate meetings and at tournaments, how war is discussed in policy debate.

POLICY DEBATE

Debate is a common co-curricular activity in colleges and universities in much of the global West (Bartanen & Littlefield, 2014). Ranging in style from various forms of parliamentary debate to Lincoln-Douglas and policy debate, it has provided a forum for students to engage with pressing philosophical and political questions (Bellon, 2000; Briscoe, 2017; Carr, 2002; Zwarensteyn, 2012). Policy debate is of particular interest to the communication discipline as it is frequently sponsored by communication departments at many of the nation's top Ph.D.-granting institutions and often revolves around questions of the impacts of various federal government policies; moreover, policy debate has been the subject of numerous research explorations in communication studies (Bartanen & Littlefield, 2014; Bellon, 2000; Gerber, 2009; Goodnight & Mitchell, 2008; Williams, McGee, & Worth, 2001) that have addressed argumentation theory in communication studies, benefits to competitors, the relationship between public policy and debate, and the relationship of debate to public discourse.

Although policy debate has changed radically in the last twenty-five years to be at the forefront of debates about identity, equality, and radical philosophy (Gerhke, 1998; Huber, 2016; Polson, 2012; Sciullo, 2019), many debates are still resolved by the weighing of high-magnitude low-probability impacts (impacts that cause many deaths but have a small likelihood of happening), first among them nuclear war (Bankey, 2013; Herbeck & Katsulas, 1992). Debates about weapons policy, foreign aid, and specific policy areas like infrastructure and agriculture are more common, but only recently have debaters been emboldened to directly address issues of identity. While previously debate teams won rounds by preserving the most life or avoiding the most deaths, norms have steadily begun to change, yet teams still seem to focus on the threat of nuclear war. Put in policy debate lingo: the bigger impact wins.

This tradition is well established in policy debate such that an observer attending any high school or college policy debate tournament in the country is likely to observe a round where the impact to a proposed affirmative team action, according to the negative team, is nuclear war. Likewise, affirmative teams proposing a federal government action often claim to be averting nuclear war in any number of ways (Briscoe, 2017). For the uninitiated, this may seem outlandish, but one of the goals of policy debate is to get students thinking like policy-makers and teach them about the advantages and disadvantages of government action in a number of contexts (Briscoe, 2017).

One of the core negative positions is the Disadvantage (DA or DisAd). This argument is that the affirmative plan causes some detrimental consequence or consequences, such as nuclear war, regional war, starvation, genocide, or some other weighty event. Because much of judging a debate is about weighing the advantages of the affirmative versus the disadvantages argued by the negative, the disadvantage is a central strategy in debate. As debate has grown to rely more on evidence and less on delivery, debate rounds often rely on a comparison of evidence presented using any number of standard theories of argumentation and rhetoric. This results in many debate rounds being decided not on who had the better idea or political proposal, but whether a specific piece of evidence (usually a scholarly source or news article) was answered by another team, argued what debaters thought it argued, or contained the warrants alleged by the team offering it. This is, of course, a hyper-specialized study in argumentation that can occasionally render debate rounds to discussions of hyper-specific questions of validity more than overarching themes of "will an affirmative plan help or hurt people?" As such, the people in debate, both debaters and those who are hypothetically harmed by affirmative or negative teams' arguments, often do not get as much attention as non-debate observers might think they should receive. Increasingly, however, policy debate rounds have centered race, gender, and indigeneity in debate to address the exclusion of marginalized debaters and persons from debate round decision-making; now, after the historic National Debate Tournament (NDT) participation of George State University Luke Floyd and John Finch, two MALs who centered their experience in their debate rounds (Sciullo, 2017), veterans have been recognized as marginalized debaters left out of debate. This exclusion or at least lack of acknowledgment results in a devaluation of people in debate. Put another way, for much of policy debate's existence, the evidence mattered more than the people making the arguments or the people about whom the arguments were made. This then puts MALs at risk of having the matters with which they may have experience discussed without their experience being valued.

There are significant implications of the constant discussion about nuclear war. First, as with other topics that are constantly discussed, there is a risk that those involved in debate and those who regularly observe it become desensitized to nuclear war and other forms of war or demonstrate philia about war (Baudrillard, 2006). Second, reliance on high-magnitude impacts averts attention from structural impacts so that debaters, judges, and coaches worry about how to address nuclear war while ignoring on-going harmful events like sexual violence, poverty, ethnic cleansing, genocide, forced labor, and other slower forms of violence. This reliance on nuclear war is likely due to the tradition of judges voting for high-magnitude impacts compared with the smaller scale impacts of low-magnitude, which rewards the use of nuclear war impacts. Third, the majority of student debaters in higher education have not experienced nuclear war. As such, this may serve to erase the experiences that military-affiliated debaters have with war by not including them, their stories, or their potential expertise in war-fighting. Ultimately, this has the potential to make it seem as if their lived experiences with war are somehow less relevant or important than nuclear war. These implications have a deleterious effect on the pedagogical value of debate because they force debaters to frame war in impersonal terms, which could potentially result in a faulty understanding of war as something nebulous and only carried out in military planning rooms or defense contractor white papers. The problem is not that nuclear war is not a catastrophe, but that its likelihood is low (Abedi, 2017; Mueller, 2010; Sheng, 2020), and one can reasonably assume discussing other scenarios for violence would be more relevant.

Nuclear war has become so common that competitors regularly joke about "X leading to nuclear war" and one might even find stickers on laptops with versions of this meme. In the same way that Godwin's law (1994) suggests that the longer an internet discussion lasts the more likely it is someone will compare someone to Nazis or Hitler, so too might an analogous law in debate state that the longer one observes a policy debate tournament, the more likely they are to hear or read a nuclear war impact. The author has heard, for example, the following nuclear war scenarios in roughly twenty years of involvement in policy debate: U.S.-China, U.S.-Russia, Russia-China, China-Japan, Israel-Iran, India-Pakistan, and no doubt more. These scenarios are repeated year after year, tournament after tournament, debate round after debate round with little to no consideration of MALs who may have experienced war and those people who have been impacted by war generally.

While debate activity participants should be able to engage in this self-critical reflection, laughing and joking about how everything leads to nuclear war ignores the fundamental reality and experiences of those who have been involved with (or may be involved with) and those who are the victims of wars. Moreover, this causal perspective of nuclear war may overlook how those directly impacted by war must grapple with its consequences. Even in today's era of drone warfare and cyberwarfare, people still make the decisions that lead up to war, engage in war-fighting efforts, and are impacted by war. MALs are often left trying to understand how to explain their involvement in war, lacking the capacity and language to do so (Alexander, 2017). This does not mean that they are necessarily "triggered" by discussions of war, are unable to discuss war, nor that discussing war is unavoidably detrimental, but rather that discussions of war in policy debate often lack the nuances that are inclusive of MALs' experiences.

War is not, one would hope, a laughing matter for these students, and seemingly should not be for the rest of the university community. Many MALs have a unique connection with the subject matter of many policy debates, including regional war, arms sales, humanitarian aid, national security, force projection, hard power, and other related concepts (Brown, 2006; Carlton, 2012). But, as much work has been completed on race and debate (Cooper, 2014; Dillard-Knox, 2014; Kraft, 2014; Loge, 1991; Reid-Brinkley, 2019; Sciullo, 2019; Warner & Bruschke, 2001), gender and debate (Bruschke & Johnson, 1994; Matz & Bruschke, 2006; Poapst & Harper, 2017; Stepp, Simerly, & Longue, 1994), and increasingly disability and debate (Dhillon, 2016; Ferguson, 2016; Nelson & Miller, 2016; Richter, 2016), very little work has been completed on MALs in debate (Sciullo, 2017). Most work indicates exclusion based on identity, especially along gender and race lines, but that work does not conclude that exclusion is inevitable nor that change is impossible. The research on disability and debate has been far better on suggesting ways to change the activity in concrete ways. Ultimately, the lack of work on MALs in debate reveals an unfortunate gap in our understanding of how policy debate participants' discussions of war and violence in the classroom and debate competitions affect MALs, as well as our understanding of diversity and difference in policy debate (Allen, Trejo, Bartanen, Schroeder, & Ulrich, 2004; Rogers, 1997; Stepp, 1997; Stepp & Gardner, 2001). Despite policy debate's ready affirmation of liberal diversity and inclusion rhetoric, the activity's discussion of nuclear war may result in lack of inclusion and alienation of MALs and their lived experiences.

Aside from the obvious criticisms made about "nuclear war impacts," including that they stem from various forms of fallacious argumentation (e.g., slippery slope, false dichotomy, etc.) (Herbeck & Katsulas, 1992; Prager, 2007), one unexplored area worthy of further study is the danger of flippantly discussing war given that some MALs have often had psychologically complex, material engagements in theaters (the place where war takes place, as well as those places in which war could take place). For these learners, war is not an abstract concept to be debated and weighed, but a lived experience that was, is, and may still be harmful for MALs, potentially producing psychological trauma that can manifest in interpersonal, intercultural, and self-communicative harms (Burnam, Meredith, Tanielian, & Jaycox, 2009; Frankfurt & Frazier, 2016; Murthy & Lakishminarayana, 2006; Pols & Oak, 2007). Although we have no data on how many MALs participate in policy debate, or any form of debate for that matter, we can assume that of the 900,000 veterans enrolled in colleges and universities (Hill, Kurzweil, Piscareta, & Schwartz, 2019), not to mention active duty service members, reservists, and relatives and partners of service members, MALs are participating directly or indirectly in aspects of debate. Yet, scholars have largely failed to study MALs in debate and have produced relatively little work on marginalized debaters throughout the lifespan of policy debate. Without continued work on MALs in debate, scholars risk reinforcing exclusionary practices and preventing MALs from having a positive experience in the activity. This is particularly true for MALs who are difficult to identify (Demers, 2011). As such, the policy debate community, in order to address MALs and become more open to their healthy engagement with higher education, must reconsider the ways it debates war to embrace their involvement in the activity.

Debating about War

The focus of policy debate is often the Resolution, which teams vote on prior to the season beginning and which is the stasis point of sorts for the entire academic year/debate season. For example, the 2015–2016 Cross Examination Debate Association/National Debate Tournament (CEDA/NDT) resolution was: "Resolved: The United States should significantly reduce its military presence in one or more of the following: the Arab states of the Persian Gulf, the Greater Horn of Africa, Northeast Asia." The affirmative or pro side would ostensibly advocate reducing military presence in one or more of these regions, and the negative or con side would oppose this government action. A judge must decide the debate based on some calculus or paradigm for decision-making based on the arguments presented. Many judges have traditionally used some version of utilitarianism to make their decisions, voting for the side that does the most good for the greatest number of people.

One of the generic criticisms of policy debate is that even though there is strong evidence that debate prepares students for business, politics, and law school, there is something nonsensical or at least less or un-efficacious about pretending to be the U.S. federal government (Zwarensteyn, 2012). Obviously, debaters are not the federal government, nor are they elected officials, nor will most debate rounds have much bearing on federal policy-making. Nonetheless, policy debate relies on this founding fiction and as a result deals with issues of national and international politics quite regularly. These issues span the range of presidential elections to international relations to the welfare of marginalized people. This makes debate interesting and engaging for students interested in politics, political theory, students who do not feel like they have much of a voice on campus, and even those who struggle to find a place in campus communities because it brings together like-minded people to discuss politics, philosophy, and student engagement in the world (Mitchell, 1998).

Regardless of who participates, debate is certainly about improving research and argumentation skills; however, there is a unique harm to people who are affected by the arguments made in debate rounds when they are not considered in the form or subject of arguments made. Shifts in policy debate, and increasingly Lincoln-Douglas and parliamentary debate, mean an increasing number of rounds are about critiquing anti-blackness and the ways in which debaters and their arguments may be anti-black (Sciullo, 2019). While this shift initially sent shockwaves through the community, it has been increasingly accepted as a positive shift in the right direction for the debate community. As such, debaters have steadily incorporated social and political issues beyond those that the federal government addresses into their discussions of public policy and philosophy. Yet when debate is about nuclear war and great power politics, debate largely fails to recognize how MALs are affected by an activity that often revolves around questions of military deployments, nuclear war, regional war, arms sales, missile treaties, genocide, humanitarian aid, and other related ideas that specifically involve military personnel (Ajello, 2013; Downs & Murtazashvili, 2012). This is because war is understood in the abstract in policy debate.

War as Metaphor

This perception of war as abstract in debate presents a problem for MALs because the wars that debaters discuss, while hypothetical, are nonetheless still about people and may even implicate the people in the debate round whose parents, siblings, and partners serve in the military, as well as those students who expect to enter the military through enlistment or other avenues. This means that debaters are challenged to be more thoughtful about what their hypothetical wars do and to whom. Wars are not mere abstractions based on relevant data. War does not just happen; it is conducted by people under the direction of other people and causes casualties among still other people. The wars that have marked United States history from the Revolutionary War to the Iraq War have resulted in nearly one million deaths (Department of Veterans Affairs, 2019). The same is true around the world where war has exacted the ultimate toll on millions (Hedges, 2003).

But what does this casual talk about war do to those closest to it? A necessary precursor to this discussion and, perhaps, an entry point into theorizing the danger of war rhetoric is the "war metaphor." Most people have heard the statement "X is war," where X is some other non-war activity from baseball to marriage to selling agricultural technology to politics. The metaphor intends to evoke the difficulty, contentiousness, stress, and struggle of the thing it describes. Metaphors are powerful rhetorically (Steuter & Wills, 2008), and war metaphors especially so (Lakoff, 2009; Steuter & Wills, 2008). Any discussion of war regardless of its specific rhetorical form is rhetorically powerful because of the gravity of war (Gross & Aoláin, 2014; Steuter & Wills, 2008). Yet, policy debaters claim nuclear war (or other war) is happening at the end of every two-hour debate round as if war was not fundamentally about people dying. The arguments roll off the tongue with such speed and are repeated round after round (some debate tournaments are eight debate rounds and two to four days long) as if no one was affected by the discussion at hand. Debate as an intellectual game that demonstrates the gross iterability of war rhetoric.

Ultimately, MALs must confront this flippant discussion of war because it is difficult to avoid and a commonplace for debaters. While many MALs would be excellent contributors to these discussions, their participation is regularly degraded (the author has heard about MALs being described as rapists, imperialists, and killers in debate rounds) (Hart and Thompson, 2016; Miller, 2019). This reckoning with war rhetoric includes being alleged to have caused a war, being ridiculed for serving in a military that is not beyond criticism, as well as having to address claims that someone causes or prevents war in the abstract in order to win a debate round (Stone-Mediatore, 2010). Then there is the uneasy way military personnel are discussed as either heroes or baby killers, which is disempowering and essentializing, and as a result often awkward at best for those who serve in the military (Moore, 2012; Sciullo, 2017; Stahl, 2009; Stone-Mediatore, 2010). This means that soldiers and veterans are often constructed as occupying one of two poles. They are either baby killers responsible for countless atrocities, violence of all types, and destruction or they are constructed as heroes constantly praised for their heroism, keeping the world safe, and making everyone's lives better. Many MALs find both poles troubling, and when one tries to bring up one's military service in a policy debate round, one gets blamed for killing civilians and sexual violence perpetrated against military personnel and civilians. The Georgia State University debate team I coached of Luke Floyd and John Finch, perhaps the most storied military-affiliated debate team who centered their military affiliation in debate, experienced just such treatment when they centered their MAL-status in policy debate rounds.

Furthermore, MALs experience war in vastly different ways. Some were so far removed from war that they resent being included in larger discussions of heroism or even the veteran identity because they served at a time when not much was asked of them or they resent being thanked because it diminishes the work they did (Haynes, 2019; Richtel, 2015; Stewart, 2015; Wood, n.d.). Others, however, may have intimate knowledge and experience with the traumas of war. It still bears considering how we discuss war, given that people have slowly come around to talking more sensitively about other forms of violence that were once more nonchalantly discussed.

For example, not too long ago it was common to compare situations to slavery, which tended to deemphasize and dehumanize Black people's experiences (Berkenwald, 2011; Eglash, 2007; Perr, 2015). Slavery became a metaphor for deprivation that erased blackness. Changing how slavery was used as a metaphor seemed reasonable to many learners and teachers because it respected the trauma of slavery. They recognized that modern usages denied the violence on which those ideas were built and the people that were a part of that violence. So, too, have similar efforts made great progress on using rape as a verb in contexts beyond sexual assault because of the ways it ignores the significance of this word for people who have undergone extreme sexual violence. We take these words more seriously now, but football games are still wars on the gridiron, dating is a minefield, one is in the trenches when they are working on a complicated work project, and the affirmative plan will still lead to some form or forms of war. The metaphors and other tropes we rely on convey much polysemic meaning, and while that makes them powerful, it also makes them dangerous.

Trivializing war is like trivializing other crises in that it denies the human impact these crises have. One need not think hard about the ways war may affect people from post-traumatic stress to depression and anxiety. While veterans are certainly not all affected by these issues, the impacts and consequences they face remain important to consider. For example, veterans' mental health issues have been widely and shockingly reported (Basu, 2013). These are not laughing matters to be trivialized. While individuals have rightfully drawn attention to Holocaust trivialization (Gerstenfeld, 2007) and genocide trivialization (Charny, 2001), not nearly as much attention has been paid to war trivialization. War causes time apart from one's home, friends, and family, death, economic hardship, loss of freedom, psychological trauma, sexual violence, and other physical and mental harm and impairments (Forrest, Edwards, & Daraganova, 2018; Levy & Sidel, 2016; Murthy & Lakishminarayana, 2006; Thompson, 1993). These are not matters to disregard with the callous use of a flippant phrase on a Saturday at the local college campus.

One might wonder what this offhand use of war looks like in an actual debate. In what ways is war discussed? In policy debate, arguments often rely on evidence that is composed of quotations from a range of sources from blogs to scholarly articles and monographs. These quotations are referred to as pieces of evidence, but they are only one part of the evidence. The three parts of evidence are the tag (headline), the cite (where the evidence came from), and the text (the paragraph or two that is being cited and read from). Policy debate has been depending on evidence constructed this way for decades, and indeed one of the major criticisms of policy debate is that it relies too much on the scripted reading of others' thoughts and not enough on public speaking skills or original thoughts by debaters (McCordick, 2017). Nonetheless, these pieces of evidence are combined to form an argument that, depending on the argument, has a particular form. The above-mentioned references to impacts are the last part of the disadvantage where a negative team explains the bad thing that will happen if the affirmative team's plan is supported.

One example of evidence, displayed in the format the evidence would take in a debate file (a Word document of evidence), about nuclear war looks like this famous piece of *Strait Times* evidence:

U.S.-China War Causes Multiple Scenarios For
Extinction; Sing Along With Me

Strait Times, 2000:
(Strait Times, June 25, 2K, Pg. l/n)

The high-intensity scenario postulates a cross-strait war escalating into *a* full-scale *war between the US and China.* If Washington were to conclude that splitting China would better serve its national interests, then a full-scale war becomes unavoidable. Conflict on such a scale *would embroil other countries* far *and* near and -horror of horrors *-raise the possibility of* a *nuclear war. Beijing* has already told the US and Japan privately that it *considers any country providing* bases and logistics *support to* any US *forces attacking China as* belligerent parties *open to* its *retaliation.* In the region, this means South Korea, Japan, the Philippines and, to a lesser extent, Singapore. If China were to retaliate, *east Asia will be set on fire.* And the conflagration may not end

> there as opportunistic powers elsewhere may try to overturn the existing world order. With the US distracted, *Russia may seek to redefine Europe*'s political landscape. *The balance of power in the Middle East may be* similarly *upset* by the likes of Iraq. In south Asia, *hostilities between India and Pakistan*, each armed with its own nuclear arsenal, *could enter a* new and *dangerous phase*. Will a full-scale Sino-US war lead to a nuclear war? According to General Matthew Ridgeway, commander of the US Eighth Army which fought against the Chinese in the Korean War, the US had at the time thought of using nuclear weapons against China to save the US from military defeat. In his book The Korean War, a personal account of the military and political aspects of the conflict and its implications on future US foreign policy, Gen Ridgeway said that US was confronted with two choices in Korea -truce or a broadened war, which could have led to the use of nuclear weapons. If the US had to resort to nuclear weaponry to defeat China long before the latter acquired a similar capability, there is little hope of winning a war against China, 50 years later, short of using nuclear weapons. The US estimates that China possesses about 20 nuclear warheads that can destroy major American cities. *Beijing* also *seems prepared to go for the nuclear option*. A Chinese military officer disclosed recently that Beijing was considering a review of its "non first use" principle regarding nuclear weapons. Major-General Pan Zhangqiang, president of the military-funded Institute for Strategic Studies, told a gathering at the Woodrow Wilson International Centre for Scholars in Washington that although the government still abided by that principle, there were strong pressures from the military to drop it. He said *military leaders considered the use of nuclear weapons mandatory if the country risked dismemberment* as a result of foreign intervention. Gen Ridgeway said that should that come to pass, we would see the destruction of civilization. There would be no victors in such a war.

At first blush, this piece of evidence, which functions as an impact to a disadvantage claiming that the affirmative team's proposed plan will lead to a conflict between the United States and China which will then lead to nuclear war, is not dissimilar from block quotations in academic writing. It is a large section of what someone else wrote proffered to prove a claim. There is a text, and a citation, and a summary of that text that leads into what the quotation supposedly does or proves. The italicized portions generally represent what the debater reads in a debate round with most competitors only reading the italicized portions of a given piece of evidence. Although this chapter does not focus on the relative merits of or processes by which evidence is italicized, this method surely has implications for argumentation and deserves further research and consideration like researching the proportion of words read to the total words in a piece of evidence, the average warrant count in evidence, and the number of times people are mentioned in war-related evidence.

But, one will note that the tag (the italicized summary at the beginning of the block quotation) makes light of the claim that nuclear war will or is likely to happen with the added phrase "sing along with me." The inclusion of this line in the tag is a nod to the ubiquity of this evidence in policy debate for the last twenty years. But, that nod suggests an alarming ambivalence about the costs of and participants in nuclear war. Nuclear weapons have killed thousands and severely polluted the environment (Listwa, 2012). Those who have worked on the technology that led to the creation of nuclear weapons have expressed regret about their involvement in developing this technology because of the tremendous toll it has exacted on life (Valiunas, 2006). Furthermore, there is no mention of the harm, pressure, and stress expressed by service members related to the control and use of nuclear weapons. The perspectives of the civilians and service members who work on nuclear weapons are nowhere to be found. There should be no singing.

This tag suggests that the text of the evidence concludes that nuclear war happens, but with obviously unclear warrants. Of course, a common criticism of policy debate is that arguments that conclude with nuclear war impacts are untrue, fallacious, or otherwise specious. Star Muir (1989, as quoted in Herbeck & Katsulas, 1992) succinctly makes such an argument:

> This takes form in two disturbing tendencies: an unwillingness to examine more real world impacts of policies, and a jaded view of global devastation. That first is apparent in the unwillingness of the debaters to argue that a recession, per se, is bad; that a regional war, not escalating to superpower conflict, would be a horrible thing. A global recession would probably not cause a nuclear war, but it doubtless would cause untold suffering and human anguish. A regional war in Africa could cost hundreds of thousands of lives, easily enough to outweigh a properly mitigated set of case scenarios. The problem is that debaters won't tell these stories, but they will take the easy way out and read a blurb on World War III (emphasis in original). (p. 11)

This chapter's criticism is not that the impact is too attenuated or not real-world, however true those criticisms may be, but rather that such a representation of war makes war seem simple and without an actor ignoring the military personnel implicated in the act. In this case, it almost seems silly, functioning as an inside joke to the debate community. Who caused the war? Where is the war? Who started the war? Who is participating in the war? Who is impacted by the war? None of these questions are answered. This might not seem tremendously important for those learners not affiliated with the military; however, for those that are, this invalidates their lived experiences. The idea that everyone understands that this is an "academic game" and that these wars are "not real" ignores the perspectives and sacrifices of MALs (Sciullo, 2017). There seems to be a long history of discussing war on news channels and in news magazines that focuses more on this idea of war, on the weapons, on the international relations or failure thereof that bring us closer to war, without nearly as much attention to those involved in war. So often we do not address who gets recruited, what racism, sexism, and other discrimination looks like in the military, where wars are conducted and who usually dies in war zones, or the larger capitalistic, racialized, and xenophobic motives that underlie war-fighting. This callousness must be criticized.

In policy debate, war is only a metaphor. This does not mean the metaphor has no meaning, nor that metaphors generally are without meaning. Rather, this means that the way debate evidence is used indicates a profound lack of appreciation for who is involved in war and the resulting effects being involved in war has on people. War will not happen because of a policy debate round, nor will any decision of a judge's ballot cause war to happen. Even the sum total of debate rounds would not have and has not had in the history of intercollegiate debates resulted in war. No matter how many pieces of evidence one reads, war never happens because of collegiate debate.

This seems obvious to many, especially those who are not participants in debate. Obviously, in most instances, what one reads does not produce an action beyond the reading one does. Any proposed plan by the affirmative team will not actually result in that plan being passed by Congress or some other government actor. But, debate judges often operate under the assumption that what debaters read will happen, so they must choose the affirmative or negative side providing for the least bad or most good situation. This assumption helps debate judges focus on the advantages and disadvantages of a proposed affirmative plan and serves the purpose of avoiding lengthy discussions about vote counts in Congress.

But what of this for MALs? This flippant debate of war does not help MALs have serious discussions about their role in war and the way it affects them. The problem is that this discourse about war ignores them, acts as if they are not present. It makes light of the risk they could die or cause others to die. It ignores their involvement in or even protest of war. It ignores the impacts deployments have on families. Since September 11, 2001, 6,967 U.S. service members have died and 52,802 have been wounded (Congressional Research Service, 2019). These numbers do not include those incidents that go unreported nor do they describe what impact these incidents have on those related to service members. When discussing nuclear war as a mere game and ignoring other violence in the world, both that which directly impacts MALs and their families, debate devalues MALs' experiences with war.

RECOMMENDATIONS FOR CHANGING DEBATES

There are several solutions that with enough work may improve policy debate for MALs. First, policy debaters should focus on more real-world impacts. The ephemeral discussions of war reduce the activity to a simulation that disorients and confuses participants about the trauma of war. This step can be achieved by talking about more plausible scenarios than the tried and accepted-as-true story of "economic decline leading to war." Rather, debaters should discuss what economic declines actually cause, such as hardship for marginalized populations, deferred maintenance on infrastructure, pressure on family food budgets, and other impacts that are real and experienced by people every day. Alternatively, if policy debaters want to discuss war, they should phrase the tags of their evidence to acknowledge the people involved. There should be mention of military personnel and civilians affected by war. Another solution is reorienting judges away from high-magnitude low-probability impacts (nuclear war) to lower magnitude high probability impacts (hunger, poverty, etc.) (Gannon, n.d.; Herbeck & Katsulas, 1992). It will not work to change how debaters think about war if we do not also change how judges think about war. This means educating judges about the effects debate arguments may have on those involved can be an important professional development strategy.

Debaters must be empowered to rethink debate's focus on war. This will not be easy. But, if coaches and judges work to encourage debaters to write cases that deal with issues that matter to them, work with MALs to discuss war in debate rounds in more meaningful and sensitive ways, work on explaining the ills of the world that actually exist, and discourage the race to nuclear war in debate rounds by not coaching these strategies and not voting for them, then the debate activity will slowly move away from the fetishistic avowal of high-magnitude impacts that make light of the complexities of war. Likewise, improvements to the way war is discussed will likely improve argument construction, refutation, and intercultural awareness, which should indirectly result in better public speaking, better writing, and improved performance in class. Indeed, increased awareness about anti-blackness among debate activity participants has at the very least begun a number of important discussions about the rules of the activity, representation on debate teams, and important racial issues across the world. Were it not for these discussions, debate would seem a much less hospitable place, and many participants would be less aware of racial injustice, discrimination, and a wide range of authors working on important questions of race and racism. That is, improvements to the debate activity should also improve participants' access to the documented benefits of debate.

Debate must also welcome MAL voices. While it is easy to criticize the military, those criticisms should include MALs, their stories, and experiences. Ignoring these voices and experiences does not help anyone understand war better and makes debate inhospitable for MALs, particularly veterans who may wish to discuss their experience in the military and have relatively few options to do so. Ignoring these debaters while focusing on the same evidence round after round and year after year prevents the inclusion of MALs and degrades our understanding of the people involved in war. One way to welcome MALs is to partner with veterans' offices on campus as well as ROTC and similar programs to both recruit debaters and to initiate conversations between debaters and MALs. Reaching out to alumni who are or were MALs and connecting them with debaters can also help. Since debate is increasingly focused on public debate, such efforts will only bolster the connections debate has across a university and include more people in important discussions about war. Furthermore, educators not involved in debate may find MALs and MAL organizations helpful in adding to classroom discussions about war and violence.

CONCLUSION

Regardless of the changes that occur, policy debate must be more attuned to how casual talk about war affects those closest to it. Regardless of whether those wars are nuclear, asymmetrical, or any other type, policy debate must discuss war in more meaningful ways that welcome those involved in it to the table. While this argument appreciates that not all military personnel experience war, nor do they all connect with any war-fighting efforts that may go on while they serve, the best pedagogical practices are those that do not make light of complex traumatic events and treat the impacts of war as tokens in an intellectual game, no matter how important or pedagogically rewarding that game might be otherwise. We have an imperative to include MALs in our pedagogical spaces (Genovese, 2013). The slow process of critical examination that has discouraged many people from using Holocaust, genocide, and rape in casual ways ought to serve as instructive reminders of the type of reforms that can be made to the language we use to describe violent events like war. Our ability to recognize the impacts language has on participants and those in close proximity to these events, especially if it ignores their relationship to war, should guide our interactions with MALs so that we do not ignore their experiences with war.

REFERENCES

Abedi, M. (2017, August 9). Reality check: Why a nuclear war likely won't break out. *Global News*. https://globalnews.ca/news/3657356/reality-check-nuclear-war-unlikely/.

Ajello, L. (2013). Soldiers, scapegoats, and the tragic demands of operation first casualty. *International Journal of Cultural Studies, 17*, 243–247.

Alexander, C. (2017). Behind the mask: Revealing the trauma of war. *National Geographic.* https://www.nationalgeographic.com/healing-soldiers/.

Allen, M., Trejo, M., Bartanen, M., Schroeder, A., & Ulrich, T. (2004). Diversity in United States forensics: A report on research conducted for the American Forensics Association. *Argumentation and Advocacy, 40*, 173–184. doi:10.1080/00028533.2004.11821605.

Bankey, B. (2013). *The "fact of blackness" does not exist: An evocative criticism of resistance rhetoric in academic policy Ddebate and its (mis)use of Frantz Fanon's* Black Skin, White Masks. Unpublished Masters Thesis, Wake Forest University, Winston-Salem, NC.

Bartanen, M. D., & Littlefield, R. S. (2014). *Forensics in America: A history.* Rowman & Littlefield.

Basu, M. (2013, November 14). Why suicide rate among veterans may be more than 22 a day. *CNN.* https://www.cnn.com/2013/09/21/us/22-veteran-suicides-a-day/index.html.

Baudrillard, J. (2006). War porn. *Journal of Visual Culture, 5*, 86–88.

Bellon, J. (2000). A research-based justification for debate across the curriculum. *Argumentation and Advocacy, 36*, 161–176.

Berkenwald, L. (2011, July 7). Abortion and the slavery analogy: Dangerous and wrong. *Ms. Magazine.* https://msmagazine.com/2011/07/07/abortion-and-the-slavery-analogy-dangerous-and-wrong/.

Briscoe, S. F. (2017). *Policy debate: A guide for high school and college debaters.* Southern Illinois University Press.

Brown, W. (2006). American nightmare: Neoliberalism, neoconservativism, and de-democratization. *Political Theory, 34*, 690–714.

Bruschke, J., & Johnson, A. (1994). An analysis of differences in success rates of male and female debaters. *Argumentation and Advocacy, 30*, 162–173. doi:10.1080/00028533.1994.11951584.

Burnam, M. A., Meredith, L. S., Tanielian, T., & Jaycox, L. H. (2009). Mental health care for Iraq and Afghanistan war veterans. *Health Affairs, 28*, 771–782. doi:10.1377/hlthaff.28.3.771.

Carlton, L. S. (2012). Together we stand, divided we fall: Building alliances with combat veterans. *Soundings: An Interdisciplinary Journal, 95*, 284–308.

Carr, J. E. (2002, January). A better investment not found on Wall Street. *Rostrum, 76*(9), 25–26.

Charny, I. W. (2001). The psychological satisfaction of denials of the Holocaust or other genocides by non-extremists or bigots, and even by known scholars. *Idea: A Journal of Social Issues, 6*. https://web.archive.org/web/20071224203627/http://www.ideajournal.com/articles.php?id=27.

Congressional Research Service. (2019, April 18). *U.S. war costs, casualties, and personnel levels since 9/11.* Congressional Research Service. https://fas.org/sgp/crs/natsec/IF11182.pdf.

Cooper, B. (2014, May 14). "I was hurt": How white elite racism invades college debate championship. *Salon*. https://www.salon.com/2014/05/13/%E2%80%9Ci_was_hurt%E2%80%9D_how_white_elite_racism_invaded_a_college_debate_championship/.

Demers, A. (2011). When veterans return: The role of community in reintegration. *Journal of Loss and Trauma, 16*, 160–179.

Department of Veterans Affairs. (2019, November). *America's wars*. https://www.va.gov/opa/publications/factsheets/fs_americas_wars.pdf.

Dhillon, K. K. (2016). Accessibility & debate camps. *National Journal of Speech & Debate, 4*, 10–11.

Dillard-Knox, T. Y. (2014). *Against the grain: The challenges of Black discourse within intercollegiate policy debate*. Unpublished Master's Thesis, University of Louisville, Louisville, KY.

Downs, D. A., & Murtazashvili, I. (2012). *Arms and the university: Military presence and the civic education of non-military students*. Cambridge University Press.

Eglash, R. (2007). Broken metaphor: The master-slave analogy in technical literature. *Technology and Culture, 48*, 360–369.

Ferguson, C. L. S. (2016). Deaf debaters. *National Journal of Speech & Debate, 4*, 8–9.

Forrest, W., Edwards, B., & Daraganova, G. (2018). The intergenerational consequences of war: Anxiety, depression, suicidality, and mental health among the children of war veterans. *International Journal of Epidemiology, 47*, 1060–1067. doi:10.1093/ije/dyy040.

Frankfurt, S., & Frazier, P. (2016). A review of research on moral injury in combat veterans. *Military Psychology, 28*, 318–330. doi:10.1037/mil0000132.

Gannon, A. (n.d.). *Comparative impact calculus*. Retrieved from https://www.forensicscommunity.com/policy/comparative-impact-calculus.

Gehrke, P. J. (1998). Critique arguments as policy analysis: Policy debate beyond the rationalist perspective. *Contemporary Argumentation and Debate, 19*, 18–39.

Genovese, J. M. (2013). *Post-Traumatic Story Disorder: Using the power of narrative to heal the invisible wounds of war*. Unpublished Master's Thesis, University of Texas Medical Branch, Galveston, TX.

Gerber, M. (2009). Toward public sphere intercollegiate policy debate: The path to participation?. *Contemporary Argumentation & Debate, 30*, 80–93.

Gerstenfeld, M. (2007, October 28). *The multiple distortions of Holocaust memory*. Jerusalem Center for Public Affairs. https://jcpa.org/article/the-multiple-distortions-of-holocaust-memory/.

Godwin, M. (1994, October 1). Meme, counter-meme. *Wired*. https://www.wired.com/1994/10/godwin-if-2/.

Goodnight, G. T., & Mitchell, G. R. (2008). Forensics as scholarship: Testing Zarefsky's bold hypothesis in a digital age. *Argumentation and Advocacy, 45*, 80–97. doi:10.1080/00028533.2008.11821699.

Gross, O., & Aoláin, F. N. (2014). The rhetoric of war: Words, conflict, and categorization post-9/11. *Cornell Journal of Law and Public Policy, 24*, 241–289.

Hart, D. A., & Thompson, R. (2016). Veterans in the writing classroom: Three programmatic approaches to facilitate the transition from the military to higher education. *College Composition & Communication, 68*, 345–371.

Haynes, M. (2019, November 12). 49% of veterans don't like to be thanked for their service, poll shows. *WUSA9*. https://www.wusa9.com/article/news/local/some-veterans-dont-like-to-be-thanked-for-their-service/65-385e06d6-eff4-4071-8b05-51cef63bf2a5.

Hedges, C. (2003, July 6). 'What every person should know about war'. *New York Times*. https://www.nytimes.com/2003/07/06/books/chapters/what-every-person-should-know-about-war.html#:~:text=How%20many%20people%20have%20died,150%20million%20to%201%20billion.

Herbeck, D. A., & Katsulas, J. P. (1992). *The use and abuse of risk analysis in policy debate*. Paper Presented at the 1992 Speech Communication Association Convention. https://files.eric.ed.gov/fulltext/ED354559.pdf.

Hill, C. B., Kurzweil, M., Pisacreta, E. D., & Schwartz, E. (2019, January 10). *Enrolling more veterans at high-graduation-rate colleges and universities*. New York, NY: ITHAKA. doi:10.18665/sr.310816.

Huber, A. (2016). *Rhetorical criticism of the use of kritik in parliamentary debate*. Unpublished Honors Thesis, Northern Illinois University, DeKalb, IL.

Kraft, J. C. (2014, April 16). Hacking traditional college debate's white-privilege problem. *The Atlantic*. http://www.theatlantic.com/education/archive/2014/04/traditional-college-debatewhite-privilege/360746/.

Lakoff, G. (2009). Metaphor and war: The metaphor system used to justify war in the Gulf. *Cognitive Semiotics, 4*, 5–19. doi:10.1515/cogsem.2009.4.2.5.

Levy, B. S., & Sidel, V. W. (2016). Documenting the effects of armed conflict on population health. *Annual Review of Public Health, 37*, 205–218. doi:10.1146/annurev-publhealth-032315-021913.

Listwa, D. (2012, August 9). *Hiroshima and Nagasaki: The long term health effects*. Columbia K 1 Project. https://k1project.columbia.edu/news/hiroshima-and-nagasaki.

Loge, P. (1991). An examination of black participation in CEDA debate. *Contemporary Argumentation and Debate, 12*, 79–87.

Matz, I., & Bruschke, J. (2006). Gender inequity in debate, legal and business professions. *Contemporary Argumentation & Debate, 27*, 29–47.

McCordick, J. (2017, September 26). The corrosion of high school debate—And how it mirrors American politics. *America Magazine*. https://www.americamagazine.org/arts-culture/2017/09/26/corrosion-high-school-debate-and-how-it-mirrors-american-politics.

Mead, W. R. (1992). Depending on the kindness of strangers. *New Perspectives Quarterly*, 9, 28–30.

Miller, K. A. (2019). *Veteran and military-connected students' experiences in college*. Doctoral Dissertation, Purdue University. https://hammer.figshare.com/articles/Veteran_and_Military-Connected_Students_Experiences_in_College/8979821.

Mitchell, G. R. (1998). Pedagogical possibilities for argumentative agency in academic debate. *Argumentation and Advocacy, 35*, 41–60.

Moore, E. (2012, September 11). *From combat to college: Student veterans in academic "contact zones"*. Institute for the Study of Societal Issues. http://www.escholarship.org/uc/item/92m9r81k.

Mueller, J. (2010). Calming our nuclear jitters. *Ideas in Science and Technology, 26*. https://issues.org/mueller/.

Murthy, R. S., & Lakishminarayana, R. (2006). Mental health consequences of war: A brief review of research findings. *World Psychiatry, 5*, 25–30.

Nelson, N., & Miller, J. H. (2016). Dyslexia & debate. *National Journal of Speech & Debate, 4*, 5–7.

Perr, J. (2015, October 12). Between Holocaust and slavery analogies, Republicans choose both. *DailyKos*. https://www.dailykos.com/stories/2015/10/12/1431205/-Between-Holocaust-and-slavery-analogies-Republicans-choose-both.

Pineda, R., & Salinas, C. (2009). Model proposal: Increasing Latina/o involvement in policy debate through summer debate workshops. *Contemporary Argumentation & Debate, 30*, 114–127.

Poapst, J., & Harper, A. (2017). Reflections on the 2014 celebration of women debate tournament at George Mason University. *Argumentation and Advocacy, 53*, 127–137. doi:10.1080/00028533.2017.1305000.

Pols, H., & Oak, S. (2007). War & military mental health: The US psychiatric response in the 20th century. *American Journal of Public Health, 97*, 2132–2142. doi:10.2105/AJPH.2006.090910.

Polson, D. (2012). *Longing for theory: Performance debate in action*. Unpublished Doctoral Dissertation, University of Maryland, Baltimore County, Baltimore, MD.

Prager, J. R. (2007). *Introduction to Lincoln Douglas Debate*. Bay City, MI: John R. Prager. https://www.boone.k12.ky.us/userfiles/781/Classes/52280/intro-ld_2007.pdf?id=558370.

Reid-Brinkley, S. R. (2019). Voice dipped in Black: The Louisville Project and the birth of Black radical argument in college policy debate. In N. S. Eidsheim & K. Meizel (Eds.), *Oxford handbook of voice studies* (pp. 215–236). doi:10.1093/oxfordhb/9780199982295.013.28.

Richtel, M. (2015, February 21). Please don't thank me for my service. *New York Times*. https://www.nytimes.com/2015/02/22/sunday-review/please-dont-thank-me-for-my-service.html.

Richter, Z. (2016). The disabled person's struggle in round & beyond: Taking back formerly ableist educational spaces in the post-ADA generation. *National Journal of Speech & Debate, 4*, 12–14.

Rogers, J. (1997). A community of unequals: An analysis of dominant and subdominant culturally linked perceptions of participation and success within intercollegiate competitive debate. *Contemporary Argumentation and Debate, 18*, 1–22.

Sciullo, N. J. (2017). Making debate more inclusive for the student-veteran debater. *Argumentation and Advocacy, 53*, 138–148. doi:10.1080/00028533.2017.1305008.

———. (2019). The racial coding of performance debate: Race, difference, and policy debate. *Argumentation and Advocacy, 55*, 303–321. doi:10.1080/10511431.2019.1672028.

Sheng, Y. (2020, March 10). China-US war unlikely despite rising hostility. *Global Times*. https://www.globaltimes.cn/content/1187947.shtml.

Stahl, R. (2009). Why we "support the troops": Rhetorical evolutions. *Rhetoric & Public Affairs, 12*, 533–577.

Stepp, P. (1997). Can we make intercollegiate debate more diverse?. *Argumentation and Advocacy, 33*, 176–191. doi:10.1080/00028533.1997.11978017.

Stepp, P., & Gardner, B. (2001). Ten years of demographics: Who debates in America. *Argumentation and Advocacy, 38*, 69–82. doi:10.1080/00028533.2001.11821558.

Stepp, P., Simerly, G., & Logue, B. (1994). Sexual harassment in CEDA debate. *Argumentation and Advocacy, 31*, 36–40. doi:10.1080/00028533.1994.11951596.

Steuter, E., & Wills, D. (2008). *At war with metaphor: Media, propaganda, and racism in the War on Terror*. Lexington Books.

Stewart, S. (2015, October 26). We need to stop saying 'Thank you for your service'. *TheTab.com*. https://thetab.com/us/syracuse/2015/10/26/thank-you-for-your-service-724.

Stone-Mediatore, S. (2010). Epistemologies of discomfort. *Studies in Social Justice, 4*, 25–45.

Valiunas, A. (2006). The agony of atomic genius. *The New Atlantis, 14*, 85–104.

Warner, E., & Bruschke, J. (2001). "Gone on debating:" Competitive academic debate as a tool of empowerment. *Contemporary Argumentation & Debate, 22*, 1–21.

Williams, D. E., McGee, B. R., & Worth, D. S. (2001). University student perceptions of the efficacy of debate participation: An empirical investigation. *Argumentation and Advocacy, 37*, 198–209.

Wood, J. (n.d.). Don't just thank me for my service, ask me about it. *Military.com*. https://www.military.com/veterans-day/dont-just-thank-me-for-my-service.html.

Zwarensteyn, E. C. (2012). *High school policy debate as an enduring pathway to political education: Evaluating possibilities for political learning*. Master's Thesis, Grand Valley State University, Allendale, MI. https://scholarworks.gvsu.edu/cgi/viewcontent.cgi?article=1034&context=theses.

Chapter 10

The Hard Sell

Supporting MALs through Campus Resource Centers

Meghan Velez, Victoria McDermott, and Kenneth W. (Broken) Marlin

To help ensure academic success, universities have developed campus resource centers from math labs to writing centers to provide additional tutoring and support for students (Heppner & Neal, 1983; Horton, 2017; Osam et al., 2017). While these campus resource centers provide an additional support network for students struggling in specific discipline areas, they remain underutilized by the majority of students (Brock, 2010). While previous research has explored the motivation behind traditional college student resource center utilization (McDermott et al., 2019), there is little information about how campus resources are serving Military-Affiliated Learners (MALs). Moreover, MALs often have layered identities and lived experiences that can present additional challenges for resource center directors tasked with developing programs to target the unique needs of this population of learners. While MALs may be resilient and self-sufficient, focusing heavily on the mission at hand (Ford & Vignare, 2015), the need for academic support to ensure program completion remains unchanged. As such, more research is needed to understand the factors that may be impacting MALs' campus resource utilization.

To better understand MALs' experiences, the present chapter explores how MALs are currently using campus resource centers. Three campus resource center staff members used an autoethnographic approach to reflect on their previous experiences serving MALs. Through the use of thematic coding, themes were identified to explain barriers MALs face to using campus resources. Through the identification of common barriers, the researchers provide recommendations campus resource centers can adopt to better serve and support MAL student populations.

LITERATURE REVIEW

Campus Resource Centers

Campus resource centers (e.g., communication centers, math labs, writing centers) provide a variety of vital support services for college students. Students who are perceived to be "at-risk" of dropping out due to personal or academic struggles are often referred to tutoring and counseling resources on campus (Capstick et al., 2019). Previous research has suggested that "campus services and key university personnel are influential in retaining and promoting positive student experiences" (Southwell et al., 2018, p. 395). Thus, campus resources seek to assist students, as well as provide a sense of commitment and community to increase academic persistence (Astin, 1993; Osborne et al., 2019; Polite, 2019). Through the assistance provided by these resources, students often find the support they need for their academic endeavors ranging from finishing an assignment to completing a degree to successfully navigating a job interview post degree completion. For example, Capstick et al. (2019) found students who participated in academic coaching (i.e., individualized and personalized coaching sessions for challenging material) had significant GPA increases and were more likely to remain at the university than their counterparts who did not participate in the academic coaching program. Furthermore, previous research has suggested that resource centers positively impact students' mental health and social lives, as well as their academic performance (Henchy, 2013; McDermott et al., 2019; Polite, 2019). Collectively, campus resource centers increase retention and improve the academic self-efficacy of students (Barefoot, 2004; Henchy, 2013; McDermott et al., 2019).

Who Does and Doesn't Use Campus Resource Centers

The researchers acknowledge research regarding the demographics of campus resource center users varies. For example, Gormley et al. (2019) found that students from diverse racial and ethnic backgrounds had higher rates of service use. Additionally, research suggests that women are more likely than men to use campus resource centers (Salem, 2016). Furthermore, Salem (2016) found that first-generation students and those with low SAT scores were more likely to use campus resource centers. However, Gormley et al. (2019) found that SAT scores were not a predictor of campus resource use. Alternatively, Olfson and colleagues (2002) argue that individuals with higher levels of education use support services to a greater degree relative to those with less education. While the specific demographics of campus resource center users remains unclear, the fact remains that a majority of students, no matter their race & ethnicity, gender identity, or previous educational history, do not use campus resource centers.

Lack of Campus Resource Use

Despite evidence of their positive impact on campus communities, campus resource centers are underutilized (Brock, 2010; Neal & Heppner, 1986; Spellman, 2007). One of the first studies of campus resource use in 1977 (Benedict et al., 1977) found that between 14 percent to 50 percent of college students lacked awareness of campus resources. Of the students who were aware of these resources, only 10 percent to 25 percent decided to seek out these services (Benedict et al., 1977). Four decades later, these trends remain stagnant. In 2017, a study of both public and private universities found that only 20 percent of undergraduate students utilized the career services center provided by their institution (Strada & Gallup, 2017). Salem (2016) explored campus resource center use over the course of students' college careers. After collecting baseline data on an incoming cohort of freshmen, over the course of four years, only 22 percent visited the writing center and 78 percent did not (Salem, 2016). Unfortunately, many similar resources, such as math labs, or communication centers at colleges and universities, continue to be underestimated and underutilized by students.

Barriers to Use

This underutilization may be due to physical or perceived barriers students have regarding campus resource centers. One of the biggest barriers to campus resource use is a lack of knowledge regarding campus resource centers. Students may not know these resources exist, how to contact them, or where they are located (Craft, 2013). Research indicates that universities with highly affluent student bodies report less students utilizing campus resources for assistance throughout the semester (Salem, 2016; Strada & Gallup, 2017). Additionally, when these students do use campus resources, they seem to rate them more unfavorably and designate resources as "unhelpful" (Strada & Gallup, 2017). Over time, this viewpoint of "unhelpful" can lead to negative word of mouth perceptions around campus. Furthermore, Salem (2016) argues that if students believe the resource center is "remedial," they will be less likely to visit even if they genuinely want or need help with an assignment.

Another reason students may not be using these campus resources is that they may not think they need help (McDermott et al., 2019). Students believe they can "do it themselves," even if this may not be the case (Craft, 2013; McDermott et al., 2019). Unless required by an instructor, many students will not visit these resource centers (Craft, 2013; McDermott et al., 2019). Other barriers to utilization may include time, convenience, and anxiety (Craft, 2013; McDermott et al., 2019). The hours a center is open can limit a student's ability to use that resource. Students have noted class scheduling conflicts as a reason for not using campus resources (Craft, 2013). Additionally, if a student is off campus, only providing face-to-face tutoring will present a challenge. Finally, a major barrier to student resource use is anxiety (McDermott et al., 2019). Students may be apprehensive about using a campus resource center because the assignment gives them too much anxiety to begin with or they believe they will be judged by the tutors at the center. For example, students may not be willing to visit a Communication Center knowing they have to practice public speaking. For some students, having to present a speech in class once already causes too much anxiety (McDermott et al., 2019). The same can be said about students who are apprehensive about getting their writing critiqued by a writing center or having their resume questioned at a career counseling center.

MAL IDENTITY

For many universities, the MAL population continues to grow with the introduction and heavy use of the Post 9/11 G.I. Bill (Ford & Vignare, 2015). Thus, universities may not have the necessary resources to support these students, as MALs may differ significantly from the traditional student populations they are used to serving (Howe & Shpeer, 2019; Iloh, 2018; see chapter 14 by Ludewig in this edited volume). The subsequent section briefly outlines some of the characteristics of MALs that can impact their campus resource use in college, such as identifying with a nontraditional student population and the impact of military values.

A Nontraditional Student Population

According to the National Center for Education Statistics (NCES), age acts a "surrogate variable that captures a large, heterogeneous population of adult students who often have family and work responsibilities as well as other life circumstances that can interfere with successful completion of educational objectives" (NCES, n.d., para. 1). In addition to age, the NCES defines nontraditional students as meeting one of seven other characteristics: delayed enrollment into postsecondary education; attends college part-time; works full-time; is financially independent for financial aid purposes; has dependents other than a spouse; is a single parent; or does not have a high school diploma (Pelletier, 2010). Alternatively, traditional students are characterized as someone who is between the ages of eighteen and twenty-two, begins college immediately after high school, enrolls full-time in college, lives on campus, and receives financial support from parents (Deil-Amen, 2011; Schmidt, 2013).

Overall, research suggests that MALs identify more closely with nontraditional learners (Southwell et al., 2018). MALs are usually over the age of twenty-four, juggle family life, work a full-time job, and may only be enrolled part-time (Ford & Vignare, 2015). Currently, a majority of top institutions have focused their infrastructure and resources design around the more traditional student learner (Iloh, 2018). This includes the way they recruit, admit, advise, and support students (Syverud, 2019). For example, universities may hold the traditional fifteen-week-long semester format. For a traditional student living on campus, attending class over this period of time three times a week may be easy. For a MAL who has a family and a job and commutes to campus, completing a fifteen-week-long course is daunting, especially if class attendance is strictly required (CCA, n.d.; Syverud, 2019).

Being a nontraditional student learner can pose a variety of challenges for degree success. Previous research shows that only 48.9 percent of adult learners graduated within six years at four-year institutions compared to a graduation rate of 64.7 percent among traditional students (Fredman, 2018). In 2013, a longitudinal study of adult learners found that 67.1 percent of part-time students were no longer enrolled in college after six years; however, only 21.9 percent of the students being tracked had completed a degree program (Shapiro et al., 2013, 2017). Thus, the other 45.2 percent of students either transferred or dropped out. Due to universities being developed to accommodate more traditional student populations (Iloh, 2018; Syverud, 2019) and the added challenges a nontraditional student can face, such as having dependents or working a full-time job, nontraditional student learners may need additional support in order to finish their college degrees successfully.

Military Values

In addition to identifying as nontraditional student learners, MALs are impacted by military values that they may bring into the classroom (see the following chapters in this edited volume: May & McDermott, chapter 2; Chiara, Chiara, & Ashlock, chapter 3; Howe, chapter 4). These military values are defined by strict codes of conduct, as well as rules and guidelines all members must follow or face severe consequences. For example, the Uniform Code of Military Justice (UCMJ) clearly outlines acceptable behaviors for members when they are both in and outside of uniform ranging from their inability to speak out against the President to the consequences of adultery (Army.mil, 2018; Howe & Shpeer, 2019; Maravilla, 2007). Even in their private lives, military members are held to a specific standard. If broken, the consequences can vary from a slap on the wrist to a loss of rank or pay and even imprisonment.

Thus, changing from the military to the classroom can present a challenge for MALs. Previous research has shown that MALs, veterans especially, tend to experience reverse culture shock (Howe & Hinderaker, 2018; Howe & Shpeer, 2019; Knight, 1990), expecting the same level of formal rules and regulations within the classroom. The values highlighted and reinforced in the military, such as a strict hierarchy, order-based decision-making, and the emphasis on acting in favor of the greater good (Army.mil, 2018; Halvorson et al., 2010; Ziff, 2017) have been shown to carry into MALs' university experiences (Howe & Shpeer, 2019; May & McDermott, 2019). These values may not be found in academic environments that tend to promote "liberal ways of thinking for constant questioning of the status quo" (May & McDermott, 2019, p. 64) and allow students to both challenge traditional institutional norms and advocate for social change. Additionally, MALs are likely to possess a "rich experiential history with exposure to organizational culture, leadership, decision-making, different management styles, and other managerial involvement and organizational engagement" (Starr-Glass, 2011, p. 150). Ultimately, a combination of their previous experience of interdependence and constant structure within the military transfer into the classroom defining the MAL identity with characteristics such as persistence, focus on the mission, and resilience (Ford & Vignare, 2015; Howe & Shpeer, 2019).

As such, identifying with a nontraditional student population, as well as the values asserted in the military culture, can make it difficult for MALs to reach out for help when they need it. Previous research has shown that nontraditional students participate less in orientation activities and utilize less university support resources than traditional students because they may believe these activities do not pertain to them or that they do not need the assistance (Wardley et al., 2013; see Chiara, Chiara, & Ashlock, chapter 3, in this edited volume). MALs may feel lost and alone trying to navigate higher education believing they must be autonomous and self-sufficient in their learning experience as both a MAL and a nontraditional student learner (Vacchi, 2012; in this edited volume, see Chiara, Ashlock, & Chiara, chapter 3). In order to better support MALs, universities and campus resource centers need to connect more effectively with this population of learners. Getting MALs to engage with and use more campus resource centers can help MALs feel more supported throughout their educational careers, as resource centers have been proven to increase student academic self-efficacy and retention (Southwell et al., 2018).

MAL Enrollment and Retention Rates

When it comes to MALs, degree completion rates are problematic both from being underreported and under-tracked by universities, as well as for being low at institutions that do track retention and completion rates (Marcus, 2017; NASPA, 2013; Schwartz & Davidson Pisacreta, 2019). Of 239 institutions surveyed, 72 percent did not track retention or degree completion rates for active-duty students, and 67 percent of institutions did not track retention or degree completion rates for veteran students (NASPA, 2013). A separate study of students receiving G.I. Bill benefits reported that only 15 percent of full-time students graduated with a two-year degree in 2014 (Marcus, 2017). This figure also includes the students who took three years to complete a two-year degree (Marcus, 2017). Not only are graduation rates problematic, but the G.I. Bill only covers a maximum of thirty-five cumulative months in college (VA, 2020). This provides MALs with enough financial aid to cover a bachelor's degree but leaves little margin for error. Moreover, Schwartz and Davidson Pisacreta (2019) found that only one in ten G.I. Bill beneficiaries are enrolled in an institution with a six-year graduation rate above 70 percent. Graduation rates are a transparent metric for holding schools accountable and can be an indicator for students about the quality of academic support previous students have received (Pairitz Morris, 2020). Moreover, veterans are overrepresented at for-profit colleges and underrepresented at non-profit high-graduation-rate institutions (Schwartz & Davidson Pisacreta, 2019). Research shows that graduation rates for non-profit institutions are marginally higher than in for-profit institutions, 65 percent compared to 28 percent (Lang & Weinstein, 2013). Although for-profit institutions may look enticing because of their open enrollment policies and flexible programs (Value Colleges, n.d.), a study by the National Bureau of Economic Research found that students who attend for-profit institutions are more likely to be unemployed, earn less money, have higher debt levels, and are more likely to default on their student loans (Lang & Weinstein, 2013). Previous research suggests that students are often dissatisfied with their educational experience at for-profit institutions and may leave with debt and without finishing a degree or rack up debt while trying to change degree or certification paths (Body, 2019). Additionally, many non-profit universities will not accept credits from for-profit institutions, meaning students have to pay to retake classes and credits at their new institution while still paying off the debt of their previous for-profit institution (Elejalde-Ruiz, 2016). Nontraditional students, and especially MALs, may not be aware of the predatory practices of for-profit institutions or the nuances of picking a university (Webley, 2012; see chapter 8 by Moullet & Belding in this volume). Overall, with the barriers MALs face from cultural differences and outside commitments to low degree completion rates, higher education is not effectively supporting MALs and their educational endeavors.

One way that universities can lessen dropout rates and increase student retention, especially for MALs, is through the use of campus resource centers (Southwell et al., 2018). Campus resource centers can provide additional instruction and support that may fill in learning gaps to increase student success and academic self-efficacy (McDermott et al., 2019; Osborne et al., 2019; Polite, 2019). Previous research by Southwell and colleagues (2018), building on research conducted by Wardely et al. (2013), found that MALs are less likely to use campus resources (i.e., registrar, financial aid office, advisers, and so on) than their civilian peers. Their research suggests that due to MALs' time restrictions and other potential outside commitments, MALs may not have the time to seek the necessary support from faculty and advisers (Southwell et al., 2018). However, when MALs do engage in these interactions with faculty and resource centers, they experienced positive benefits, such as increased academic persistence. Moreover, frequent visits to faculty members and academic advisers have positive implications for MALs, such as feeling better supported and connected to their campus communities (Southwell et al., 2018).

Currently, there is a dearth of research on the experiences of MALs when they do use campus resource centers and how campus resource center staff sensemake their experiences working with MALs. Some research to date has explored the traditional students' experience using campus resources (McDermott et al., 2019), but none focus specifically on the MAL experience. MALs may have different perspectives and barriers regarding campus resource use. Furthermore, since these campus resource centers were developed to assist traditional student learners in successfully completing their degrees, it is crucial for center directors and staff to fully understand how to better support their MAL student population specifically. Thus, the dearth of research and understanding of MALs campus resource use informed the following research question:

RQ1: How can student resource centers better support MALs?

METHODS

Through the critical autoethnographic reflections of three campus resource center staff members, the researchers were able to identify common themes for supporting MALs in campus resource centers. The present section overviews the literature of autoethnography, the authors' background, and the step-by-step autoethnographic process the researchers used to compile their results.

Autoethnography

Autoethnography has been used in a variety of studies to sensemake, describe, and interpret cultural texts, experiences, beliefs, and practices (Adams et al., 2017; Ellis & Bochner, 2000; Mendez, 2013). Autoethnography requires working at the intersection of autobiography and ethnography. First, we must reflect on past experiences, talk with others about the past, and examine archives, such as personal journals, recordings, news stories, or blogs. Next, we observe, participate in, and write about a culture to connect personal lived experiences to larger formal research and theoretical framework (Adams et al., 2017). McIlveen (2008) argues that autoethnography is not just writing about oneself but about "being critical about personal experiences in the development of the research being undertaken, or about experiences of the topic being investigated" (p. 3). Reed-Danahay (1997) assigned three main characteristics to autoethnography:

> (1) The role of the autoethnographer in the narrative: is the autoethnographer an insider or an outsider of the phenomenon being described? (2) Whose voice is being heard: who is speaking, the people under investigation or the researcher? (3) Cultural displacement: some realities are being described by people who have been displaced from their natural environment due to political or social issues. (p. 3–4)

Overarchingly, the researcher must move back-and-forth between "experiencing and examining a vulnerable self and observing and revealing the broader context of that experience" (Ellis, 2007, p. 14).

Ultimately, autoethnography serves a variety of purposes in research. First and foremost, it provides a voice to personal experiences and provides alternatives to dominant or even harmful cultural scripts, stories, and stereotypes (Boylorn, 2014). These personal experiences can "complement or fill in gaps for existing research . . . [and] may show how the desire for, and practice of, generalization in research can mask important nuances of cultural issues" (Adams et al., 2017, p. 3). Secondly, autoethnography allows researchers to articulate insider knowledge of cultural experience. Autoethnography can describe moments of everyday life that cannot be captured through traditional research methods (Ellis, 2009; Boylorn, 2011). Thirdly, autoethnography shows how researchers are implicated by their observations and conclusions and encourages others to share counternarratives (Crawford, 1996). Plummer (2001) states, "What matters is the way in which the story enables the reader to enter the subjective world of the teller—to see the world from her or his point of view, even if this world does not 'match reality'" (p. 401). Finally, autoethnography allows for greater accessibility. Through the use of autoethnography, researchers can create texts that can be used outside of academic settings. Autoethnography allows the researchers to provide a point of connection with the audience. Previous research has noted that much of academic writing suppresses the "sensory, emotional, and bodily life" of research (Tamas, 2011, p. 73). With its emphasis and value on personal experience Adams (2012) explains:

> Much of what is said happened in concrete contexts, there-by offering lessons applicable to everyday life. Autoethnography thus embodies the concept of praxis—a melding of theory and practice, form and content—and, in so doing, offers engaging and accessible texts for others to easily comprehend and use. (p. 189)

Ultimately, autoethnography allows researchers to reflect on and connect their lived experiences to the larger cultural narrative to explore the nuances of their experiences. Through this self-reflection, researchers are able to engage with both the academic and nonacademic communities simultaneously while developing new counternarratives to give voice to potentially underrepresented different lived experiences.

Background of the Researchers

Meghan Velez oversees the Virtual Communication Lab (VCL) at Embry-Riddle Aeronautical University (ERAU) Worldwide. The VCL opened in 2019 and was designed to support ERAU-Worldwide's predominantly online, nontraditional population, which includes a high number of active duty and recently retired military members, military spouses, and other MALs. The lab provides one-on-one tutoring and workshops, both of which are hosted in a synchronous online format, as well as a communication resource repository students can access any time. In addition to running the VCL, Velez teaches courses in English composition, rhetoric, and technical writing. She has spent more than ten years working in writing centers, and although she has no military affiliation herself, she has previously worked with MAL populations in writing centers.

Kenneth Marlin has served as the lead tutor for the VCL since it opened 2019, offering workshops, creating resources, and working one-on-one with students to help them grow their confidence and skills in the area of communication. A military veteran and a longtime instructor at ERAU, Marlin has become the most sought-after tutor at the center and is currently developing training materials for the lab's other staff. In addition to tutoring, Marlin teaches a variety of courses at ERAU-Worldwide, including first-year composition, literature, and technical writing.

Alex Rister provided a third perspective as Director of ERAU's Virtual Environment for Communication: Teaching, Outreach, and Research

(VECTOR).[1] As an overall center, VECTOR comprises the Virtual Communication Lab run by Velez, a Communication Across the Curriculum faculty development program, and a communication research support hub. Rister teaches a variety of courses in communication, including general education speech courses as well as courses within the communication major. She does not identify as a MAL but has worked with MALs throughout her history at ERAU-Worldwide.

The Process

In order to conduct this autoethnographic research, researchers Velez and Marlin first developed five questions to prompt personal reflection about how our center and other resource centers can support MALs. After developing the questions, all three participants sat down over the course of five days to think about, reflect on, analyze, and answer the posed questions. The questions used for reflection were:

1. What are the biggest complaints/barriers we hear from MAL students for using campus resource centers (not just our center)?
2. How does our center support MALs differently than other parts of the university (their classes, other resources or departments, and so on)?
3. With unlimited resources, time, and support, what would our center look like? What would we want to provide specifically for our MAL population?
4. What do we know about supporting the MAL population that we want others to know?
5. What do we still need to learn about supporting MALs based on our experiences so far?

Once all the participants had answered these questions, they came together to compare answers, discuss similarities, and dive more in-depth into their reflections through the use of thematic coding (Geisler, 2003; Richards & Hemphill, 2017). To begin the coding process, the three participants shared an electronic version of their answers with each other. During this quiet read through, the participants did not take notes or highlight important ideas. After reading through everyone's responses, the three participants then began a second read through, looking for common and differentiating themes, motifs, and emotions (Thornberg & Charmaz, 2012). At this point the researchers began discussing their responses, highlighting frequent and recurring themes, as well as pulling quotes to support the themes presented to answer the research question. Through this use of analytic coding (Richards & Hemphill, 2017), the researchers were able to create categories and themes to answer the research question.

This inductive method is a way for researchers to focus on what the data is telling them rather than imposing conventions or themes from external sources. Geisler (2003) claimed that inductive approaches to coding "let each segment of the data suggest appropriate categories to describe what's happening with the phenomenon of interest" (p. 60). Adopting an inductive approach to the analytic coding process allowed Velez, Marlin, and Rister to identify moments of interest in their separate responses, follow up, and reflect on those responses. Moreover, it allowed them to analyze and determine the meaning and frequency of words or phrases before having to determine whether or how to categorize the data into specific themes.

RESULTS

Although the participants responded to these questions separately at first, answers were further developed in the subsequent reading and discussion. The inductive coding process resulted in five thematic codes: Resilience, Flexibility, Mission, Trust, and Representation/Relatability. "Resilience" refers to the characteristics of MALs described above, including their persistence and self-reliance. "Flexibility" refers to MALs' needs for a flexible approach to factors like time, modality (online vs. face-to-face), and pedagogy in tutoring sessions. "Mission" refers to MAL perceptions of an assignment or course as a mission. "Trust" refers to barriers as well as strategies to advise and support MALs through earning confidence or respect. Finally, "Representation/Relatability" refers to the notion that MALs may be drawn to resource centers that have MAL representation or that make a visible effort to relate to MAL needs and identities. Table 10.1 represents the appearance of each thematic code in the researchers' individual and group responses to the five discussion questions.

Because the written responses and subsequent discussion often drew on more than one question at a time, the remainder of this section is organized by theme in response to the overarching research question: How can student resource centers better support MALs?

Flexibility: Time, and Accessibility

The kinds of barriers that MALs often face in education in general, and in using resource centers in particular, fall into two categories: (1) the practical or pragmatic, and (2) the attitudinal or affective. In both the individual responses and group discussion, time emerged as the most significant of the practical barriers. As described earlier in this chapter, MALs, like many non-traditional students, often balance school with work and family commitments.

Table 10.1 Thematic Codes

Discussion Question (DQ)	*Themes Present in Responses*
1. What are the biggest complaints/barriers we hear from MAL students for using campus resource centers (not just our center)?	Flexibility, Resilience
2. How does our center support MALs in ways that they might not be supported by other parts of the university (their classes, other resources or departments, and so on)?	Flexibility, Representation/ Relatability
3. With unlimited resources, time, and support, what would our center look like? What would we want to provide specifically for our MAL population?	Flexibility, Representation/ Relatability, Mission
4. What do we know about supporting the MAL population that we want others to know?	Resilience, Representation/ Reliability, Flexibility, Trust
5. What do we still need to learn about supporting MALs based on our experiences so far?	Trust, Mission

Active-duty military students face additional time-based challenges when they are deployed to or stationed in a time zone that is different from that of their campus.

Both of these factors influence MALs' ability to practically access resource centers. They may not have enough time to work with a tutor or consultant in addition to their other responsibilities, or (more commonly, in the researchers' experiences), the resource center hours may not be compatible with the hours that MALs are working on their assignments (e.g., nights and weekends). If MALs are able to visit a campus resource center, they may not be able to visit frequently. MALs may not have the option to leave the center after a tutoring session and come back another day with a new draft of an assignment. As a result, they may want to remain at the center for long periods to get their work done in the presence of the tutor. This can pose a challenge when many centers may have time limits on tutoring sessions to increase how many students they are able to assist or tutor in a day. All three researchers noted that our center provides and prioritizes flexibility in its programming (Discussion Question 2; DQ2), but that one of our main objectives with additional funding and resources would be to further increase tutoring hours and formats (DQ3). For instance, all three researchers listed more tutoring hours as an outcome of increased funding. Two researchers, Rister and Velez, also mentioned the addition of more tutoring modalities, such as more asynchronous feedback options.

An additional barrier identified was accessibility. This was characterized as having multiple options for tutoring (e.g., virtual, face-to-face). Whether MALs are deployed and taken away from the campus environment or are otherwise unable to visit campus resource centers in person, accessing resource centers may be a challenge if the services are limited to the face-to-face modality. Ultimately, MALs may be more inclined to take online classes to optimize their potential for class success due to the easy access and flexibility they afford. MALs who choose online course options for flexibility need a similar flexibility from campus resource centers. Providing strictly face-to-face tutoring use options can highly limit MALs' ability to use these campus resources, particularly if they are working full time, balancing school and family life, or (as is often the case at ERAU), they are active-duty military and are deployed or reassigned to new locations mid-semester. Two of the researchers wrote about the importance of flexibility with MALs, calling it the "first thing" we must think about with this student population. At the same time, we noted that not all of the faculty and staff on campus understand this importance (DQ2). Resource centers may not be able to convince faculty to be more flexible with elements of their courses, such as assignment deadlines, but centers can be there as a source of support to help keep students on track.

Resilience: The Mission and the Hard Sell

MALs at our institutions are often self-sufficient and reluctant to seek outside help because they are used to solving problems on their own. This result aligns with the previous research on MAL identity that characterized MALs as persistent and resilient (Ford & Vignare, 2015). In his individual response, Marlin wrote that one of our center's biggest challenges is convincing MALs that they both need and want the help that a communication center can provide: 'They push hard, and they rarely ask for help except as a last resort. How do we reach them before then?' This theme recurred in multiple responses from all participants and was a focal point of our reflective discussion. Notably, we characterized resilience as both a reason that MALs may be reluctant to seek out center services (DQ1) and as a concept that we understand about supporting MALs that others may not realize (DQ4).

While their resilience may prevent them from asking for help, MALs may also be reluctant to utilize resource centers if the center is billed as remedial or supplemental. Faculty and advisers who mean well in recommending the center but frame the center in a remedial way may do more harm than good, and the researchers discussed this knowledge of "framing" as something that we wanted others at our own institution to know about (DQ 4). The historical framing of writing centers as remedial "fix it shops" or clinics for poor writers, for instance, is well-documented (Carino, 1996; Bouquet, 1999; Lee, Caronia, & Beltran, 2014). While many resource centers must constantly work against these perceptions to convince students and faculty alike that their services can benefit all students, regardless of ability or performance, this framing may pose a more significant challenge for gaining the trust of

MALs. MALs are often high achievers in their careers, bringing a wealth of experiential knowledge to their education, and the idea of "needing extra help" may be unappealing or antithetical to the persistent attitude MALs apply to their work.

Finally, MALs bring a mission-driven focus to their education, as noted by Ford and Vignare (2015). The idea of the mission came up frequently in the researchers' individual responses and group discussion. The researchers agreed that resource centers may sometimes seem "external" to the primary goal of pushing through the course or assignment because they exist outside of the classroom and thus, sometimes, outside of students' consciousness. This may be due in part to the ways centers are framed administratively, in boilerplate language about student success, rather than in consistent messaging to students. Regardless of the cause, if the resource center is not seen or understood as a necessary or vital aspect of the "mission" of completing an assignment, course, or degree, MALs may not make resource center visits part of their practice. Since resource centers are external to courses, often operating on the sidelines offering support for those who want it, their services may be seen as unrelated to the mission at hand. The researchers noted that some faculty at our institution are interested in embedding communication center visits into assignments or course requirements and that this may help make the practice of visiting the center seem more central to the mission of academic success. However, most centers, including our own, cannot sustain this embedded requirement on a large scale without increased funding and resources (DQ3).

Trust and Relatability

While it can be difficult to break through the barrier of the resilient and self-sufficient attitude described above, the researchers all noted that trust was a key element in successfully appealing to MALs. Several factors contribute to building MALs' trust, but representation emerged as a closely related theme to trust-building. Center staff who have military affiliations themselves, or who purposely work to foster relatability with MALs through their language and pedagogical choices, can further build trust between MALs and the center. Building relationships with other MAL-trusted leaders across campus also featured heavily in the researchers' discussion, particularly in DQs 4 and 5.

Relatability through Language

The individual and group responses to DQ 4 focused on both administrative (i.e., staffing) and pedagogical choices centers can make to improve the experience for MALs. Center staff, particularly tutors, who themselves have military backgrounds can tap into what Marlin and Velez referred to as "military as a second language" to facilitate "lightbulb" moments with MALs. Marlin described his method of speaking with students about their communication projects and finding common ground through military-focused analogies, such as comparing an outlining process to an Army SOP. However, this is not something that every tutor can or should do. Velez described her own self-awareness as a civilian and noted that she would not feel comfortable attempting to leverage these communication strategies when working with students: "I would feel disingenuous trying to speak that language." However, she acknowledged that she *could* invite students to put concepts in terms they are familiar with, to create those analogies themselves, and to explain them to her. All tutors, whether they have a military background or not, can learn to ask questions that prompt MALs to share their language and culture. These questions can begin simply, asking students about their jobs and work lives, or they can invite comparisons between communication and education tasks in the military and in higher education, such as: "how would you present these findings to your superior officer?" or "How do you usually interpret or apply feedback from instructors or SOs?" Making space for military language, whether the student or tutor is using this language, allows students to leverage their experiences in the educational setting.

Representation and the Trusted Leader

As a MAL himself, Marlin acknowledged his appeal to students who may be drawn to a tutor "who is like me." In addition to being a sought-after tutor, Marlin's ethos as a veteran contributes to the trustworthiness of the center as a whole. As a faculty member, he is also able to recommend the center to his students. Rister described the role of faculty members in recommending the center to students: "We don't always know how it is framed. Faculty who heavily criticize students' work and then recommend the center have damaged the center's credibility and trustworthiness." As discussed previously in this chapter, framing the center as a space of remediation can make MALs less likely to utilize its services. On the other hand, Marlin described faculty and other key stakeholders who have "earned the privilege to recommend" support services like communication centers by establishing themselves as sympathetic toward and knowledgeable about issues MALs face. Moving forward, we will continue to leverage these relationships to shape the institution's knowledge about supporting MALs through the communication center (DQ4).

Finally, in our responses to discussion question 5, the researchers indicated that there is still much we do not know about supporting MALs through our communication center, such as the specific hours and modalities that MALs prefer for communication tutoring and the specific challenges that academic communication practices pose for MALs used to military communication. However, the researchers agreed that trust is an important component of our plan to continue learning. The common theme in this area was that our first and continued step is to provide MALs with opportunities to tell us what they need. Marlin noted, "we know what we think they need," and Rister added that "there may be gaps between our perception and theirs." As we continue to earn their trust, soliciting feedback from our MALs (both those who do and those who do not currently use the center) is our most important task. We hope to provide multiple opportunities for that feedback as our center grows.

DISCUSSION

When students use campus resource centers there is an increase in student retention, as well as positive effects on students' academic success, mental health, and social lives (Brock, 2010; Henchy, 2013; McDermott et al., 2019; Osborne et al., 2019; Polite, 2019). Even though campus resource centers have been developed to increase student success and retention in college, many resources remain underused (Benedict et al., 1977; Brock, 2010; Strada & Gallup, 2017). When thinking about the MAL population, resource centers may not provide the services and accessibility necessary to garner enough interest for these students to use the resources at their disposal (Southwell et al., 2018). Through the autoethnographic reflections of three campus resource staff (Adams et al., 2017), this chapter was able to identify the key barriers MALs may face to using campus resources. These reflections also informed the creation of recommendations for improving and increasing campus resource centers use among the MAL population.

While previous research has briefly explored the reasons why traditional students visit campus resource centers, as well as the barriers they face to using campus resource centers (McDermott et al., 2019), no research to date has specifically looked into the MAL student experience. As previously mentioned, many MALs identify as nontraditional student learners, meaning their schedule and time outside of class may be filled with additional obligations such as work and family (Southwell et al., 2018). Our findings indicate that both the platform for tutoring (e.g., only face-to-face, only virtual) and hours of operation present a large challenge for MALs accessibility, thus, making campus resource centers more accessible through the use of multi-modal platforms, as well as altering times of operation may increase MAL student resource use.

Furthermore, MALs have been characterized as persistent, resilient, and mission-focused (Ford & Vignare, 2015). While these traits may serve MALs well in both the military and in university life, they can also prevent

MALs from utilizing campus resources. Our findings indicate that MALs may not feel comfortable asking for help or using campus resources. MALs may believe they have the skills and capabilities to accomplish tasks without needing outside assistance potentially leading to lower grades, frustration, and even failure. Student failure has been known to lead to a lack of self-worth, decreased self-efficacy, and depression (Heimerdinger & Hinsz, 2008). Additionally, MALs may view campus resources as irrelevant to the "mission" of completing their course. Since the military has strict order and regulations for how to behave both inside and outside of uniform, students may treat a syllabus and instructor directions with the same adherence (Howe & Shpeer, 2019; Maravilla, 2007). If campus resource centers are seen as outside the scope of the class, MALs may be less likely to use them.

Finally, MALs may not feel connected with their campus communities (see May & McDermott, chapter 2). As the military and academia have distinct organizational cultures (May & McDermott, 2019), students transitioning from the military to higher education may feel uncomfortable with the lack of structure and newfound independence (Howe & Shpeer, 2019). Oftentimes, there are limited MAL staff or faculty to relate to, which may make MALs even warier of using campus resources. Since trust and loyalty are pillars of the military community, the present research has found that building trust among the MAL community is a cornerstone for attracting MALs to use campus resources (Army.mil, 2018; Howe & Shpeer, 2019).

RECOMMENDATIONS FOR RESOURCE CENTERS

Although campus resource centers have been historically underutilized and undervalued, there are changes campus resource centers can make to increase their MAL attendance rates and better support the MAL community. These recommendations were determined based on the above autoethnographic reflections and current literature. The researchers recommend the following strategies for resource centers serving MALs.

Offer Flexible and Online Options

While all students can benefit from online support services, MALs may not be able to access resource centers at all unless virtual options are available (see chapter 12 by Cobos in this edited volume). Since MALs often prefer to take online classes, resource centers must offer support in a similar modality. Synchronous virtual tutoring services can be offered through software like Zoom or WebEx. Asynchronous feedback options can be made available for students who are unable to meet with tutors during scheduled hours through options such as Google document comments or Kaltura screencast videos. Additional software such as GoReact can be used to embed comments throughout a students' video to further engage students with feedback asynchronously.

However, we also recommend offering live support, if at all possible, during later evening and weekend hours when MALs are most likely to be hard at work. Offering a variety of access options and service times will help address the time and access-related barriers that MALs face. These options must be flexible not only in scheduling, but in formality within tutoring, advising, or coaching sessions. For instance, access to childcare may be a barrier for military spouses seeking support from resource centers. Therefore, centers should consider inviting parents to bring their child with them to the center, creating a welcome, supportive environment for MALs juggling school and parenting commitments.

Partnerships and Representation

Centers should seek out partnerships with faculty, staff, and students that have established themselves as trustworthy to MALs. If at all possible, centers should have representation from the MAL population on staff (e.g., a faculty or student tutor). For the researchers' center, this representation has been crucial to getting buy-in from the MAL population. Centers can also partner with military-affiliated faculty, veteran student services offices, and other trusted members of the campus community for targeted outreach opportunities. The key to this strategy is locating advocates who have earned what Marlin called "the privilege to recommend" a visit to a resource center. These may be military-affiliated faculty who leverage this background to connect with students or other faculty who have taken the time to understand this population's unique needs. These trusted leaders can recommend the center by framing it as a strategy for completing assignments or course missions successfully. Additionally, Green Zone (Nichols-Casebolt, 2012) is a successful nationwide program for training faculty, students, and staff on creating military safe environments on campus (see May & McDermott, chapter 2 in this edited volume). Having a center become Green Zone approved can both provide staff with the necessary knowledge for better supporting MALs and make MALs feel more comfortable in the tutoring space. Moreover, inviting VA Center directors or ROTC leaders can help to enhance the credibility of the resource center. These additional MAL-centered campus resources may also be willing to provide trainings or workshops for tutors and resource center leaders to increase their military culture knowledge and sensitivity. Partnering with these additional MAL-centric resources will help to encourage MALs to visit campus resource centers, as well as build community networks among a variety of other campus resources.

Resource centers can also become campus-wide advocates for MAL needs, educating faculty about this population to make the classroom a more hospitable place for MALs. For example, resource center staff can attend faculty meetings across departments and disciplines to share information about MAL-friendly pedagogy. Currently, the researchers are collecting data on MAL needs (in communication specifically, but also in general support) as they are reported during tutoring sessions and during interviews/focus groups. This data will inform the creation of faculty and professional development programming to help faculty better understand and support MALs' needs.

Consider Embedded Supports

As our results indicated, MALs may not seek out resource centers if the center visit doesn't seem integral to the mission of completing an assignment or course. Because MALs are used to doing things on their own, developing embedded tutoring programs within select courses, or asking trusted faculty to integrate center visits as part of an assignment or class requirement (i.e., part of the mission) may go a long way toward getting buy-in from this population (as well as bringing more students to the center overall). The authors acknowledge that due to their campus being predominantly MAL-populated, the VA/military support is embedded throughout all campus resources (e.g., financial aid, advising) rather than having a standalone VA center. Since this might not be the case for all universities, explicitly partnering with VA Centers on campus can help embed resources in support systems MALs might already be using. Additionally, the credibility behind VA Centers can enhance the validity of these resources from the MAL perspective. Although this strategy may not be appropriate for every center, and it should not be adopted as a blanket strategy across large courses that would overtax the center's capacity, we recommend considering this course of action, particularly at institutions with large MAL populations.

CONCLUSION

The above recommendations represent small, yet concrete actions resource centers can take to increase support for MALs. However, this process is a gradual one. As more research is conducted on MAL needs in various academic support areas, institutions must continue to collaborate and share their knowledge from working with MALs. The remainder of this chapter acknowledges the limitations of the present study, suggests directions for future research and offers some final takeaways for resource centers seeking to better support MALs on their campuses.

Limitations of the Study

While this is just the beginning of research identifying the key components of supporting MALs in campus resource centers, the authors acknowledge that the present chapter has some limitations. Ultimately, autoethnography has been criticized for evoking feelings in readers that may be unpleasant since the connections readers make to narratives cannot be predicted (Bochner & Ellis, 1996). Other limitations include that it requires researchers to honestly and willingly self-disclose inner feelings and thoughts. Ethically, autoethnography has been criticized for its strong emphasis on the self (Atkinson, 1997; Coffey, 1999). Walford (2004) questioned the research method wondering how much of the accounts presented represent reality as they happened or how much they are inventions of the authors. Moreover, there were only three researchers who responded to the questions and came together to answer them and analyze the responses. Additionally, all three researchers hold positions at the same university. To mitigate some of these criticisms the researchers used the same five questions to prompt specific reflection. The researchers also chose to first reflect and answer the questions individually, before coming together to talk about common response themes. This was to enhance interrater reliability during the coding process.

Directions for Future Research

In order to continue this area of study, future research should survey and interview MALs—both those who do and those who do not use campus resources—to understand their motivations for visiting the centers or not. Rister and Velez are currently working on such a study at their institution; however, more institutions should also explore this research for a more comprehensive understanding of MAL motivations. Exploring the benefits MALs perceive from using campus resources will help to inform strategies for marketing campus resources to more MALs.

Furthermore, surveying and interviewing campus resource staff on their knowledge of MALs and how to better support MALs may help to highlight areas of improvement in staff training. Exploring the perception of campus resources from VA Center staff and Reserve Officer Training Corps (ROTC) faculty will help to inform new approaches for developing relationships between campus resource centers. Overall, this is an area of research that should be explored in more depth as many campus resources may not be fully supporting MALs and MALs may not be using all the resources they have at their disposal.

Eliminating Barriers to MAL Success

This chapter has presented results and recommendations for resource centers to help increase the amount of support they offer MAL populations. As MAL and other nontraditional student populations continue to increase on college campuses, all resource centers should consider the kinds of outreach, partnerships, and programming they employ to allow them to better serve a more diverse group of learners. As is clear from the previous literature and the current study, simply "existing" is not enough. MALs are not guaranteed to use a center's services simply because they are there. MALs are resilient, reluctant to ask for help, and face considerable material barriers to accessing resource center services. Centers must respond with changes to programming, including virtual and flexible face-to-face options, that circumvent those obstacles, but they must also respond to the perception MALs may have that the center is unhelpful or unrelated to their educational missions. Ultimately, campus resource centers need to continue to build trust with their MAL campus community to truly support their academic success.

NOTE

1. Due to the nature of autoethnography, Velez, Marlin, and Rister all contributed to the data collection and analysis, which resulted in the themes described later in this chapter, but Rister did not participate in writing the results. The chapter authors are grateful for her contributions and voice which helped to inform this chapter's recommendations.

REFERENCES

Adams, T. E. (2012). The joys of autoethnography: Possibilities for communication research. *Qualitative Communication Research, 1*(2), 181–194.

Adams, T., Ellis, C., & Jones, S. H. (2017). Autoethnography. In *The international encyclopedia of communication research methods* (pp. 1–11). Wiley-Blackwell.

Allen, I. E., & Seaman, J. (2011). *Going the distance: Online education in the United States.* Babson Survey Research Group and Quahog Research Group. https://files.eric.ed.gov/fulltext/ED529948.pdf.

Allen, I. E., & Seaman, J. (2013). *Changing course: Ten years of tracking online education in the United States.* Babson Survey Research Group and Quahog Research Group. http://www.onlinelearningsurvey.com/reports/changingcourse.pdf.

Army.mil. (2018). *Features-Values.* https://www.army.mil/values/.

Astin, A. W. (1993). *What matters in college? Four critical years revisited.* Jossey-Bass.

Atkinson, P. (1997). Narrative turn or blind alley?. *Qualitative Health Research, 7*(3), 325–344.

Barefoot, B. O. (2004). Higher education's revolving door: Confronting the problem of student drop out in US colleges and universities. *Open Learning, 19*(1), 9–18.

Benedict, A. R., Apsler, R., & Mordson, S. (1977). Student views of their counseling needs and counseling services. *Journal of College Student Personnel, 18*(1), 110–114.

Body, D. (2019, March 19). *Worse off than when they enrolled: The consequences of for-profit colleges for people of color.* The Aspen Institute. https://www.aspeninstitute.org/blog-posts/worse-off-than-when-they-enrolled-the-consequence-of-for-profit-colleges-for-people-of-color/.

Bouquet, E. (1999). Our little secret: A history of writing centers, pre-to post-open admissions. *College Composition and Communication, 50*(3), 463–482.

Boylorn, R. M. (2011). Gray or for colored girls who are tired of chasing rainbows: Race and reflexivity. *Cultural Studies Critical Methodologies, 11*, 178–186.

Brock, T. (2010). Young adults and higher education: Barriers and breakthroughs to success. *The Future of Children, 20*(1), 109–132.

Capstick, M. K., Harrell-Williams, L. M., Cockrum, C. D., & West, S. L. (2019). Exploring the effectiveness of academic coaching for academically at-risk college students. *Innovative Higher Education, 44*, 219–231. doi:10.1007/s10755-019-9459-1.

Carino, P. (1996). Early writing centers: Toward a history. *The Writing Center Journal, 15*(2), 103–115.

CCA. (n.d.). *Promoting college success in nontraditional students.* https://completecollege.org/wp-content/uploads/2017/11/TPMA-Nontraditional-Student-Success-White-Paper_TPMA-V2.pdf.

Chiara, N. A., Chiara, M., & Ashlock, M. Z. (2020). What is impacting the military-affiliated learner experience in Higher Education. In V. McDermott, L. H. Hernández, & A. May (Eds.), *Supporting the military-affiliated learner: Communication approaches to military pedagogy & education* (pp. 47–65). Lexington Books.

Cobos, A. (2020). Two-way communication as a constructivist pedagogy tool for enhancing meaning making in digital learning classrooms with military affiliated learners. In V. McDermott, L. H. Hernández, & A. May (Eds.), *Supporting the military-affiliated learner: Communication approaches to military pedagogy & education* (pp. 257–278). Lexington Books.

Coffey, A. (1999). *The Ethnographic Self.* Sage.

Cooper, B. C., Morris, S. M., & Boylorn, R. M. (Eds.) (2017). *The crunk feminist collection.* The Feminist Press.

Craft, R. W. (2013). *Underutilization of student-centered resources: Understanding student response to supports for improved institutional practice* (Publication No.

3562373). Doctoral Dissertation, Wilmington University, ProQuest Dissertations & Theses Global.

Crawford, L. (1996). Personal ethnography. *Communication Monographs, 63*, 158.

Deil-Amen, R. (2011, November). *The "traditional" college student: A smaller and smaller minority and its implications for diversity and access institutions* (Paper Presentation). Mapping Broad-Access Higher Education Conference, Stanford University, Stanford, CA, United States. https://cepa.stanford.edu/sites/default/files/2011%20Deil-Amen%2011_11_11.pdf.

Denny, H., Nordlof, J., & Salem, L. (2018). 'Tell me exactly what it was that I was doing that was so bad': Understanding the needs and expectations of working-class students in writing centers. *Writing Center Journal, 37*(1), 67–98.

DiRamio, D., Ackerman, R., & Mitchell, R. L. (2008). From combat to campus: Voices of student-veterans. *Journal of Student Affairs Research and Practice, 45*, 73–102.

Elejalde-Ruiz, A. (2016, October 7). Why lower-income students are drawn to for-profit schools. *Chicago Tribune*. https://www.chicagotribune.com/business/ct-black-youth-for-profit-trade-schools-1009-biz-20161007-story.html.

Ellis, C., & Bochner, A. P. (1996). *Composing ethnography: Alternative forms of qualitative writing*. Scholar Commons.

Ellis, C., & Bochner, A. P. (2000). Autoethnography, personal narrative, reflexivity: Researcher as subject. In N. K. Denzin & Y. S. Lincoln (Eds.), *Handbook of Qualitative Research* (pp. 733–768). Sage.

Ellis, C. (2007). Telling secrets, revealing lives: Relational ethics in research with intimate others. *Qualitative Inquiry, 13*, 3–29.

Ellis, C. (2009). *Revision: Autoethnographic reflections on life and work*. LeftCoast Press.

Ford, K., & Vignare, K. (2015). Evolving military learner population: A review of literature. *Online Learning, 19*(1), 7–30.

Fredman, J. (2018, March 2). Report finds lower completion rates for students in two-year institutions, adult learners. *NASFAA*. https://www.nasfaa.org/news-item/14541/Report_Finds_Lower_Completion_Rates_for_Students_in_Two-Year_Institutions_Adult_Learners.

Geisler, C. (2003). *Analyzing streams of language: Twelve steps to the systematic coding of text, talk, and other verbal data*. Longman.

Gormley, M. J., DuPaul, G. J., Weyandt, L. L., & Anastopoulos, A. D. (2019). First-year GPA and academic service use among college students with and without ADHD. *Journal of Attention Disorders, 23*(14), 1766–1779. doi:10.1177/1087054715623046.

Heimerdinger, S. R., & Hinsz, V. B. (2008). Failure avoidance motivation in a goal-setting situation. *Human Performance, 21*, 383–395.

Henchy, A. (2013). The perceived benefits of participating in campus recreation programs and facilities: A comparison between undergraduate and graduate students. *Recreational Sports Journal, 37*(2), 97–105.

Heppner, P. P., & Neal, G. W. (1983). Holding up the mirror: Research on the roles and functions of counseling centers in higher education. *The Counseling Psychologist, 11*(1), 81–89.

Horton, J. (2017). Identifying at-risk factors that affect college student success. *International Journal of Process Education, 7*(1), 83–101.

Howe, W. T., & Hinderaker, A. (2018). "The rule was the rule": New member socialization in rigidly structured totalistic organizations. *Atlantic Journal of Communication, 26*(3), 180–195.

Howe, W. T., Jr., & Shpeer, M. (2019). From military member to student: An examination of the communicative challenges of veterans to perform communication accommodation in the university. *Journal of Intercultural Communication Research, 48*(3), 203–220. doi:10.1080/17475759.2019.1592770.

Howe, W. T. (2020). Supporting military veterans in the graduate classroom: A case study of political beliefs and discussion. In V. McDermott, L. H. Hernández, & A. May (Eds.), *Supporting the military-affiliated learner: Communication approaches to military pedagogy* & education (pp. 67–86). Lexington Books.

Iloh, C. (2018). Not non-traditional, the new normal: Adult learners and the role of student affairs in supporting older college students. *Journal of Student Affairs, 27*, 25–29.

Knight, J. P. (1990). Literature as equipment for killing: Performance as rhetoric in military training camps. *Text and Performance Quarterly, 10*(2), 157–168.

Lang, K., & Weinstein, R. (2013). The wage effect of not-for-profit and for-profit certifications: Better data, somewhat different results. In *Labor economics* (pp. 240–243). National Bureau of Economic Research.

Lee, J. C., Caronia, N., & Beltran, D. Q. (2014). Interdisciplinary writing center collaborations. *Academic Exchange Quarterly, 18*(4), 1–6.

Ludewig, A. (2020). Collaborative apprenticeship approaches to writing instruction in the military. In V. McDermott, L. H. Hernández, & A. May (Eds.), *Supporting the military-affiliated learner: Communication approaches to military pedagogy & education* (pp. 311–336). Lexington Books.

Marcus, J. (2017, Marcus 21). Community colleges rarely graduate the veterans they recruit. *The Atlantic.* https://www.theatlantic.com/education/archive/2017/04/why-is-the-student-veteran-graduation-rate-so-low/523779/.

May, A., & McDermott, V. (2019). A cultural collision: A review of literature. In L. H. Hernández & J. Belding (Eds.), *Navigating life with a graduate degree as a military spouse: Interdisciplinary approaches to challenges, lessons learned, and thriving amidst uncertainty* (pp. 59–85). Lexington Books.

May, A., & McDermott, V. (2020). Understanding dual identity formation: Identifying commonalities and resolving "differentness" for military affiliated learners. In V. McDermott, L. H. Hernández, & A. May (Eds.), *Supporting the military-affiliated learner: Communication approaches to military pedagogy & education* (pp. 11–45). Lexington Books.

McDermott, T., May, A., & Bahr, N. (2019, November). *Two sides of the coin: Understanding perception of campus resources in student life* (Paper Presentation). National Communication Association Convention, Baltimore, MD.

McIlveen, P. (2008). Autoethnography as a method for reflexive research and practice invocational psychology. *Australian Journal of Career Development, 17*, 13–20.

Mendez, M. (2013). Autoethnography as a research method: Advantages, limitations and criticisms. *Colombian Applied Linguistics Journal, 15*(2), 279–287.

Moullet, Z., & Belding, J. N. (2020). Supporting the military-affiliated learner's medical school aspirations. In V. McDermott, L. H. Hernández, & A. May (Eds.), *Supporting the military-affiliated learner: Communication approaches to military pedagogy & education* (pp. 159–180). Lexington Books.

NASPA. (2012). *Measuring the success of student veterans and active duty military Students.* https://www.naspa.org/images/uploads/main/NASPA_vets_13(1).pdf.

NCES. (n.d.). *Nontraditional undergraduates: Definitions and data.* https://nces.ed.gov/pubs/web/97578e.asp.

Neal, G. W., & Heppner, P. P. (1986). Problem-solving self-appraisal, awareness, and utilization of campus helping resources. *Journal of Counseling Psychology, 33*(1), 39–44.

Nichols-Casebolt, A. (2012). The Green Zone: A program to support military students on campus. *About Campus*, pp. 26–29. doi:10.1002/abc.21070.

NSSE. (2010). *Major differences: Examining student engagement by field of Ssudy.* https://nsse.indiana.edu/NSSE_2010_Results/pdf/NSSE_2010_AnnualResults.pdf.

Olfson, M., Marcus, S. C., Druss, B., & Pincus, H. A. (2002). National trends in the use of outpatient psychotherapy. *American Journal of Psychiatry, 159*, 1914–1920.

Osam, E. K., Bergman, M., & Cumberland, D. M. (2017). An integrative literature review on the barriers impacting adult learners' return to college. *Adult Learning, 28*(2), 54–60. doi:10.1177/1045159516658013.

Osborne, J., Parlier, R., & Adams, T. (2019). Assessing impact of academic interventions through student perceptions of academic success. *Learning Assistance Review, 24*(1), 9–26.

Pairitz Morris, C. (2020, June 18). Why school graduation and retention rates are important to consider. *Earnest.* https://www.earnest.com/blog/graduation-and-retention-rates/.

Pelletier, S. G. (2011). Success for adult students. *Public Purpose.* https://www.aascu.org/uploadedFiles/AASCU/Content/Root/MediaAndPublications/PublicPurposeMagazines/Issue/10fall_adultstudents.pdf.

Plummer, K. (2001). The call of life stories in ethnographic research. In P. Atkinson, A. Coffey, S. Delamont, J. Lofland, & L. Lofland (Eds.), *Handbook of ethnography* (pp. 395–406). Sage.

Polite, D. (2019). *First-generation college students' perceptions regarding the impact of tutoring and mentoring on persistence* (Publication No. 2312835233). Doctoral Dissertation, Capella University, ProQuest Dissertations Publishing.

Reed-Danahay, D. E. (1997). *Auto/Ethnography: Rewriting the self and the social.* Berg.

Richards, K. A. R., & Hemphill, M. A. (2017). A practical guide to collaborative qualitative data analysis. *The Journal of Teaching in Physical Education.* doi:10.1123/jtpe.2017-0084.

Salem, L. (2016). Decisions … decisions: Who chooses to use the writing center?. *The Writing Center Journal, 35*(2), 147–171.

Schmidt, S. W. (2009). Understanding the online learner. In V. X. Wang (Ed.), *Handbook of research on e-learning applications for career and technical education: Technologies for vocational training* (pp. 482–494). doi:10.4018/978-1-60566-739-3.ch038.

Schwartz, E., & Davidson Pisacreta, E. (2019, January 10). New report: Enrolling more veterans at high-graduation-rate colleges and universities. *Ithaka S+R*. https://sr.ithaka.org/blog/new-report-enrolling-more-veterans-at-high-graduation-rate-colleges-and-universities/.

Shapiro, D., Dundar, A., Ziskin, M., Yuan, X., & Harrell, A. (2013). *Completing college: A national view of student attainment rates-fall 2007 cohort* (Signature Report No. 6). National Student Clearinghouse Research Center.

Shapiro, D., Dundar, A., Huie, F., Wakhungu, P. K., Yuan, X., Nathan, A., & Bhimdiwali, A. (2017). *Completing college: A national view of student completion rates – fall 2011 cohort* (Signature Report No. 14). National Student Clearinghouse Research Center.

Spellman, N. (2007). Enrollment and retention barriers adult students encounter. *The Community College Enterprise, 13*(1), 63–79.

Southwell, K. H., Whiteman, S. D., MacDermid Wadsworth, K. M., & Barry, A. E. (2018). The use of university services and student retention: Differential links for student service members or veterans and civilian students. *Journal of College Student Retention: Research, Theory, & Practice, 19*(4), 394–412. doi:10.1177/1521025116636133.

Starr-Glass, D. (2011). Military-learners: Experience in the design and management of online learning environments. *MERLOT Journal of Online Learning and Teaching, 7*(1), 147–158.

Strada-Gallup. (2017). Crisis of confidence: Current college students do not feel prepared for the workforce. *Gallup*. http://news.gallup.com/reports/225161/2017-strada-gallup-college-student-survey.aspx.

Syverud, K. (2019, June 21). Commentary: Universities need to better serve military students and veterans. *U.S. News*. https://www.usnews.com/news/education-news/articles/2019-06-21/commentary-higher-education-must-better-serve-military-students-and-veterans.

Tamas, S. (2011). *Life after leaving: The remains of spousal abuse*. Left Coast Press, Inc.

Thornberg, R., & Charmaz, K. (2012). Grounded theory. In S. D. Lapan, M. T. Quartaroli, & F. J. Riemer (Eds.), *Qualitative research: An introduction to methods and designs* (pp. 41–67). Jossey-Bass.

VA. (2020, February 11). *Education and training*. https://www.benefits.va.gov/GIBILL/resources/benefits_resources/rate_tables.asp#ch33.

Vacchi, D. T. (2012). Considering student veterans on the twenty-first-century college campus. *About Campus, 17*(2), 15–21. doi:10.1002/abc.21075.

Value Colleges. (n.d.). *For-profit vs. non-profit cchools. What's the difference?* https://www.valuecolleges.com/resources/faqs/for-profit-vs-non-profit-schools-whats-the-difference/.

Walford, G. (2004). Finding the limits: Autoethnography and being and Oxford University proctor. *Qualitative Research, 4*, 403–417.

Wardley, L. J., Bélanger, C. H., & Leonard, V. M. (2013). Institutional commitment of traditional and non-traditional-aged students: A potential brand measurement? *Journal of Marketing for Higher Education, 23*, 90–112. doi:10.1080/08841241.2013.810691.

Webely, K. (2012, January 9). For-profit schools: 'Agile predators' or just business savvy? *Time.* https://business.time.com/2012/01/09/for-profit-schools-agile-predators-or-just-business-savvy/.

Ziff, E. (2017). "The mommy deployment": Military spouses and surrogacy in the United States. *Sociological Forum, 32*(2), 406–425.

Part III

BUILDING INCLUSIVE LEARNING COMMUNITIES VIRTUALLY AND BEYOND

Chapter 11

Best Practices in Supporting Military-Affiliated Learners

Suggestions from Communication and Psychological Sciences

Leandra Hinojosa Hernández
and Jennifer N. Belding

In an era where the value of a college degree is increasing, the demographic characteristics of the undergraduate population in the United States are changing greatly (e.g., Anderson, 2003; Carnevale et al., 2013). Due in part to a plethora of scholarship opportunities available, one population that has grown in particular over recent years is veterans (see Barry et al., 2014). Estimates suggest that approximately 68 percent of post-9/11 veterans do not have a bachelor's degree, yet veterans are overrepresented in two- and four-year for-profit institutions and underrepresented in public four-year institutions; moreover, veterans are half as likely to enroll in high-graduation-rate institutions (Hill, Kurzweil, Pisacreta, & Schwartz, 2019). These findings suggest that veterans are indeed pursuing higher education at both traditional brick-and-mortar and online institutions, but may face a variety of unique barriers to student success compared to traditional college students (e.g., Barry et al., 2012; Rumann & Hamrick, 2009).

While veterans presumably represent the largest proportion of military-affiliated learners (MALs), it is important to note that the phrase MALs refers to the combination of active duty service members, reservists, veterans, military spouses, and military children, and that each of these groups are pursuing higher education at increasing rates as well. Indeed, estimates suggest that military spouses are more highly educated than their civilian peers, but are more likely to be underemployed (e.g., Lim & Schulker, 2010). Military spouses, along with children from military families and reservists, are also regularly present in the college classroom (Ott et al., 2018).

Ultimately, both the military and academia are "greedy institutions" that require much time, effort, energy, and emotional labor from its service members, employees, and family members (Belding & Hernández, 2019; Coser, 1974; May & McDermott, 2019; Segal, 1986). As we have noted in past research (Belding, 2019; Belding & Hernández, 2019; Hernández, 2019; Moullet & Belding, chapter 8 in this volume), strides have been made to improve social support, mental health, and career and educational opportunities for service members, reservists, veterans, military spouses, and military children. Nonetheless, in educational contexts, MALs may face unique barriers, such as lack of knowledge about available programs and institutions, how to navigate the college system (including funding sources and cultural fit), and how to transfer institutions when it is in their best interest (Drummet et al., 2003; Hill et al., 2019). This uncertainty about the college environment, along with the uncertainty that is an unavoidable element of the military lifestyle, can present a challenge to student success (e.g., Sollitto & Cole, 2019). We believe that concentrated efforts to support MALs have the potential to greatly improve their experiences in higher education and ultimately increase graduation rates.

Whereas traditional college students matriculate to college promptly following graduation from high school, MALs overwhelmingly do not. As a result, the life frame of reference for a MAL is undoubtedly different than that of a traditional college student, which further illustrates the need for a MAL-sensitive approach to higher education pedagogical and administrative training and readiness preparation. The diversity of life experiences that MALs bring to the classroom has also provided opportunities for them to grow specific skills, such as the ability to work independently and to guide themselves toward the completion of their goals. For example, they are often more mature as a result of their time in the military, and they have diverse life experience and perspectives that can benefit course curricula and assignments (Bricker, 2017).

In this chapter, we argue that MALs are more likely to experience an effective learning environment when higher education educators and student support specialists understand the unique experiences and challenges that may influence educational pursuits of MALs. We argue that the use of appropriate pedagogical best practices that have been shown to be effective for undergraduate students in general will allow educators and student support specialists to create a welcoming and supportive environment for MALs in particular. In support of this view, we briefly summarize some of the common challenges that MALs may face as a result of their military affiliation. We then identify a selection of empirically supported best practices from the scholarship of teaching and learning (SOTL) to attenuate such challenges and promote student success (Brown & Gross, 2011; Semer & Harmening, 2015).

Based upon our review and synthesis of MAL pedagogies in communication and psychology disciplines, we provide a tangible list of recommendations, strategies, and best practices to assist current and future instructors to work with MALs in the higher education classroom.

THE MILITARY-ACADEMIA CULTURAL COLLISION

Military-affiliated individuals occupy a unique space in the American imaginary. On one end of the spectrum, military-affiliated individuals garner reverence and respect because of their service to the United States; on the other end of the spectrum, military-affiliated individuals are also stereotyped in problematic ways (See chapter 2 by May & McDermott, chapter 3 by Chiara, Chiara, & Ashlock, and chapter 4 by Howe in this volume). For example, research by Lieberman and Stewart (2014) found that "Americans believe that veterans are more likely to carry negative attributes than an average civilian, in opposition to reality" (p. 4). Furthermore, according to Lieberman and Stewart (2014), the larger American consciousness views veterans as both damaged and heroic, as well as strong and flawed. These perceptions have the potential to impact the higher education classroom from the perspective of instructors and students alike, particularly if they have little to no experience interacting with MALs.

Research that explores the experiences of MALs in higher education often situates this phenomenon within a deficit-based versus assets-based approach. A MAL deficit-based approach in higher education, as Ludewig describes in chapter 14, positions MALs' success within the framework of their shortcomings that result from their military service and experiences. In general, deficits could include diagnostic, medical, and relational concerns that may influence the way in which MALs learn, such as traumatic brain injury (TBI), post-traumatic stress (PTS), traumatic experiences, disability, substance abuse, transition issues, lack of stability, lack of connection with others, increased levels of stress, and lack of preparation for college (Hart & Thompson, 2016).

One specific domain in which MALs may differ from their traditional student peers is with regard to mental health. College is an inherently stressful time for students for a variety of reasons. In addition to the normal stressors associated with academic pursuits, MALs also bring unique stressors associated with their military experience. For example, service members and veterans face disproportionately higher rates of psychological symptoms, risky health behaviors, and personal and educational adjustment difficulties (Barry et al., 2014). They may also struggle to connect with peers and faculty members on campus when compared to their civilian counterparts. Previous research has also demonstrated that service members and veterans struggle to discuss their mental health concerns with family members for several reasons, including what falls within societal constructions of "normalcy," "weakness," and "strength." These concerns along with fear of implications raised by acknowledging and disclosing mental health issues are situated within concerns associated with trauma, military culture, and risk/benefit assessments (Wilson et al., 2015; Wilson et al., 2019).

Indeed, service members and veterans are hesitant to initiate conversations about their mental health and wellbeing with close others (Gettings et al., 2019), which suggests that it is likely that MALs are even more hesitant to initiate such conversations with individuals in higher education. We posit that if MALs already find it difficult to disclose mental health concerns to close family members and friends, they are even less likely to share these feelings with classmates, educators, and administrators with whom they do not have pre-developed relationships. Some support for this idea comes from research that demonstrated that veterans viewed their classmates as immature and unfavorable, which led to further feelings of isolation from their civilian counterparts. As a result of these feelings of isolation and mental health-related struggles, veterans may struggle with the re-/integration process (Elliott et al., 2011; Livingston et al., 2011).

When framed this way, these potential "deficits" suggest that MALs can benefit from intentional consideration and support from higher education instructors, staff members, and MAL support centers. Even so, viewing MALs as having "deficits" can be problematic because it can contribute to instructors' and administrators' preconceived notions of MAL abilities and competencies or lack thereof. Weiss (2015) elaborates that such preconceived notions can do unintended, nonconscious damage because students are likely to respond accordingly, regardless of their intellectual competencies or individual identities that are separate from their military identities and experiences. Ultimately, while the military experience is indeed likely a significant component of a MAL's larger worldview, it should not comprise or primarily constitute the veteran higher education experience (Hart & Thompson, 2016; Weiss, 2015).

Furthermore, while MALs may have unique experiences and characteristics that can present a challenge in higher education environments, it is also important to note that they also bring many assets to the classroom. One asset in particular is the remarkable diversity of MALs. Not only do MALs represent a fascinating cross-section of the population (Barroso, 2019), they also have likely lived in several different places and have a plethora of cultural experiences that may be rare among more traditional college students. For example, as a MAL and as a result of her husband's military service, Leandra has lived both on and off base in two states in the United States and also in Japan. Similarly, Jennifer has had to move five times in four years and lived both on base and off base while her husband served with units in the United States Navy and Marine Corps. Furthermore, MALs often experience a great number of demands and personal responsibilities and may demonstrate more grit than their civilian counterparts as we have argued elsewhere (e.g., Belding, 2019).

Taken together, we believe that there is a delicate balance between viewing MALs with empathy regarding the unique challenges they may face while simultaneously respecting and affirming their service and right to belong in a college classroom. The question of how to support MALs without adopting a potentially reductionist perspective is challenging, even for those with direct experience with the military lifestyle. To help those who wish to support MALs more effectively in higher education, we acknowledge some of the "deficits" per se that shape the MAL experience in higher education (or, rather, perceptions of MALs in higher education; e.g., Bricker, 2017; Jenner, 2017). These barriers include geographic mobility and frequent transfers (see chapter 12 by Cobos in this edited volume); financial challenges associated with the responsibilities of providing for families, working full time, attending school in at least a part-time capacity (see chapter 2 by May & McDermott in this edited volume), and often surviving on one income because of struggles military spouses face with unemployment & underemployment (see Hernández & Belding, 2019); a lack of understanding on how to study effectively or lack of preparation for higher education (Callahan & Jarrat, 2014); mental health issues; and barriers to retention and success as a result of institutional shortcomings (see chapter 10 by Velez, McDermott, & Marlin in this edited volume).

As we will discuss below, viewing veterans, active duty service members, reservists, military spouses, and military children from an asset perspective enhances the MAL learning experience and creates a more inclusive learning environment. We argue that this positive construal of MALs, when used in conjunction with pedagogical best practices identified within the SOTL, can promote MAL success in higher education. We discuss specific best practices regarding how to improve experiences associated with the culture shift, pedagogical training to improve the development of learning objectives, how to help MALs study more effectively, and how to provide support for MALs specifically in terms of geographic mobility and transfers.

SUPPORTING THE CULTURAL SHIFT FROM MILITARY TO ACADEMIC SETTINGS

A Cultural Collision

At a broad macro-level, there is a military-civilian divide (Knobloch & Wehrman, 2014; Rahbek-Clemmensen et al., 2012; Wilson, 2020) within which the cultural collision between military and academia is located. The military-civilian divide, according to Rahbek-Clemmensen and colleagues (2012), is comprised of four sectors: cultural gaps, demographic gaps, policy preference gaps, and institutional gaps. First, the cultural gap refers to "whether the attitudes and values of civilian and military populations differ" (Rahbek-Clemmensen et al., 2012, p. 671). The military as an institution stresses discipline, conformity, and self-sacrifice, whereas the U.S. civilian culture is firmly entrenched within individualist cultural values and prioritizes individualism, self-gain, and independence (Knoblock & Wehrman, 2014; Wilson, 2020). Such cultural gaps can influence service members and other military-affiliated individuals to feel like cultural outsiders. Additionally, civilians may have incorrect or limited perceptions of military-affiliated individuals (Knoblock & Wehrman, 2014; Wilson, 2020). The second and third gaps, demographic and policy preferences, explore whether the military population represents the larger U.S. population politically and demographically and whether policy gaps exist between military and civilian elites (Rahbek-Clemmensen et al., 2012). The fourth gap, institutional relations, refers to "whether the relationship between military and civilian institutions such as the media, the courts, and the education system can be characterized in terms of harmony or conflict" (Rahbek-Clemmensen et al., 2012, p. 673).

Of most relevance to our analysis is the role of the education system. As Wilson (2020) describes, academia and the military are well aligned when universities establish centers for veteran support. However, this alignment is limited when barriers exist pertaining to awareness and utilization of center services and when there is no support for other MALs, such as spouses and children. Moreover, there is rich diversity among student veterans and other MALs, illustrating the areas for improvement faced by "one size fits all" veteran campus initiatives (Vaccaro, 2015).

Unfortunately, there is a cultural collision between academia and the military, which are two distinct cultures with different norms, values, and belief systems (May & McDermott, 2019). On the one hand, the military has traditionally been defined by various conservative and liberal ideologies (May & McDermott, 2019; Malka & Lelkes, 2010). Military training emphasizes uniformity and conformity, which are practices that train service members to follow orders, defend one's country and fellow service members, and reinforce guiding military cultural practices and principles (Bricker, 2017; May & McDermott, 2019; Tuchik & Wilson, 2010). On the other hand, while academia is also steeped in tradition and requires much of its students, faculty, staff, and administrators, there is more freedom and more space for liberal modes of thinking and for questioning the status quo that do not fall within the typical military chain of command parameters and expectations (May & McDermott, 2019).[1] Ultimately, May and McDermott (2019) conclude:

> Collectively, the academy strives to further society's understanding of social issues to bring about change. As with the military, clearly defined values reflect both history and a changing landscape in which the academy must evolve, adapt, and address significant issues that plague espoused values. With a focus on academic freedom, inclusion, and changing the narrative on violence, the academy has a different structural core that is, at times, diametrically opposed to the military. Moreover, members of both cultures often are deeply entrenched, committing to a career that requires loyalty and full immersion to survive and thrive. (pp. 69–70)

Although some elements of the two cultures are very similar, others could not be more different. One could argue that both the military and academia share a sense of greediness in that they both demand remarkable amounts of time and attention. Indeed, the military has garnered the label of "greedy institution" (Segal, 1986). A similar sentiment could be shared for institutions of higher education, which often require students to be enrolled full time in order to obtain maximum financial aid.

As MALs transition from the military to academic settings, they are subject to this stark cultural shift. One of the many ways in which military and academic settings differ is in perceived control or autonomy. Whereas military service is associated with a lack of autonomy or perceived control over one's life choices (Leahy, 2013), academic institutions often pride themselves on the opposite (e.g., Romo de la Rosa, 2007). Navigating this remarkable difference can thus present a dramatic culture shift for MALs. For example, during military service, MALs might reasonably expect their military leader or training program to dictate virtually all aspects of their life including what time to wake, shower, eat, sleep, study, and more (Leahy, 2013). Academic institutions, on the other hand, rarely offer such mandatory structure; indeed, relatively few undergraduate courses require in-person attendance or activities at specific times (St. Clair, 1999). The freedom thus granted by transitioning between military service and undergraduate education can understandably be a challenge for students. Nonetheless, the flexibility and personal freedom available in academic settings is beneficial and ought not be compromised.

Learning Mindsets

Although the military and academia differ in these and many other ways, it is clear that the people living and working within these systems are goal-directed in that they are regularly behaving in a way that helps them achieve their goals. As MALs transition from military to academic contexts, we believe that there are two particular types of mindsets that can help ensure educational success: mastery goals and growth mindsets. Broadly speaking, research on the SOTL suggests that instructors that promote both mastery goals and growth mindsets can be effective at promoting student learning (e.g., Dweck, 1986; Ehrlinger & Shain, 2014; Van Doorn & Van Doorn, 2014). Although growth mindsets and mastery goals are often considered trait-level differences in that some people are more likely to endorse them than others, both can be affected by the presence of others and thus have the potential to be influenced by higher education instructors and student support specialists. We review both of these mindsets in turn and then offer concrete suggestions for how to promote these mindsets as a way of alleviating the challenges associated with the military-academic cultural clash.

Mastery versus Performance Goals

One frequently discussed distinction in the scientific literature on goal pursuit, particularly that associated with the pursuit of academic goals, is the distinction between mastery and performance goals (e.g., Dweck, 1992). On the one hand, mastery goals in academic settings correspond to a desire to acquire knowledge and subject matter expertise. Performance goals, on the other hand, are concerned with demonstrating competence, outperforming classmates, and earning high marks (Benita et al., 2014). Although both types of goals are associated with academic success in different ways, mastery goals have been shown to be associated with continued interest in a given academic course and subsequent enrollment in related courses (Harackiewicz et al., 2000). Additionally, mastery goals have been associated with encouraging more deep thinking, facilitating student help-seeking, and ultimately greater learning over the academic term (see Ehrlinger & Shain, 2014).

Growth versus Fixed Mindsets

Students who embrace a growth mindset believe that they can learn and improve over time with sufficient effort, whereas students with a fixed mindset believe that they have an innate ability that cannot be improved regardless of the amount of effort exerted (see Hochanadel & Finamore, 2015). Students with growth mindsets tend to be more resilient in the face of poor performance and respond by redoubling their efforts and improving over time, whereas those with fixed mindsets are more inclined to disengage from the course. As a result, previous research has shown that growth mindsets tend to be associated with greater learning and better performance over time (Yeager et al., 2019).

Strategies to Promote Mastery Goals and Growth Mindsets

In order to help promote both mastery goals and growth mindsets in their students, instructors can thoughtfully build these concepts into their course design and structure through both the clear articulation of relevant learning objectives and thoughtfully designed assignments. Specifically, instructors can emphasize learning, understanding, and improvement over time as such foci have been associated with mastery goals and subsequent beneficial outcomes (Linnenbrink, 2005; Darnon et al., 2009). The use of specific strategies, such as low-stakes quizzing (e.g., Hinze & Rapp, 2014) and scaffolded assignments that allow students to engage in revision and improvement over time (Cardinale & Johnson, 2017; Coe, 2011), are other avenues to produce such outcomes. Such assignments can work particularly well for MALs who might be juggling family life, deployment, military training, and family separation, given that they may have less bandwidth to dedicate to certain assignments at any given time. Furthermore, because emotional support from one's instructor has been associated with significant increases in student engagement and mastery motivation (Ruzek et al., 2016), instructors who attempt to build strong rapport with their students may also reap the benefits associated with mastery goals and growth mindsets. As Hernández et al. discuss in chapter 7 in this edited volume, instructor support can both increase student engagement and validate student lived experiences in ways that help MALs feel like an integral part of the classroom community.

EFFECTIVE LEARNING OBJECTIVES

Learning Objectives Clarify Expectations

As a result of the cultural clash between military and academic settings, many MALs may struggle to understand the expectations associated with individual courses. Although military training and college courses are similar in that both should be designed with clear objectives in mind, completion of certain tasks tends to be heavily regulated and monitored in military settings, but not necessarily academic settings. For example, a military leader may regularly check in with their subordinates to ensure tasks are completed daily and may refuse to release the member from work for the day until one's task is complete. In contrast, it is rare for a college instructor to hound students to complete a given assignment for a course. Again, a cultural clash can emerge due to a difference in expectations of one's behavior or performance.

One way that instructors can help MALs understand the expectations associated with their course is to use effective learning objectives. To do this, we must first clarify what we mean by "effective learning." In college, learning means being able to remember, retrieve, and apply information, which requires being able to distinguish critical concepts from ancillary concepts and being able to generate and use schemas to organize and structure knowledge. Effectively learning a specific subject matter sets the foundation for future understanding of that material and an ability to think critically about it when presented with new information. As an illustration, imagine that your colleague said they taught their dog to play the piano. Your initial reaction might be a combination of surprise, amazement, and potentially disbelief. However, your colleague quickly clarifies by saying something along the lines of, "I said I taught my dog to play the piano. I didn't say he learned how to do it." Just because we teach our students something doesn't mean they have effectively learned it.

In order to promote student learning, instructors should begin by establishing effective learning objectives, which are each around one to two sentences in length. Learning objectives are an essential element of a course syllabus and provide a clear summary of the knowledge, skills, and abilities (KSAs) that students should be able to demonstrate at the conclusion of the course (Chatterjee & Corral, 2017). If an instructor is not able to directly articulate these learning objectives, it is unlikely that students will be able to recognize them or the purpose of individual tasks within a course, especially for MALs who may be adjusting to increased flexibility and freedom. As a result, a plethora of pedagogical research suggests that instructors should begin by identifying what their learning objectives are and design their course in such a way that specific activities and assignments directly contribute to those specific objectives in a process called "backwards design" (Wiggins & McTighe, 2005). In essence, instructors should ask themselves what KSAs may differ between a student at the beginning and end of the course and use that response to craft coherent and straightforward learning objectives. Instructors teaching courses with large MAL populations can create effective learning objectives and a course syllabus that privilege flexibility, empathy, and understanding of the MAL experience as one strategy to enhance MAL success.

How to Write Learning Objectives

Bloom's Taxonomy

Once instructors have identified the relevant KSAs, they can begin to draft their learning objectives. In general, many instructors find it helpful to leverage the framework articulated in Bloom's taxonomy for this purpose. Briefly, Bloom's taxonomy is a prominent framework in education that suggests that students progress in hierarchical fashion through six general stages of learning: remembering, understanding, applying, analyzing, evaluating, and creating (Adams, 2015; Chatterjee & Corral, 2017). For example, before students can truly understand a given subject matter, they must first remember it; before students can apply a specific subject matter, they must first understand it. If instructors are able to identify the level of depth with which students should master the course's subject matter, they can pick effective action verbs that correspond to each level to help them generate concise and unambiguous single sentence learning objectives for the course. For example, introductory courses that might focus on having students remember and understand information can use terms such as "recall" or "summarize," respectively. In mid-level courses focused on application and analysis, instructors could use terms such as "utilize" or "distinguish," respectively. In upper-level or senior capstone courses targeted toward evaluation and creation, instructors could use phrases such as "justify" or "formulate," respectively. Instructors looking for inspiration with such tasks can find multiple tables of suggestions on the internet with a quick Google search. Additionally, when working toward application-based learning objectives, instructors can design assignments in which students must relate relevant subject matter to their own experiences, including military-affiliated experiences, which could promote a feeling of inclusion and deeper processing of course material.

SMART Goals

Because learning objectives are goals for student learning, they can also benefit from being written in such a way that follows the "SMART goals" framework. Briefly, SMART goals are goals that are specific, measurable, attainable, realistic, and time-bound (Latham, 2003; Locke & Latham, 1990). In this sense, instructors should write their learning objectives in a way that highlights the specific subject matter students are expected to learn. Their ability to learn such information should be measured by performance on activities and assessments used in the course (e.g., exams, written assignments). Because learning objectives available on a syllabus are inherently time-bound by the length of the academic term (e.g., a ten-week quarter, a fifteen-week semester), it is important to ensure that the instructor's expectations are attainable and realistic. For example, it is not reasonable to expect that students will be able to move from remembering a specific psychological theory to designing and conducting a scientific study testing that theory in their very first psychology course. In general, such activities and assessments should be designed with careful regard for the bandwidth that students have to contribute to a class, which would allow the flexibility for MALs in particular to succeed despite the plethora of demands on their time.

HELPING STUDENTS STUDY EFFECTIVELY

Academic Self-Efficacy

Another component of maximizing the MAL higher education experience is helping them learn to study effectively. At the outset, many student veterans and other MALs are unaware of the unwritten rules needed to navigate higher education, such as registering for classes, completing admissions paperwork, managing G.I. Bill and service-related paperwork, and interacting with college staff (Jones, 2017; Wyner, 2014). MALs also struggle with the transition to higher education, citing academic self-efficacy as a barrier to academic success. Academic self-efficacy is defined as students' beliefs and confidence in their ability to succeed in academic tasks and has been shown to be a reliable predictor of academic outcomes (Eakman et al., 2019; Gore, 2006; Putwain et al., 2013). Research suggests that veterans and service members "with greater academic self-efficacy and fewer academic problems may, therefore, possess a critical constellation of resilience supports capable of enhancing academic success" (Eakman et al., 2019, p. 1016).

In other words, how can instructors enhance MAL academic self-efficacy, given that service members and veterans are socialized to write, speak, and study in a particular manner (as Ludewig discusses in chapter 14 and St. Pierre discusses in chapter 15) that may be at odds with academic self-directed modes of study and inquiry? Additionally, military spouses and children may face academic challenges associated with frequent and permanent relocations, which may disrupt their ability to enroll in online and face-to-face courses and make it difficult to develop and maintain relationships with faculty members and advisers.

Common Misconceptions about Learning

Many people, both MALs and civilians in general, believe that they have a given "learning style," such as being a visual or auditory learner, and that learning information in their preferred modality is more effective and efficient than other modalities. However, these perceptions have remained largely unsupported in the scientific study of cognition and memory (Pashler, McDaniel, Rohrer, & Bjork, 2008) and thus is a common misperception about how to study effectively for college courses. Alas, it is not the only misconception that students have about studying. Indeed, many undergraduate students fail to understand how to study effectively, which can impede their performance (Nonis & Hudson, 2010). Although there is limited research to date on the misconceptions that MALs have about learning in particular, it is plausible that MALs may be more adversely affected by misconceptions about learning than their traditional counterparts as we described in our summary of the deficits-versus-assets framework.

One of the most straightforward findings from decades of research on cognition and memory is that memory is better for material that people have processed deeply (see Carpenter, 2014). In general, the more people think about a given topic or idea, the more likely they are to remember it over time, with deep, meaningful rehearsal being more effective than shallow, meaningless rehearsal (Hyde & Jenkins, 1969). This is true for personal experiences as well as undergraduate coursework. As a result, students are more likely to retain information and thus perform better on assessments when they have processed the material deeply (see Carpenter, 2014). Instead of focusing on one's "learning style" or other potential misconceptions about learning, one way to clearly support MALs is to help them understand that in order to understand and retain material from one's course, they must think deeply about it and do so repeatedly over time.

Encouraging Deep Processing of Course Content

Active Learning

One way to promote deep thinking and studying is to use as much active learning in the classroom as possible. In general, when students are active participants in their learning rather than passive recipients of knowledge, they are more likely to retain information over time (Bertsch & Pesta, 2014). Thoroughly describing what active learning is and how to promote it in the college classroom has been discussed ad nauseum elsewhere and is beyond the scope of this chapter. Strategies to promote active learning include but are not limited to encouraging students to self-generate test questions (e.g., Van Blerkom et al., 2006; Bertsch & Pesta, 2014), providing guided but not full notes so that students must take their own notes (see Russell et al., 1983), encouraging students to take notes by hand versus on a laptop (e.g., Mueller & Oppenheimer, 2014), promoting problem-based learning (e.g., Schmidt et al., 2011), and using flipped classrooms (e.g., Gilboy et al., 2015). Problem-based learning as one strategy can be particularly effective for MALs, especially if the learning activities pertain to military and cross-cultural experiences. Communication instructors, for example, could devise problem-based learning activities that draw in military policies, politics, experiences, and cases to provide students an opportunity to engage in group reasoning about the case to solve it and defend their stances collectively.

Encouraging Studying Over Time

Beyond active learning, for many undergraduate courses, students must study outside of the classroom and must use effective self-directed strategies to do so (Pintrich, 2004). Unfortunately, many students opt to "cram" the night before an exam rather than spacing out their studying over time, which is suboptimal (e.g., Kornell, 2009). Although there may be very valid barriers hampering students' ability to engage in spaced studying, particularly for MALs (e.g., increased family demands for military spouses when a service member is deployed), scientific research has clearly demonstrated that spaced learning is more effective than the alternative, which is often referred to as "massed learning" (see Cepeda et al., 2006). An instructor who finds ways to promote distributed practice or studying (i.e., studying that occurs over a period of time rather than in a single "cram" session) can thus directly promote effective student learning.

Instructors who wish to promote spaced learning can attempt to do so by implementing a combination of regular quizzing and/or cumulative exams (see Cepeda et al., 2008). If students are responsible for material learned throughout the entire academic term, they are more likely to devote time to studying that material over time. Students may even default to interleaved practice in which they return to material covered earlier in the academic term after more recently presented information. This can serve to promote the ability to compare and contrast between similar topics (see Carpenter, 2014). Thus, instructors who continue to assess material covered cumulatively throughout a course are more likely to promote effective studying habits among their students, though it is plausible that this strategy can simultaneously cause more stress for students.

Metacognitive Awareness

One additional benefit of spaced learning is that it can also improve metacognitive awareness of what one has learned (Ehrlinger & Shain, 2014). Ehrlinger and Shain (2014) stated the value of metacognition beautifully:

> Effective learning requires metacognition—the process of reflecting upon which learning concepts one has already mastered, what still needs to be learned, and how to best approach the task of learning. Within educational contexts, metacognitive perceptions are defined as students' beliefs about their learning process, including the degree to which they have mastered learning tasks. (p. 142)

In this sense, metacognition involves both an understanding or awareness of how one thinks and the ability to engage in self-regulation of that thinking process (Girash, 2014). In short, students who are able to explicitly identify and articulate when they truly do or do not understand a concept have metacognitive awareness of their learning, whereas those who fail to know that which they don't know lack metacognitive awareness and as a result may perform more poorly (Girash, 2014).

Although teaching students metacognitive awareness is not routinely part of standard course content and is neither necessarily easy or natural for college students to develop (see Ambrose, et al., 2010), several activities have been shown to be effective, particularly if they are built into the class itself (Veenman et al., 2006). Tanner (2012) reviewed several such strategies that can easily be integrated into college coursework. For example, prior to covering a new topic in class, the instructor can set aside time to have a discussion with students about what they already know about a given topic. This can be particularly effective for MAL student populations, given potential misconceptions or lack of knowledge about MALs and the MAL experience. Moreover, at the end of each class period, the instructor can ask students to write on a piece of scrap paper what the "muddiest point" from that day's material was. After collecting these responses, the instructor can then review the topics in the subsequent class. If the instructor has allotted time for students to study in class with each other (even in just ten-minute increments), the instructor can intentionally interrupt the studying process after a few minutes by explicitly asking students to identify what specific actions students are taking to study, why they are engaging in those specific actions rather than others, and how such specific studying activities are helping the students (see Girash, 2014). Each of these activities have been shown to increase metacognitive awareness of learning.

Promoting Self-Reflection

Another effective strategy that can be both built into the class or offered outside of class are activities that promote self-reflection after one has performed poorly on an exam. One possible option is to assign students a "reflective journal" assignment in which they must write a short (one page) paper summarizing how they studied and the barriers and facilitators that influenced their studying process (see Tanner, 2012). Alternatively, instructors can invite students to complete an "exam wrapper" activity in which they prompt students to engage in a similar reflection in a more guided and faster approach. For example, Edlund (2020) assessed the effectiveness of a simple 5-question exam wrapper in which students indicated the amount of time they spent studying, the percentage of time they spent engaging in a variety of studying tasks (e.g., rereading the textbook, reading notes from class), the number of points lost on a graded exam for a variety of reasons (e.g., studying the wrong topics, trouble with terminology), ways to improve their study plan for the next exam, and what the students believed the instructor could do to support them on the next exam. (A full copy of the exam wrapper is available in the Appendix of Edlund's 2020 article.) When the exam wrapper was completed one class period after an exam and was returned to the students one week before their next exam, students showed statistically significant increases in scores compared to a control condition that simply provided information about study strategies (Edlund, 2020). Another more thorough example of an exam wrapper was developed by Samford University's Dr. Stephen Chew, a renowned expert on teaching effectiveness in psychology, and is available in Appendix 2 of a packet of helpful tips and strategies for instructors developed as part of a workshop he led at Belmont University in 2013.[2] Such an activity could also provide MALs the opportunity to compare, contrast, and evaluate military and academic preparation, time management, and study skills.

While these strategies have been empirically supported, they are often conducted over time (e.g., after the first exam), which means that a student's grade may already have been negatively affected by poor performance. Proactive strategies to promote an accurate understanding of how-to study are crucial, particularly for MALs, who may otherwise struggle to study effectively as we have previously described. Unfortunately, college courses are often packed with essential course content that may not allow for sufficient in-class dedicated time to review effective studying strategies. Even if such time is available, instructors who dedicate too much time to this topic run the risk of alienating, boring, or otherwise disappointing students who already have this metacognitive awareness. As such, providing resources that are short, easily accessible, informative, and available without an instructor's involvement or assistance is helpful.

One such resource is a series of five YouTube videos created by Dr. Chew; a playlist can be easily accessed by searching for "Stephen Chew How to Get the Most Out of Studying." The videos include: (1) "Beliefs that make you fail . . . or succeed," (2) "What students should understand about how people learn," (3) "Cognitive principles for optimizing learning," (4) "Putting the principles for optimizing learning into practice," and (5) "I blew the exam, now what?" Each video is less than ten minutes long, provides concrete suggestions and examples for students, and has even embedded the empirical evidence supporting the strategies suggested. To promote the likelihood that students actually watch the videos early in the term, instructors can tell students in advance that there will be a low-stakes quiz in the second class period on which several questions will assess whether students watched and understood the material. For additional information on how to encourage students to get the most out of studying, interested readers are referred to Chew (2014). Collectively, these different strategies may be particularly helpful for MALs returning to the college classroom after having been out of school for a prolonged period of time.

STRUCTURAL CHALLENGES

MAL Geographic Mobility

While much of our chapter thus far has focused on alleviating challenges that can result from the cultural clash between military and academic environments, there is one additional challenge that is more structural rather than cultural in nature. As a function of their connection to the military, MALs (particularly those serving on active duty and/or their dependents) may experience geographic mobility during their pursuit of higher education (see chapter 12 by Cobos in this edited volume). Frequent moves as a result of permanent changes of station or temporary duty assignments can affect schooling in that they may necessitate institutional transfer and adjusted schedules (Callahan & Jarrat, 2014). As a result, MALs may be forced to navigate the transfer process more frequently than their traditional counterparts. Military spouses, for example, experience educational dissatisfaction resulting from permanent changes of station that disrupt their programs of study. Hernández (2019) found that military spouses pursuing graduate studies either had to temporarily pause their studies or transfer to new programs that exclusively offered online educational programs because of their role as a trailing spouse (Bouchard, 2019). Although most instructors of higher education courses may not be aware of a student's transfer status, there are several actions instructors can take to support students over time.

Supporting Students Undergoing Institutional Transfers

First and foremost, instructors should design their course syllabus to be as specific and detailed as possible with regard to the learning objectives and assignments required in a given course. Because the syllabus serves as both a contract between students and the instructor as well as a permanent record of the course (see Parkes & Harris, 2002), students may use a copy of the course syllabus to apply for transfer credit for the course from one institution to the next. Syllabi that are clear and detailed can assist the student in making a stronger argument that their previous coursework achieves the learning and curriculum objectives required at their new institution.

Second, several actions that have been articulated as advantageous in the pursuit of supporting nontraditional students would also be helpful in supporting MALs facing geographic mobility (e.g., Cross, Mandernach, & Huslig, 2014). For example, because students transferring from one institution to the next may not be aware of university-specific resources like a campus Writing Center or Financial Aid Office, instructors can serve as a critical source of information by creating and disseminating short flyers about such resources on their course websites or in the course syllabus. In many cases, the relevant resources on campus may already have flyers prepared and available upon request, resulting in minimal effort on the part of the instructor. Additionally, instructors can help nontraditional students understand the language and culture of the university (e.g., specific terminology for specific procedures or offices on campus) through a variety of means (e.g., by providing information on university resources on the course syllabus or website) and thereby help attenuate the risk that a perceived lack of fit may result in attrition (see Monroe, 2006).

Third, instructors can help support students by making a concentrated effort to be empathetic to students' situations and experiences. Moving, in general, is challenging and comes with a remarkable number of stressors, particularly for military families (Tong et al., 2018). Instructors have the potential to help offset these challenges by providing appropriate emotional and social support, which can serve to foster academic engagement (Monroe, 2006). For example, if a student asks for an extension on an assignment, instructors who engage in "teaching empathy" listen to the student's perspective, make an attempt to understand the nature and scope of the problem, and offer appropriate accommodations after considering how an extension may relate to the learning objectives of the course. This sort of empathetic response can facilitate the student's learning and performance in the class as a whole (Meyers et al., 2019). We call for more teaching empathy for those interacting with MALs because many may encounter challenges through no fault of their own as a result of their military service and are simply trying their best to succeed in higher education.

CONCLUDING THOUGHTS

In this chapter, we have argued that the increasing presence of MALs, including active duty service members, reservists, veterans, military spouses, and military children, warrant additional consideration and support in the college classroom. Utilizing a deficits-versus-assets framework, we have described some of the unique life experiences that MALs may have, which can present both barriers to and simultaneously facilitate their undergraduate education. It is our belief that while MALs may experience some challenges in the pursuit of higher education as a result of their military service, they greatly enhance the diversity of and are welcome additions to the college classroom. We have highlighted four broad categories of empirically supported pedagogical practices that college instructors can use to create a more inclusive learning environment for MALs including supporting the cultural shift from military to academic settings, promoting student learning through mastery and growth mindsets, encouraging the use of effective study strategies, and easing the challenges associated with geographic mobility.

While the direct support provided by college instructors is perhaps the most visible weapon in the arsenal of institutional support, there are undoubtedly other means to effectively welcome these valuable learners into the classroom environment and facilitate their educational pursuits. For example, university administrators and support centers can emphasize the diversity in MAL backgrounds and experiences (Jenner, 2017). Instead of focusing solely on veterans and active duty service members, administrators and support centers can utilize MAL diversity—nontraditional status, first-generation status, identity diversity, and political diversity—as a basis for outreach programs and educator training. Moreover, administrators and support centers can create and sustain developed educator training to assist educators with knowledge about the culture clash detailed in this chapter, mental health resources that can be provided to MALs, and best pedagogical practices to inform classroom construction and instruction. Additionally, co-identity organizations can be created and sustained administratively to serve as a "transformative intervention" because they can serve a crucial social identity theory function in creating a safe, familiar, and welcoming space for MALs (Jenner, 2017).[3] Military leaders and policy makers, on the other hand, can continue to advocate for funding for resources that support MALs and ensure that such information (e.g., scholarships, resources) is widely disseminated. Additionally, this volume's conclusion provides further suggestions for recommendations at the administrative, institutional, and policy level.

The value of a college education is tremendous. We believe that when college instructors, administrators, military leaders, and policymakers unite to advocate for the educational pursuits of this valuable population of students, many of the barriers to student success can be easily surmounted. Indeed, we believe that such improvements would likely pay remarkable dividends in the form of increased employment for veterans after service, reduced military spouse unemployment and underemployment, reduced healthcare costs, and more. In short, effectively supporting MALs in the college classroom should be a priority for all.

NOTES

1. It is important to note that there are still many limitations hindering valid and valuable attempts to challenge the inadequacies of the status quo, particularly for minoritized individuals of diverse racial, ethnic, sexual identity, and gender identity backgrounds.

2. See http://www.belmont.edu/tcenter/pdf/schew_video_resource.pdf for pedagogical tips and strategies.

3. Jenner (2017) describes co-identity organizations as "any organization, institutional safe-space, or student group that is centered around one or more aspects of identity, such as race/ethnicity, religion, sexuality, gender, ability/disability, marriage and parental status, age, or veteran status. From an institutional perspective, such organizations and activities may represent easy-to-implement and cost-effective interventions" (p. 27).

REFERENCES

Adams, N. E. (2015). Bloom's taxonomy of cognitive learning objectives. *Journal of the Medical Library Association, 103*(3), 152–153. doi:10.3163/1536-5050.103.3.010.

Ambrose, S. A., Bridges, M. W., DiPietro, M., Lovett, M. C., & Norman, M. K. (2010). *How learning works: Seven research-based principles for smart teaching.* Jossey-Bass.

Anderson, E. L. (2003). Changing US demographics and American higher education. *New Directions for Higher Education, 2003*(121), 3–12.

Barroso, A. (2019). *The Changing profile of the U.S. military: Smaller in size, more diverse, more women in leadership.* Pew Research Center. https://www.pewresearch.org/fact-tank/2019/09/10/the-changing-profile-of-the-u-s-military/.

Barry, A. E., Whiteman, S. D., & MacDermid Wadsworth, S. (2014). Student service members/veterans in higher education: A systematic review. *Journal of Student Affairs Research and Practice, 51*(1), 30–42. doi:10.1515/jsarp-2014-0003.

Barry, A. E., Whiteman, S. D., MacDermid Wadsworth, S., & Hitt, S. (2012). The alcohol use and associated mental health problems of student service members/veterans in higher education. *Drugs: Education, Prevention, and Policy, 19*(5), 415–425. doi:10.3109/09687637.2011.647123.

Belding, J. N. (2019). Military spouses with advanced degrees: A unique subpopulation to study for the science of motivation and goal pursuit. In L. H. Hernández & J. N. Belding (Eds.), *Military spouses with graduate degrees: Interdisciplinary approaches to thriving amidst uncertainty* (pp. 217–238). Lexington Books.

Belding, J. N., & Hernández, L. H. (2019). Military spouses with advanced degrees: Current findings and pathways moving forward. In L. H. Hernández & J. N. Belding (Eds.), *Military spouses with graduate degrees: Interdisciplinary approaches to thriving amidst uncertainty* (pp. 1–14). Lexington Books.

Benita, M., Roth, G., & Deci, E. L. (2014). When are mastery goals more adaptive? It depends on experiences of autonomy support and autonomy. *Journal of Educational Psychology, 106*(1), 258–267. doi:10.1037/a0034007.

Bertsch, S., & Pesta, B. (2014). Generating active learning. In *Applying science of learning in education: Infusing psychological science into the curriculum* (pp. 71–77). Society for the Teaching of Psychology.

Bricker, S. (2017). *An instructor's guide to teaching military students: Simple steps to integrate the military learner into your classroom*. Lexington Books.

Brown, P. A., & Gross, C. (2011). Serving those who have served – Managing veteran and military student best practices. *The Journal of Continuing Higher Education, 59*(1), 45–49. doi:10.1080/07377363.2011.544982.

Callahan, R., & Jarrat, D. (2014). Helping student service members and veterans succeed. *Change: The Magazine of Higher Learning, 46*(2), 36–41.

Cardinale, J. A., & Johnson, B. C. (2017). Metacognition modules: A scaffolded series of online assignments designed to improve students' study skills. *Journal of Microbiology & Biology Education, 18*(1), 18.11.13. doi:10.1128/jmbe.v18i1.1212.

Carnevale, A. P., Strohl, J., & Melton, M. (2013). *What's it worth?: The economic value of college majors*. Georgetown University Center on Education and the Workforce. https://repository.library.georgetown.edu/bitstream/handle/10822/559309/whatsitworth-complete.pdf?sequence=1.

Carpenter, S. K. (2014). Spacing and interleaving of study and practice. In *Applying science of learning in education: Infusing psychological science into the curriculum* (pp. 131–141). Society for the Teaching of Psychology.

Cepeda, N. J., Pashler, H., Vul, E., Wixted, J. T., & Rohrer, D. (2006). Distributed practice in verbal recall tasks: A review and quantitative synthesis. *Psychological Bulletin, 132*(3), 354–380. doi:10.1037/0033-2909.132.3.354.

Chatterjee, D., & Corral, J. (2017). How to write well-defined learning objectives. *The Journal of Education in Perioperative Medicine, 19*(4), E610.

Chew, S. L. (2014). Helping students to get the most out of studying. In *Applying science of learning in education: Infusing psychological science into the curriculum* (pp. 215–223). Society for the Teaching of Psychology.

Chiara, N. A., Chiara, M., & Ashlock, M. Z. (2020). What is impacting the military-affiliated learner experience in Higher Education. In V. McDermott, L. H. Hernández, & A. May (Eds.), *Supporting the military-affiliated learner: Communication approaches to military pedagogy & education* (pp. 47–65). Lexington Books.

Cobos, A. (2020). Two-way communication as a constructivist pedagogy tool for enhancing meaning making in digital learning classrooms with military affiliated learners. In V. McDermott, L. H. Hernández, & A. May (Eds.), *Supporting the military-affiliated learner: Communication approaches to military pedagogy & education* (pp. 257–278). Lexington Books.

Coe, C. D. (2011). Scaffolded writing as a tool for critical thinking: Teaching beginning students how to write arguments. *Teaching Philosophy, 34*(1), 33–50. doi:10.5840/teachphil20113413.

Coser, L. A. (1974). *Greedy institutions: Patterns of undivided commitment*. Free Press.

Cross, T., Mandernach, B. J., & Huslig, S. (2014). Academic advising for online students. In R. L. Miller & J. G. Irons (Eds.), *Academic advising: A handbook for advisors and students. Volume 1: Models, students, topics, and issues*. Society for the Teaching of Psychology (pp 104–111). http://teachpsych.org/ebooks/academic-advising-2014-vol1.

Darnon, C., Dompnier, B., Delmas, F., Pulfrey, C., & Butera, F. (2009). Achievement goal promotion at university: Social desirability and social utility of mastery and performance goals. *Journal of Personality and Social Psychology, 96*(1), 119–134. doi:10.1037/a0012824.

Drummet, A. R., Coleman, M., & Cable, S. (2003). Military families under stress: Implications for family life education. *Family Relations, 52*(3), 279–287. doi:10.1111/j.1741-3729.2003.00279.x.

Dweck, C. S. (1986). Motivational processes affecting learning. *American Psychologist, 41*(10), 1040–1048. doi:10.1037/0003-066X.41.10.1040.

Dweck, C. S. (1992). Article commentary: The study of goals in psychology. *Psychological Science, 3*(3), 165–167. doi:10.1111/j.1467-9280.1992.tb00019.x.

Eakman, A. M., Kinney, A. R., Schierl, M. L., & Henry, K. L. (2019). Academic performance in student service members/veterans: Effects of instructor autonomy support, academic self-efficacy and academic problems. *Educational Psychology, 39*(8), 1005–1026.

Edlund, J. E. (2020). Exam wrappers in psychology. *Teaching of Psychology, 47*(2), 156–161. doi:10.1177/0098628320901385.

Ehrlinger, J., & Shain, E. A. (2014). How accuracy in students' self-perceptions relates to success in learning. In *Applying science of learning in education: Infusing psychological science into the curriculum* (pp. 142–151). Society for the Teaching of Psychology.

Elliott, M., Gonzalez, C., & Larsen, B. (2011). U.S. military veterans transition to college: Combat, PTSD, and alienation on campus. *Journal of Student Affairs Research & Practice, 48*(3), 279–296.

Gettings, P. E., Hall, E. D., Wilson, S. R., Kamal, D. M., Inderstrodt-Stephens, J., & Hughes-Kirchubel, L. (2019). Effects of reintegration difficulties, perceived message acceptance and perceived autonomy support on US military veterans' evaluations of messages encouraging them to seek behavioral health care. *Communication Monographs, 86*(2), 205–228.

Gilboy, M. B., Heinerichs, S., & Pazzaglia, G. (2015). Enhancing student engagement using the flipped classroom. *Journal of Nutrition Education and Behavior, 47*(1), 109–114. doi:10.1016/j.jneb.2014.08.008.

Girash, J. (2014). Metacognition and instruction. In V. A. Benassi, C. E. Overson, & C. M. Hakala (Eds.), *Applying science of learning in education: Infusing psychological science into the curriculum* (pp. 152–168). Society for the Teaching of Psychology.

Gore, P. A. (2006). Academic self-efficacy as a predictor of college outcomes: Two incremental validity studies. *Journal of Career Assessment, 14*, 92–115.

Harackiewicz, J. M., Barron, K. E., Tauer, J. M., Carter, S. M., & Elliot, A. J. (2000). Short-term and long-term consequences of achievement goals: Predicting interest and performance over time. *Journal of Educational Psychology, 92*(2), 316–330. doi:10.1037/0022-0663.92.2.316.

Hart, D. A., & Thompson, R. (2016). Veterans in the writing classroom: Three programmatic approaches to facilitate the transition from the military to higher education. *College Composition and Communication, 68*, 345–371.

Hernández, L. H. (2019). "We get it done because we have to:" Military spouses with advanced degrees, career and educational experiences, and grit amongst uncertainty. In L. H. Hernández & J. N. Belding (Eds.), *Military spouses with graduate degrees: Interdisciplinary approaches to thriving amidst uncertainty* (pp. 239–258). Lexington Books.

Hernández, L. H., De Los Santos Upton, S., & Aguirre, A. A. (2020). Nepantla, liminality, and teaching and learning in the space between: A borderlands approach to military-affiliated learner pedagogy. In V. McDermott, L. H. Hernández, & A. May (Eds.), *Supporting the military-affiliated learner: Communication approaches to military pedagogy & education*. (pp. 133–157). Lexington Books.

Hill, C. B., Kurzweil, M., Pisacreta, E. D., & Schwarz, E. (2019). *Enrolling more veterans at high-graduation-rate colleges and universities*. https://sr.ithaka.org/publications/enrolling-more-veterans-at-high-graduation-rate-colleges-and-universities/.

Hinze, S. R., & Rapp, D. N. (2014). Retrieval (sometimes) enhances learning: Performance pressure reduces the benefits of retrieval practice. *Applied Cognitive Psychology, 28*(4), 597–606. doi:10.1002/acp.3032.

Hochanadel, A., & Finamore, D. (2015). Fixed and growth mindset in education and how grit helps students persist in the face of adversity. *Journal of International Education Research, 11*(1), 47–50.

Howe, W. T. (2020). Supporting military veterans in the graduate classroom: A case study of political beliefs and discussion. In V. McDermott, L. H. Hernández, & A. May (Eds.), *Supporting the military-affiliated learner: Communication approaches to military pedagogy & education* (pp. 67–86). Lexington Books.

Hyde, T. S., & Jenkins, J. J. (1969). Differential effects of incidental tasks on the organization of recall of a list of highly associated words. *Journal of Experimental Psychology, 82*, 472–481. doi:10.1037/h0028372.

Jenner, B. N. (2017). Student veterans and the transition to higher education: Integrating existing literatures. *Journal of Veterans Studies*, *2*(2), 26–44.

Jones, K. C. (2017). Understanding transition experiences of combat veterans attending community college. *Community College Journal of Research & Practice, 41*(2), 107–123.

Knobloch, L. K., & Wehrman, E. C. (2014). Family relationships embedded in United States military culture. In C. R. Agnew (Ed.), *Advances in personal relationships. Social influences on romantic relationships: Beyond the dyad* (pp. 58–82). Cambridge University Press.

Kornell, N. (2009). Optimising learning using flashcards: Spacing is more effective than cramming. *Applied Cognitive Psychology, 23*(9), 1297–1317. doi:10.1002/acp.1537.

Latham, G. P. (2003). Goal setting: A five-step approach to behavior change. *Organizational Dynamics, 32*(3), 309–318. doi:10.1016/S0090-2616(03)00028-7.

Leahy, J. F. (2013). *Honor, courage, commitment: Navy boot camp*. Naval Institute Press.

Lieberman, D., & Stewart, K. (2014). *Strengthening perceptions of America's Post-9/11 veterans survey analysis report*. https://www.dillonconsult.com/wp-content/uploads/2013/03/Strengthening-Perceptions-of-Americas-Post-911-Veterans-Survey-Analysis-Report-Got-Your-6-June-2014.pdf.

Lim, N., & Schulker, D. (2010). *Measuring underemployment among military spouses*. RAND Corporation. http://www.rand.org/pubs/monographs/MG918.

Linnenbrink, E. A. (2005). The dilemma of performance-approach goals: The use of multiple goal contexts to promote students' motivation and learning. *Journal of Educational Psychology, 97*(2), 197–213. doi:10.1037/0022-0663.97.2.197.

Livingston, W. G., Havice, P. A., Cawthon, T W., & Fleming, D. S. (2011). Coming home: Student veterans' articulation of college re-enrollment. *Journal of Student Affairs Research & Practice, 48*(3), 315–331.

Locke, E. A., & Latham, G. P. (1990). *A theory of goal setting and task performance*. Prentice Hall.

Ludewig, A. (2020). Collaborative apprenticeship approaches to writing instruction in the military. In V. McDermott, L. H. Hernández, & A. May (Eds.), *Supporting the military-affiliated learner: Communication approaches to military pedagogy & education* (pp. 309–334). Lexington Books.

Malka, A., & Lelkes, Y. (2010). More than ideology: Conservative-liberal identity and receptivity to political cues. *Social Justice Research, 23*, 156–188.

May, A., & McDermott, V. (2019). Cultural collision: A review of literature. In L. H. Hernández & J. Belding (Eds.), *Military spouses with graduate degrees: Interdisciplinary approaches to thriving amidst uncertainty* (pp. 59–85). Lexington Books.

May, A., & McDermott, V. (2020). Understanding dual identity formation: Identifying commonalities and resolving "differentness" for military affiliated learners. In V. McDermott, L. H. Hernández, & A. May (Eds.), *Supporting the military-affiliated learner: Communication approaches to military pedagogy & education* (pp. 11–45.) Lexington Books.

Meyers, S., Rowell, K., Wells, M., & Smith, B. C. (2019). Teacher empathy: A model of empathy for teaching for student success. *College Teaching, 67*(3), 160–168. doi:10.1080/87567555.2019.1579699.

Monroe, A. (2006). Non-traditional transfer student attrition. *The Community College Enterprise*, 12(2), 33–54.

Moullet, Z., & Belding, J. N. (In Press). Supporting the military-affiliated learner's medical aspirations. In V. McDermott, L. H. Hernández, & A. May (Eds.), *Supporting the military-affiliated learner: Communication approaches to military pedagogy & education* (pp. 159–180). Lexington Books.

Mueller, P. A., & Oppenheimer, D. M. (2014). The pen is mightier than the keyboard: Advantages of longhand over laptop note taking. *Psychological Science, 25*(6), 1159–1168. doi:10.1177/0956797614524581.

Nonis, S. A., & Hudson, G. I. (2010). Performance of college students: Impact of study time and study habits. *Journal of Education for Business, 85*(4), 229–238. doi:10.1080/08832320903449550.

Ott, L. E., Morgan, J. K., & Akroyd, H. D. (2018). Impact of military lifestyle on military spouses' educational and career goals. *Journal of Research in Education, 28*(1), 30–61.

Parkes, J., & Harris, M. B. (2002). The purposes of a syllabus. *College Teaching, 50*(2), 55–61. doi:10.1080/87567550209595875.

Pashler, H., McDaniel, M., Rohrer, D., & Bjork, R. (2008). Learning styles: Concepts and evidence. *Psychological Science in the Public Interest, 9*(3), 105–119. doi:10.1111/j.1539-6053.2009.01038.x.

Pintrich, P. R. (2004). A conceptual framework for assessing motivation and self-regulated learning in college students. *Educational Psychology Review, 16*(4), 385–407. doi:10.1007/s10648-004-0006-x.

Putwain, D., Sander, P., & Larkin, D. (2013). Academic self-efficacy in study-related skills and behaviours: Relations with learning-related emotions and academic success. *British Journal of Educational Psychology, 83*, 633–650.

Rahbek-Clemmensen, J., Archer, E. M., Barr, J., Belkin, A., Guerrero, M., Hall, C., & Swain, K. E. O. (2012). Conceptualizing the civil–military gap: A research note. *Armed Forces & Society, 38*(4), 669–678.

Romo de la Rosa, A. (2007). Institutional autonomy and academic freedom: A perspective from the American continent. *Higher Education Policy, 20*(3), 275–288. doi:10.1057/palgrave.hep.8300160.

Rumann, C. B., & Hamrick, F. A. (2009). Supporting student veterans in transition. *New Directions for Student Services, 126*, 25–34. doi:10.1002/ss.313.

Ruzek, E. A., Hafen, C. A., Allen, J. P., Gregory, A., Mikami, A. Y., & Pianta, R. C. (2016). How teacher emotional support motivates students: The mediating roles of perceived peer relatedness, autonomy support, and competence. *Learning and Instruction, 42*, 95–103. doi:10.1016/j.learninstruc.2016.01.004.

Schmidt, H. G., Rotgans, J. I., & Yew, E. H. J. (2011). The process of problem-based learning: What works and why. *Medical Education, 45*(8), 792–806. doi:10.1111/j.1365-2923.2011.04035.x.

Segal, M. W. (1986). The military and the family as greedy institutions. *Armed Forces & Society, 13*(1), 9–38. doi:10.1177/0095327X8601300101.

Semer, C., & Harmening, D. S. (2015). Exploring significant factors that impact academic success of student veterans in higher education. *Journal of Higher Education Theory and Practice, 15*(7), 31–43.

Sollitto, M., & Cole, C. (2019). Military spouses' uncertainty management: Navigating academic goals and military needs. In L. H. Hernández & J. Belding (Eds.), *Military spouses with graduate degrees: Interdisciplinary approaches to thriving amidst uncertainty* (pp. 125–142). Lexington Books.

St. Clair, K. L. (1999). A case against compulsory class attendance policies in higher education. *Innovative Higher Education, 23*(3), 171–180. doi:10.1023/A:1022942400812.

St. Pierre, C. (2020). Adapt and overcome: Military-affiliated learners and transfer of learning. In V. McDermott, L. H. Hernández, & A. May (Eds.), *Supporting the military-affiliated learner: Communication approaches to military pedagogy & education* (pp. 335–369). Lexington Books.

Tanner, K. D. (2012). Promoting student metacognition. *CBE—Life Sciences Education, 11*(2), 113–120. doi:10.1187/cbe.12-03-0033.

Tong, P. K., Payne, L. A., Bond, C. A., Meadows, S. O., Lewis, J. L., Friedman, E. M., & Hernandez, E. J. M. (2018). *Enhancing family stability during a permanent change of station: A review of disruptions and policies*. RAND Corporation. https://www.rand.org/pubs/research_reports/RR2304.html.

Turchik, J. A., & Wilson, S. M. (2010). Sexual assault in the U.S. military: A review of the literature and recommendations for the future. *Aggression & Violent Behavior, 15*, 267–277.

Vaccaro, A. (2015). "It's not one size fits all": Diversity among student veterans. *Journal of Student Affairs Research and Practice, 52*(4), 347–358.

Van Blerkom, D. L., Van Blerkom, M. L., & Bertsch, S. (2006). Study strategies and generative learning: What works?. *Journal of College Reading and Learning, 37*, 7–18.

Van Doorn, J. R., & Van Doorn, J. D. (2014). The quest for knowledge transfer efficacy: Blended teaching, online and in-class, with consideration of learning typologies for non-traditional and traditional students. *Frontiers in Psychology, 5*, 1–14. doi:10.3389/fpsyg.2014.00324.

Veenman, M. V. J., Van Hout-Wolters, B. H. A. M., & Afflerbach, P. (2006). Metacognition and learning: Conceptual and methodological considerations. *Metacognition and Learning, 1*(1), 3–14. doi:10.1007/s11409-006-6893-0.

Velez, M., McDermott, V., & Marlin, K. W. (2020). The hard sell: Supporting MALs through resource centers. In V. McDermott, L. H. Hernández, & A. May (Eds.), *Supporting the military-affiliated learner: Communication approaches to military pedagogy & education* (pp. 201–228). Lexington Books.

Weiss, E. L. (2015). Building community in a diverse college classroom. In J. E. Coll & E. L. Weiss (Eds.), *Supporting veterans in higher education: A primer for administrators, faculty, and advisors* (pp. 104–129). Lyceum.

Wiggins, G., & McTighe, J. (2005). *Understanding by design* (2nd ed.). Association for Supervision and Curriculum Development.

Wilson, S. R. (2020, January 16). Four military-civilian gaps. *Military-civilian dialogue*. https://militaryciviliandialogue.blogspot.com/2020/.

Wilson, S. R., Gettings, P. E., Hall, E. D., & Pastor, R. G. (2015). Dilemmas families face in talking with returning US military service members about seeking professional help for mental health issues. *Health Communication, 30*(8), 772–783.

Wilson, S. R., Hintz, E. A., MacDermid Wadsworth, S. M., Topp, D. B., Southwell, K. H., & Spoont, M. (2019). Female US military veterans' (non) disclosure of mental health issues with family and friends: Privacy rules and boundary management. *Health Communication*, 1–12. Online ahead of print.

Wyner, J. S. (2014). *What excellent community colleges do: Preparing all students for Success.* Harvard Education Press.

Yeager, D. S., Hanselman, P., Walton, G. M., Murray, J. S., Crosnoe, R., Muller, C., … Dweck, C. S. (2019). A national experiment reveals where a growth mindset improves achievement. *Nature, 573*(7774), 364–369. doi:10.1038/s41586-019-1466-y.

Chapter 12

Two-Way Communication as a Constructivist Pedagogy Tool for Enhancing Meaning-Making in Digital Learning Classrooms with Military-Affiliated Learners

April Cobos

According to the Department of Defense (DOD)'s census survey, there were 2,917,825 active duty service members and their dependents enrolled in higher education in 2017 (Office of Deputy Assistant Secretary, 2017). Despite this large population of Military-Affiliated Learners (MALs) in post-secondary education systems across the globe, pedagogical research focused on them and their learning experiences is substantially lacking (NASPA, 2013). There is very little information available about the success rates of MALs and a dearth of research available regarding pedagogies that positively impact their experiences, given their unique challenges to education due to military lifestyle and work constraints. While there is some research related to active duty and veteran-affiliated learners in the classroom (Hart & Thompson, 2016; Molina & Morse, 2017; Falkey, 2017), typically this research does not focus on other MALs, specifically dependents (spouses, children, or dependent parents), who are also affected by circumstances related to military service (Ott, Morgan, & Akroyd, 2018).

In conjunction with this, Starr-Glass (2011) asserts that MALs are more likely to choose distance learning given the unpredictable nature of military service (deployments, transfer of duty stations, and so on). Alongside these findings, the *Military Times* reports a heavy reliance on programs that offer fast track and flexible learning options for active duty service members to earn associates, bachelors, and graduate degrees, which includes distance, hybrid, or live streaming course options (Altman, 2019). Given this, there is a need for additional research that specifically points toward MALs' growing reliance on distance learning and the ways in which they can thrive best in these learning environments. Despite a lack of research specifically related to MALs and distance learning, there is, however, a plethora of research available on the growing numbers of university distance learning programs and options, as well as research on the best practices for teaching and learning in online and blended learning environments. Thus, this chapter will use the emerging research on the best practices for distance and online learning to offer suggestions to better support the academic success of MALs in online and blended classroom environments. In particular, I advocate for using a student-centered, constructivist theory of learning to consider the lived experiences of MALs in creating successful distance learning opportunities for this population of learners. More specifically, two-way communication models have been successfully used and modeled in military communities in the past few decades both in professional communications, as well as in connecting military service members with their families from a distance. While not all MALs have experience with distance learning platforms, most have some knowledge or experience utilizing two-way communication tools; thus, using these foundational communication principles in distance learning classrooms where MALs are present can help them to transfer the knowledge they have obtained from their personal and professional lives and apply it to their work in online and hybrid classrooms. After reviewing the research on distancing learning, challenges for MALs in higher education, and pedagogies related to constructivism and students' lived experiences, this chapter will provide specific two-way communication examples that can aid educators in creating and structuring digital learning assignments and syllabi for courses involving MALs.

My professional interest in the successes of MALs stems from being a military spouse, which naturally led me to teaching in different universities and community colleges throughout the country that served MALs in some regard. I have worked in community colleges and universities with large surrounding military populations teaching face-to-face, hybrid, and online courses where spouses and children of active duty service members, as well as current active duty service members, and recently retired veterans have attended classes. However, at our most recent duty station in Rota, Spain, I worked for the Navy College office where all of the students in my classes were MALs from two nearby bases and a remote base in Italy. I had the unique opportunity to see and understand the experiences and challenges of these students, especially living overseas where they have restricted access to opportunities for higher education and related resources. These restrictions include limitations on class offerings and frequent class cancellations due to limited enrollment being a small base with a high turnover rate of active duty service members deploying or transferring duty stations. With this in mind, the Navy College office would try to accommodate students as much as possible by offering a variety of fast-paced, distance, hybrid, or live streaming course options, all of which come with their own challenges, which will be considered throughout this chapter.

While the focus of this chapter remains on ways to mitigate challenges prevalent with MALs in distance learning, in the midst of writing this chapter, postsecondary educators nationwide were faced with educating during a global pandemic due to Covid-19, which has brought to light the need to revisit best practices for teaching when faced with challenges like transforming traditional university classrooms into hybrid, hyflex, or completely online spaces. According to Wilkerson (2020), due to the pandemic, "students are quickly realizing that online and blended learning will feature more prominently in their college experience than they had initially bargained for" (para. 7); however, most students and professors were "generally dissatisfied with their emergency remote learning experience last spring" for several reasons, including personal struggles related to the pandemic, professors who had never taught online and/or had little time to make a transition, and a lack of interaction and strong sense of isolation by students and professors alike (Lederman, 2020, para. 14). These abrupt changes to teaching and learning have caused a great deal of concern and even chaos in higher education, and require reconsidering fundamental best practices for future distance and hybrid learning in these strained, constantly changing environments, especially with the reality that distance learning will feature a more prominent role in all higher education. This chapter is timely in calling for revisiting those foundational pedagogical best practices of using two-way communication tools and constructivist pedagogies, not just with MALs in distance learning education, but more broadly for all higher education settings in order to mitigate the disparities between in-person and distance learning. While the challenges educators and students face during a global pandemic are not exact mirrors of the challenges faced by MALs, the same pedagogical principles can be applied; most especially that as educators, it is important to revisit these foundational theories of two-way communication in order to better connect and engage with all of our students in virtual learning environments.

LITERATURE REVIEW

As technology capabilities for distance learning have rapidly evolved over the past two decades, so have opportunities for revising and updating related teaching pedagogies to include the use of more two-way communication tools, which pull from foundational communication theories about the most effective ways to create meaning-making and maintain student-centered learning despite the physical distance in online classrooms. As noted in the introduction, while the focus in this article remains on the needs of MALs in distance learning, these same principles can be applied to a broader range of students and educators who have had to cope with abrupt interruptions in their traditional learning and teaching pedagogies. Specifically, for supporting MALs, however, two-way communication tools can be effective in order to mitigate the challenges of physical space, scheduling, and accessibility that may negatively impact their educational success. In particular, despite online learning affording MALs more opportunities to be successful in higher education, educators still need to consider that MALs might be impacted by other issues, such as access to technology due to remote or overseas locations, or the impact of demanding schedules, which limits their capabilities to work effectively or maintain a consistent, presence in an online classroom. I experienced many of these issues with MALs while living abroad due to the remote locations of both of the bases (rural, southern Spain and rural, mountainous Italy), but also because many countries like this do not have the same access or dependability of the Internet as the majority of America does. Thus, while implementing constructivist pedagogy and two-way technologies in the classroom can aid this demographic of learners, it is still imperative for educators to keep in mind the issues of physical space, scheduling, and accessibility in creating flexible, dynamic distance learning models for MALs.

In short, the move into using more proficient twenty-first century technologies has afforded these learners more opportunities for accessibility in spite of their circumstances, but this also requires the need to revisit and revise the pedagogical tools and practices that are beneficial to MALs. These same technologies are ones many MALs are familiar with due to professional and personal circumstances that have required them to use two-way communication in an effort to effectively communicate in a variety of different circumstances. Thus, transferring those lived experiences to the classroom may be an effective way to help this population of learners succeed in distance learning environments.

Distance Learning and MALs

According to Starr-Glass (2011), MALs represent a growing number of non-traditional students, and while some may attend class on campus, military lifestyle constraints encourage them to enroll in distance ed classes to include onsite, hybrid, or live streaming classes at military installations and at a variety of different private, public, and for-profit universities. The number of MALs is on the rise due to the introduction of the post 9/11 G.I. Bill, which was reinstated in August 2009 to compensate those who served on active duty after September 11, 2001, during the Iraq and Afghanistan conflicts (Bill, 2009). An article titled "Who Are Today's Veterans" (n.d.) published on the Veterans Affairs website reports that 79 percent of student veterans using G.I. benefits are attending public schools. The remaining 21 percent are equally split between nonprofit private schools and for-profit schools. Alongside the use of the G.I. Bill to fund their higher education endeavors, in 2018, the following percentages of active duty military students used their tuition assistance (TA) funds to take classes: 39.7 percent at public, not-for-profit institutions, 34.7 percent at for-profit institutions, and 25.7 percent at private nonprofit institutions (Altman, 2018). While many service members have used the G.I. bill themselves, several others have passed their funds along to eligible dependent family members. Overall, this has increased family members' desires and accessibility for higher education; however, there is little research that explores the school choices of dependents and their subsequent impacts on educational success (Ott, Morgan, & Akroyd, 2018).

Challenges of MALs

MAL Physical Space

The 2017 report on the demographics of the military by the Office of the Deputy Assistant Secretary of Defense for Military Community and Family Policy lists active duty bases by state and distance to major cities. This report indicates the need, and likely reliance, on distance education opportunities for both active duty and their dependents, no matter the funding source. For example, Fort Riley in Kansas, which serves 36,333 active duty service and family members, is 50 miles from Topeka. Fort Campbell in Kentucky, which serves 65,973 active duty service and family members, is 50 miles from the nearest major city. As well, in 2017, there was an international American military population of 351,249, to include all seven continents and ships afloat (Office of the Deputy Assistant Secretary, 2017), although there are no reports that indicate exactly how many of these service members or affiliated family members are pursuing higher education while living or deployed abroad. While many of the major installations in the United States and abroad offer post-secondary education options on base with opportunities for both face-to-face and distance education, these opportunities are not guaranteed or consistent across the different branches of service and the different international bases. Moreover, for military members serving overseas, opportunities to pursue a degree may be more limited or come with concerns of having sufficient classes offered or enough faculty to teach the courses. During my time living abroad in Rota, Spain, for example, I taught through the University of Maryland-University College, which operated through the Navy College office on base. This office also offered several other universities' courses face-to-face, hybrid, and online to include Central Texas College and the University of Oklahoma. However, several classes I was scheduled to teach were cancelled because not enough students were enrolled due to the small base population, the high deployment rates, and the high relocation rates of active duty, contractors, and their dependents. Many of the students I encountered were also stationed at the base for less than one year. This often left students unable to complete their necessary coursework in a timely manner and uncertain if they would have the same opportunities at a follow-up duty station.

While these issues undoubtedly impact active duty service members (Ford & Vignare, 2018), they also impact their family members or other dependents, such as spouses and children, whose lives are significantly uprooted by the same variables regarding deployments and relocations. Ultimately, this has the potential to disrupt their postsecondary education as well (Ott, Morgan & Akroyd, 2018). For these reasons, distance and hybrid education models provide accessibility to MALs who are stationed in remote areas or overseas locations or who cannot maintain attendance in a physical classroom due to lifestyle and workforce challenges. The prior barriers to the physical aspects of learning for MALs have been aided over the past two decades by the movement toward distance education programs in various postsecondary institutions, affording MALs access to a wider selection of programs and opportunities to attend from any physical location. The intentions of such programs are to facilitate learning opportunities to those with physical and lifestyle constraints and stresses that can restrict their access to traditional higher education programs (Sumner, 2000).

MAL Accessibility

For MALs who are currently active duty or activated reservists, they face accessibility conditions that must be taken into consideration by educators working with MALs in distance learning programs. While the need for many of them to be deployed to Iraq and Afghanistan has subsided, there are similar concerns with near-future conflicts that could send additional troops into deployment status. For example, in December 2019, 750 soldiers from 82nd Airborne deployed with little notice to Iraq, and the indication in the related *Military Times* article is that an additional 4,000 troops would be sent shortly after (Snow, Altman, & Athey, 2019). Whether or not a direct conflict is occurring, there are still troops deployed to more remote locations for standard deployments or training cycles. As Starr-Glass (2011) notes "Military learners in these locations ('in-country' or 'down range') often find that Internet access is limited, intermittent, or simply impossible. Restrictions on Internet access may be for several days or longer" (p. 153). The same can be true for those active duty or active reservist individuals who are serving on ships, depending on the locality and the conditions on the ship and the missions they are participating in.

While active duty dependents may not be placed in a situation listed above, they might still be living in remote or overseas locations without regular Internet access, where Internet access is not as easy to obtain, or is not as reliable. During my time living as an active duty dependent spouse in Spain, we had Wi-Fi included in our rent payment, but the Internet and the power would regularly go out, causing connectivity issues. I was also informed by my students living in the barracks on base that Wi-Fi was an additional monthly charge that many could not afford to pay. Furthermore, "even when Internet connections are available for academic work, connection speed and bandwidth may present limiting factors. These limitations should be taken into account when considering the technical design and bandwidth requirements of the virtual classroom" (Starr-Glass, 2011, p. 153). In short, while the remainder of this chapter will provide suggestions for using two-way communication as an effective model for engaging MALs in distance learning, it is important to keep in mind that not all of these models should be focused on synchronous or high bandwidth methods, as this will severely limit the accessibility for some MALs or preclude them from participating in the course at all.

MAL Scheduling

Another consideration for the majority of MALs is that they are entering the university as adult learners, which significantly impacts their learning experience. A majority of MALs will have alternative duties and responsibilities they are juggling outside of the virtual or physical classroom to include their work duties, family, and other related commitments (Molina & Morse, 2017; Ford & Vignare, 2015). Work schedules may include travel, contract requirements, longer or atypical work hours, deployments, or Temporary Assigned Duty (TAD) (Molina & Morse, 2017). Military-affiliated spouses might be managing the home as a single parent if their active duty spouse is currently deployed, TAD, or on travel (Smucny & Stover, 2013). According to the survey conducted by Starr-Glass (2011), one of the biggest stressors for MALs is "concern regarding unanticipated scheduling issues," which can include abrupt changes in workload or the location of the work (p. 153). These changes are always outside of the hands of the students and are not negotiable. While the military welcomes and encourages higher education, they also rank it as a second priority with the mission of service being the highest priority (Starr-Glass, 2011). In my own personal experience with MALs, they admitted to being frustrated with the lack of consistent understanding from instructors when broaching the subject of unexpected schedule interruptions, location changes for work, or even abrupt duty station moves. As well, these unanticipated course interruptions impact the MAL's ability to participate and interact, specifically with collaborative types of activities required in courses. And while active engagement with the learning process is most beneficial to productive learning, as discussed in the section below, when working with MALs, there needs to be careful consideration about creating group projects where other non-MAL teammates might be impacted by the military personnel who have an unanticipated interruption (Starr-Glass, 2011).

LEARNING SPACES AND MEANING-MAKING

Student-centered learning has been described in the past as only truly being experienced in face-to-face environments because it requires students to be producers or active agents in their learning, working in the role of collaborators in the production of meaning-making (Spellman-Cann et al., 2016). These learning models also emphasize the focus on developing relationships and producing work for authentic audiences, reinforcing why active, student-centered learning was not available in the first few generations of distance education. Interestingly enough, the origins of distance education lie with the United States military, as leaders created correspondence courses to train their service members while they were serving at a distance as far back as WWI and WWII (Duncan, 2005). In fact, the U.S. Army Training Support Center has existed in Virginia since 1976 and first included training manuals, phone conferences, and video lectures. These correspondence courses were replaced by technological tools such as email, online course platforms, DVDs, and even video conferences as advanced technologies became available that allowed for distance students to learn in more effective manners. When distance education was primarily dominated by one-way communication, the concerns over whether or not distance education could facilitate the same kind of meaning-making were obvious; however, technological advances over the past decade have changed the way distance education can be facilitated with far more opportunities for active student engagement in a variety of different settings.

In the present context, there are now strong advocates of distance education. In particular, Memic-Fisic and Bijedic (2017) argue that the tools and technologies used in distance education will help prepare learners for the future workforce needs, which only strengthens the argument for using them in classrooms with MALs. The Internet has provided the capacity for "delivering knowledge that even the futurists of twenty-five years ago would not have envisioned. It has created an innovative environment where technology has started to 'spin on its own axis,' often irrespective of learning theory and application" (Duncan, 2005, p. 404). These methods of educational delivery are especially beneficial to MALs, who are primarily made up of non-traditional learners already in the twenty-first-century workforce or seeking to move into the workforce at more rapid rates. In 2005, Duncan argued that the main concerns over acceptance or implementation of distance programs included a lack of institutional and administrative support, a lack of instructor competence in the use of technology, and students' limited abilities to navigate the technology while having a meaningful learning experience. While some of this lack of institutional or administrative support or lack of knowledge by certain faculty still exists, which became particularly evident during the spring of 2020 with the Covid-19 pandemic, the evolving technology and increased acceptance of online learning has certainly helped to mitigate some of those concerns or fears (Duncan, 2005).

Changes in physical space from traditional classrooms to online learning affect learning. These changes are not negative, however, just different and demand educators to consider them differently and with different practices and tools in mind. In particular, Oblinger (2006) argued that three important changes affect the way learning spaces should be designed in contemporary classrooms: (1) the students have changed, (2) the availability of instructional technology has advanced, and (3) what educators and researchers know about learning and how it is acquired has changed. All of these impact meaning-making, and as Knight (2013) argues, "our understanding of meaning-making is being reworked, in step with the changing times," which includes how students are making meaning through more visual and interactive means of communication available (p. 148). This also includes needing to teach students how their audiences in the real world will create this meaning from new and evolving technologies. Students are learning to operate where the "physical and virtual worlds intermingle," both inside and outside of the classroom, and educators have to reconsider the ways in which they facilitate that learning in all of their classrooms (Cheers et al., 2012, p. 266).

These new spaces and technologies require educators, and universities as a whole, to reconceptualize how online and blended learning environments will operate in order for students to more adequately participate and engage in more meaningful learning. The reality is that the digital world is now part of all students' lives, and as previously noted, a very important and beneficial part of the educational opportunities for MALs; thus, related teaching pedagogies must be readdressed and re-employed (Cheers et al., 2012; Knight, 2013). Even in a traditional classroom, most educators are moving toward using technologies that have changed the way learning occurs. In short, those like Wahlstrom (2004) argue that because of the changes in technology, a serious restructuring of American education is required as we move away from the prior technology of the book to the technology of the digital world. Sumner (2000) underscores the importance of rethinking our education models and reinforces that technology has always had an "intimate relationship" with distance education to mediate the separation between the teacher and the learner, but now its importance is heightened because of technology's "growing range and accessibility, lower costs, greater ease of use, expanding pedagogical power, and increasing political and social cachet" (p. 271).

As these technologies continue to adapt and advance, educators also need to adapt to this environment. Since online environments can be isolating for students, particularly MALs who are working in environments that do not allow for traditional learning, this can decrease motivation and increase attrition, according to Spellman-Cann et al. (2016). Thus, a major component of the instructor's role is to ensure that learning remains socially constructed, considering the ways in which students have socialized in the world, including their world views and values. Some educators might want to just leave things as they were for a variety of reasons, including their own fears of the technology or rationalizing to themselves that their students are capable of learning through traditional models of teaching in online spaces (Memic-Fisic & Bijedic, 2017); however, in reality, teachers, administrators, and researchers need "to cope with and shape the changes underway" (Wahlstrom, 2004, p. 306).

Constructivist Pedagogy and Students' Lived Experiences

A general understanding of constructivist pedagogy recognizes the importance of students' experiences in the classroom as a way to engage and create connections with them, and ultimately for them to create additional knowledge and meaning from these experiences in combination with what they learn in the classroom (Arghode, Brieger, & McLean, 2017). This has been cited as a particularly effective pedagogy for engaging adult learners, since it pulls on the knowledge they have already acquired from their professional and personal life experiences (Arghode, Brieger, & McLean, 2017). Steen, Mackenzie, and Buechner (2019) argue that "an important consideration for us, and for others working with adult populations, is the significant professional experience students bring to the 'learning table'" (p. 406). In particular, for MALs who are more likely to come into higher education with more substantial life experiences than their traditional peers, or those students who come straight from high school into college, pulling on their prior experiences and expertise is an effective way for them to make meaning. It is also beneficial for helping them find connections between their professional and personal lives and the work they are being tasked with in the classroom, which heightens the knowledge and meaning-making transfer potential.

Effective and strategic communication are essential parts of working within the military, and finding ways to increase the effectiveness of communication is always at the forefront of military strategy (Yarnell, Dullea, & Grunberg, 2018). One of the most essential components of effective military communication is the two-way communication model, in particular, the "feedback" model, or the "receiver's response to the sender's message, which can indicate understanding, lack of understanding, misunderstanding, agreement, disagreement, desire for more information, etc." as a primary way of ensuring that everyone within the communication act effectively understands how to accomplish the task at hand (University of Akron, 2014, p. 7). The increased technology over the past two decades for the U.S. military has primarily focused on these two-way models, especially as they have an ever-expanding global presence. This includes synchronous online training and video teleconferences (VTCs) as main methods of instructional technology and facilitating meetings from varied physical spaces. However, the demand for two-way communication began long before the past few decades, as two-way radio is cited as a necessity in military communications dating back to World War II (Clarkson, n.d.). Two-way communication used in the military is not just about the tools service members are taught to work with; it is about how service members are taught to communicate, particularly with the "feedback" model being ingrained in them from the very early days of their entrance into service. For example, phrases such as "repeat back," "copy," or "roger," which are derived from the days of the two-way radio, are used to acknowledge whether or not the receiver has effectively understood the message of the sender. Thus, a method of meaning and knowledge-making in the virtual classroom for MALs is to pull from their lived experiences with the two-way communication approach that they already use and understand.

While feedback can be given effectively through one-way communication models like an instructor's feedback left on a paper draft, Sumner (2000) argues for the importance of distinguishing between the different kinds of technologies used in the classroom to make sure educators understand which ones more effectively facilitate student-centered learning and growth. She argues that technology can either serve as a creator of public spheres through collaborative action or can serve as a destroyer of that public space by creating isolation and decreased interest in learning (Sumner, 2000). In short, "interactivity involves the ability of the learner to respond in some way to the teaching material, and to obtain comment or feedback on the response, which considerably increases learning effectiveness" (Sumner, 2000, p. 272). In order to promote the creation of communicative action, she argues for returning to a focus on the basic fundamental principles of two-way communication and related tools, allowing communication to flow back and forth between the two individuals in a communication act. Implementing these fundamental practices is imperative in place of one-way communication for interactivity between teachers and learners, as well as among the distance students themselves. For example, one-way communication in the distance learning classroom includes the instructor simply posting instructions and materials without the ability for the learner to respond with questions or concerns. Alternatively, two-way communication involves implementing methods for a student to respond back to the instructor and ask questions. This can be done through the use of ideas such as online office hours with a video or text chat feature or using a shareable document like Google Docs or Google Classroom where the students can post assignment questions and comments and have them answered later by the instructor. While Sumner's article was written in 2000 when two-way communication technologies were just becoming available in distance and online learning classrooms, her main premise of using two-way communication becomes even more imperative and effective in the year 2020 where educators have a plethora of digital technologies available. Particularly with MALs, who have experience using these kinds of learning and working models and who are adept at quick, efficient, but clear communication, two-way communication can be used to help them have more successful learning experiences in spite of the challenges they may face in postsecondary education, as discussed above. Implementing these strategies in virtual classrooms helps to enact constructivist pedagogy by integrating "their own lived experience as it relates to the lesson, content, and discussion" (Steen, Mackenzie, & Buechner, p. 407)

DISCUSSION: APPLYING LIVED EXPERIENCES THROUGH TWO-WAY COMMUNICATION IN CLASSROOMS WITH MALS

As distance education becomes more prominent in the higher education setting, and as MALs continue to choose this as a more reasonable option for their postsecondary education, educators should consider ways to more deliberately reinforce and implement the use of foundational two-way communication theories into their online, hybrid, and live streaming classrooms in order to make the distance learning experience more effective in shared meaning-making and communication contexts in a way that enacts constructivist pedagogy by using MALs' lived experiences (Starr-Glass, 2011). I argue that these two-way models of communication and technologies are effective tools for re-envisioning digital classrooms, specifically in terms of facilitating learning for MALs since these students have lived experiences that encompass using these kinds of technologies in their professional and personal lives. As well, two-way models of communication fall in line with active learning pedagogies like social and student-centered learning that help to build trust and mutual respect among the educator, the learner, and their peers. Offering more ways for shared meaning-making and pulling from their lived experiences can aid in making MALs feel connected in classrooms that are often physically distant and where they are likely to be a non-traditional student with differing life experiences. As noted in the introduction, while the focus of this chapter is on mitigating these challenges for MALs, this fundamental theory and related tools are imperative to include in higher education classrooms that have been impacted, perhaps temporarily, perhaps more long-term, due to the global pandemic of 2020.

The realities of distance learning classrooms in various forms including totally online, hybrid or blended, and live streaming means that one-way communication cannot be altogether avoided, but even one-way communication can be improved by putting into place differentiated teaching practices that more readily reach various kinds of learners and use techniques from student-centered and constructivist pedagogies that consider students' prior and current lived experiences. This can include a combination of "asynchronous and synchronous activities that allowed participants from various time zones to connect and receive timely feedback on their ideas" (Spellman-Cann, 2016, para. 28). Memic-Fisic and Bijedic (2017) suggest that providing a variety and flexibility of electronic learning materials can help in working with students who have different learning styles or have different preferences and needs for support. In the case of MALs, this variety and flexibility are very important. In essence, not all one-way communications are negative if they are integrated with other kinds of two-way communication that make the course more effective overall for all types of learners, in particular for MALs who might have different time and schedule constraints. It can be difficult to keep in mind that some MALs are still active duty and might have constraints related to synchronous, two-way communication, or to accessing the Internet on a regular basis. On a side note, it is imperative to establish guidelines, expectations, and exceptions with MALs regarding attendance and participation or making up irregular attendance at the beginning of the course in order to avoid confusion or frustration over the duration of the course. Having these statements clearly indicated on the syllabus would be beneficial, although it would be substantially more effective if they were implemented at the institutional or departmental level (Wilkes, 2017). This might need to be an individualized plan with a particular student who discloses that they are planning to be gone for field exercises or a Temporary Duty (TDY) during the course or who might be going out on a ship or to a forward operating base (FOB) and not have regular access to the Internet due to sensitive maneuvers.

Despite the needed variety and flexibility in the assignment design and the needed flexibility on the part of both the instructor and the students, it is essential that teaching materials for "distance education provide clear and specific instructions and guidelines for work" (Memic-Fisic & Bijedic, 2017, p. 45). These guidelines and instructions often take the place of what would happen more naturally in a traditional classroom setting, where the instructor would spend a portion of the course talking about the upcoming assignments, overviewing assignment requirements, and accepting questions the students might have regarding those assignments. While some of this can be done through one-way communication, like the instructor posting a short video that explains to the students the steps for the next upcoming assignment or writing out a very detailed assignment sheet and rubric, the literature indicates that two-way communication has more potential in eliminating some of these challenges (Memic-Fisic & Bijedic, 2017). This could include short group or class meetings online or even individual conferences periodically throughout the semester to ensure students understand the assignments sufficiently. Two-way communication can be implemented in daily or weekly coursework assignments, such as asking students to use a social media applications like a chat feature of a Twitter hashtag to collaborate with one another about assignment questions or for contributing to a low-stakes assignment that will help them better understand the context of the larger assignment to follow.

As a reminder, in differentiating two-way communication from one-way communication, it does not necessitate that two-way communication happens instantaneously, but that the feedback is given to the giver or receiver depending on the flow of communication. This means that communications like e-mails or comments left in a discussion board post are considered two-way communication, but they still need to be differentiated with more synchronous types of two-way communication to make distance learning the most effective for varied types of learners. In particular, with MALs, while it cannot be generalized across the board, Starr-Glass's (2011) survey found several characteristics of MALs that can be beneficial for keeping in mind when planning the curriculum for an online class: maturity, international awareness, high motivate, and familiarity with teamwork and collaboration. However, these beneficial characteristics and lived experiences do not always make up for the substantial challenges listed above in terms of accessibility, scheduling, and physical space that certainly complicate more traditional teaching models, their inability to participate in a group project because of their schedules, or their inability to participate regularly in synchronous activities or even social media online activities due to restrictions from leadership.

Listed below are various kinds of two-way communication that can be effective in a digital classroom with a short description of each tool and how it may potentially be used by instructors. Using these different multimodal technologies, tools, and communications also prepares students for the different professional communities they will be part of in the future, whether remaining military-affiliated or not. Additionally, many of my prior students inside and outside the government sector have reported the ways in which they use many of these technologies on a daily basis in the office setting.

Different Options for Two-Way Communication

- *Video conferencing software.* There are many variations of this software used and available throughout a majority of universities. Some of them are already embedded into the university's Learning Management System (LMS), such as Blackboard Collaborate Ultra, or with instructors given an account connected to the university that can be used for individual, group, or class meetings with their students. Examples include Webex, Zoom, or Adobe Connect. Prior to the past decade, asynchronous communication would have been preferred to this kind of video conferencing because of issues of affordability, bandwidth, and accessibility, but over the past decade it has now become much more common in distance education programs and is a viable, affordable option (Murphy & Ciszewska-Carr, 2007). Issues of accessibility for MALs should be kept in mind, however, as not all students will be capable of participating in online conferencing; thus making these kinds of conferences mandatory or for a grade should be avoided. However, using video conferencing software for office hours and scheduled conferences can be beneficial. These can be effective for MALs with unpredictable schedules, as they can come to the online office hours when they have time or can request an appointment outside of the instructor's regular schedule.
- *Online Discussion Software.* While discussion boards are often built into the university's LMS, many students and professors find them uninviting, restrictive, or not beneficial for two-way communication or for receiving helpful feedback (Lieberman, 2019). Depending on the kinds of classroom discussion desired, there are a variety of different options that allow for synchronous and asynchronous discussions to happen during or after class time to include Flipgrid, Backchannel Chat, Kialo, and NowComment. While it depends on the instructor's preference, it can be beneficial to use these discussion platforms in the midst of an online synchronous or hyflex class to help the students ask questions to one another like a backchannel, in case they missed part of a conversation or lecture, or to put questions down to ask the instructor once they are done with a lecture or demonstration. This can be especially beneficial for MALs, who might be in different locations, those who have issues with strong internet access, or those who cannot attend the entire class due to work restrictions.

- *Cloud-Based Storage Services and Platforms.* These platforms offer many different opportunities for educators and students to collaborate synchronously and asynchronously by sharing documents, PowerPoint slides, and even surveys and sheets (like Excel spreadsheets). While there are several options available like Dropbox, Microsoft One Drive, and Google Drive, the latter has become the most popular in K-12 learning due to the collaborative capabilities. Due to this popularity among K-12 instructors, Google Drive can establish continuity and familiarity in postsecondary education, although currently, Google Drive is not particularly popular in postsecondary classrooms, likely because of the already embedded Learning Management Systems (Kim, 2017). However, some of the additional features of Google Drive include the ability for students to take notes in Keep, to video chat or text chat through Meet, and to create websites through Sites. Google Docs allows for collectively writing while making comments that can easily be responded to in a quick, efficient manner, as well as assigning tasks to different students who might be working on a group assignment or project. This program is also especially beneficial for saving various drafts since it allows for returning to see a prior version, and the drafts are saved instantaneously.

  In the case of a live streaming class where some students are present with the instructor and others are at a remote location, a disconnect can occur. In place of having these students only participate in group work with other students in the same location, using a group Google Doc can allow them to collaborate with all students in the class or to form a variety of different groups for low-stakes in-class assignments, for peer reviews, or for larger group assignments. Google Docs can also be used for students to take collective notes from the class period that can help for MALs who were unable to attend that class session or for the remote students in case of a technology issue that keeps them from hearing all of the course content. The instructor and students can, of course, go back to these documents after class and make additional notes or ask questions, which continues to facilitate two-way communication after the physical class meeting.
- *Chat features.* These are available in many different software programs and applications to include those available in Google Drive through Chats, in Google Docs so students and the instructor can chat with one another while they are working through a collaborative assignment or sharing feedback on an assignment, Facebook Messenger, Slack, GroupMe (a phone app), and Google Groups. From my personal experience as a doctoral student in a program that held synchronous classes with both distance and campus-based students, allowing students the capability to use a chat feature such as Facebook Messenger or Google Chat can help to create connectivity among the students. This is also helpful for distance MALs in a live streaming

class to let the instructor or students at the physical location know when they have a question, when they have completed an activity, or when they have lost connectivity or are having technical difficulties.

- *Slack.* Slack is also a good application to use with a blended learning course to bridge the gap between what happens during the physical classroom time and the online course time. It likely mimics a similar model used by active duty service members in communicating within their own commands. Slack is a software application that is considered a "collaboration hub" that "can replace email" to help a group of people working together communicate more effectively (Slack, 2020). The instructor can set up a workspace specifically for the class and different channels may be created that help direct conversations related to different class themes, topics, assignments, or even group projects. This application also allows for voice and video calls. There might be low-stakes assignments, group projects, or other larger assignments started in class that need to be completed the remainder of the week in an online format. Having a digital archive of the discussions that began in class can help students better navigate the shift from in-class to online. Using Slack in the online classroom setting can help MALs prepare for their current or future workplaces, as many workplaces have moved toward using a model such as this to communicate inside and outside of the office.
- *Social Media Applications.* While some instructors might frown upon using social media communication in their classroom, it does offer distance learning students the ability to communicate instantaneously with one another and can be used for inside- or outside-of-classroom time with the instructor or just among the students. For instance, creating a Twitter hashtag with the course name would allow for students to write out questions or concerns about the course (Bista, 2015). It could also help students who were not in attendance to get caught up by asking questions about what they missed or by looking at some tweets other students posted about what the instructor said during class time for a hybrid course. Having the students participate in a Twitter hashtag allows for student-centered learning and engagement using two-way communication capabilities but with the freedom and flexibility to answer when they are capable. These kinds of digital archives can also be beneficial for MALs who were without Internet access for a period of time. They can revisit the conversation asked by other students and the instructor to maintain connectivity with the course material. This tool is also likely one that MALs are extremely familiar with for staying connected to family members around the world during their time in service, which will help to pull from their lived experiences; however, it needs to be kept in mind that there might be some restrictions to their abilities to post certain things that could fall under Operational Security (OPSEC) violation.

- *Blogging or Vlogging.* Both the students and the instructor can partake in blogging or vlogging (video blogging) in order to share ideas in a similar manner as they would with a journal entry, a discussion board post, or to reflect on a major assignment recently completed (Blackstone & Wilkinson, 2011). This is a way of providing feedback that can be turned into a two-way communication, where the educator and the learner both create short videos or blog posts on a website like YouTube, Tumblr, Flipgrid, or a Wordpress site they create, and where both the educator and students can respond to one another; however, it still allows for versatility and scheduling variations that would be especially beneficial to MALs who are managing challenges of accessibility and scheduling. Making these weekly assignments a regular part of the course with consistency but flexibility in the deadlines provides a more dynamic version of a discussion board post where students feel like the learning is more centered on them and their agency (Blackstone & Wilkinson, 2011). It is likely that MALs are familiar with creating or watching some kind of video presentations through their personal or professional lives, so showing them how to adapt these to the classroom objectives would be beneficial in their overall knowledge-making that builds from their lived experiences.
- *Voice and Video Feedback.* These applications allow instructors to create a "digital recording or any instructional activity performed on a computer screen" (Ruffini, 2012, para. 4). In addition to this, they can be used to review an assignment while the instructor provides feedback orally so that the student can see the instructor scrolling through the assignment and talking through points for revision or rationale for points off an assignment. As DeWaard (2016) notes, feedback can improve outcomes within teaching and learning. Ruffini (2012) specifically applies this to screencasting by arguing that "a screencast can provide learners a student-centered and engaging learning experience in both distance and traditional learning settings" (para. 1). Options for screencasting include Voicethread, Jing, Screenr, and CamStudio, although there are many more apps and software programs available. In particular, for MALs with physical restrictions, accessibility concerns, or scheduling issues, using voice and video feedback allows them to still receive viable feedback on their assignments with the essence of two-way communication in mind, but the ability to review the feedback at their discretion.

CONCLUSION

The varied demographics and challenges of MALs, and the many variations available of distance and online learning, make it challenging to provide a one-size-fits-all best practice for this demographic of learners in distance and online education. The same is true for the new generation of educators and students who have found themselves in the midst of varied distance and hybrid teaching and learning due to Covid-19, which will inevitably impact the way learning is delivered in higher education in the future. Thus, the technologies, tools, and suggestions listed in this chapter are intended to provide opportunities for variation and adaptation depending on the specific demographic(s) of MALs or other student populations and the kind of online learning space the educator is working with. Educators will also need to take into consideration other factors, such as whether there are other non-MALs in the classroom and whether the course takes place during a traditional semester or is condensed, fast-paced, or a mini session.

As technologies continue to evolve, the capacity for two-way communication tools in distance and online learning classrooms have substantially increased, but it is imperative for educators to reconsider how to implement these in mindful, beneficial ways that focus on constructivist, student-centered learning to help them make meaning more effective inside and outside of the classroom. While these two-way communication theories seem rudimentary, the struggle to quickly and effectively implement them into virtual learning spaces in Spring 2020 shows how beneficial it can be to spend time reconsidering ways to more effectively integrate them into distance learning classrooms. Particularly, for MALs, this also requires educators to keep in mind the challenges they face in their personal and professional lives, using these tools in order to alleviate concerns of accessibility, scheduling and physical space. Ultimately this requires flexibility on the part of the instructor, but with clearly defined rules, regulations, and alternatives that might still incorporate one-way communication if MALs are unable to participate using the two-way technologies provided. These different tools and technologies can and should be adapted to help better facilitate alternative scenarios, which will increasingly be the case as universities look for new and inventive ways to aid both MALs and non-MALs in completing their coursework while maintaining active and constructivist pedagogies to engage students in meaningful learning.

REFERENCES

Altman, G. (2019, July 23). Fewer troops are using tuition assistance benefits. *Military Times*. https://www.militarytimes.com/education-transition/2019/07/23/fewer-troops-are-using-tuition-assistance-benefits/.

Army ROTC. (2014). Introduction to tactical leadership: Effective Army communication. *The University of Akron*. https://www.uakron.edu/armyrotc/MS1/36.pdf.

Beery, T., Shell, D., Gillespie, G., & Werdman, E. (2013). The impact of learning space on teaching behavior. *Nurse Education in Practice, 13*, 382–387.

Bergamin, B., Werlen, E., Siegenthaler, E., & Ziska, S. (2012). The relationship between flexible and self-regulated learning in open and distance universities. *The International Review of Research in Open and Distance Learning, 13*(2), 101–123.

Bista, K. (2015). Is Twitter an effective pedagogical tool in higher education? Perspectives of education graduate students. *Journal of the Scholarship of Teaching and Learning, 15*(2), 83–102.

Blackstone, B., & Wilkinson, M. (2011). Blogging for pedagogical purposes: A transformational methodology for post-secondary education. In L. M. Ho, A. P. Leong, & K. T. Anderson (Eds.), *Transforming literacies and language* (pp. 187–206). Bloomsbury.

Borsari, B., Yurasek, A., & Carey, K. (2017). Student service members/veterans on campus: Challenges for reintegration. *The American Journal of Orthopsychiatry, 87*(2), 166–175.

Boyd, D. (2004). The characteristics of successful online students. *New Horizons in Adult Education and Human Resource Development, 18*(2), 31–39. doi:10.1002/nha3.10184.

Cheers, C., Eng, C. S., & Postle, G. (2012). Experiential space. In M. Keppell, K. Souter, & M. Riddle (Eds.), *Physical and virtual learning spaces in higher education: Concepts for the modern learning environment* (pp. 266–277). Information Science Index.

Clarkson University. (n.p.) *Wireless: Two-way radio technology*. Clarkson University. https://people.clarkson.edu/~collinme/wireless.

Department of Veterans Affairs. (n.d.) *Who are today's student veterans?* https://www.mentalhealth.va.gov.

DeWaard, H. (2016). Voice and video feedback to enhance instructor presence. In W. Kilgore (Ed.), *Humanizing online teaching and learning* (Ch. 4). Pressbooks.

Duncan, S. (2005). The U.S. Army's impact on the history of distance education. *The Quarterly Review of Distance Education, 6*(4), 397–404.

Falloon, G. (2012). Inside the virtual classroom: Student perspectives on affordances and limitations. *Journal of Open, Flexible, and Distance Learning, 16*(1), 108–126.

Ford, K., & Vignare, K. (2015). The evolving military learner population: A review of the literature. *Online Learning, 19*(1), 7–30.

Hart, A., & Thompson, R. (2016). Three approaches to veterans in the writing classroom. *CCC, 68*(2), 345–371.

Harting, K., & Erthal, M. (2005, Spring). History of distance learning. *Information Technology, Learning, and Performance Journal, 23*(1), 35–44.

Hrastinski, S. (2019, September). What do we mean by blended learning? *Tech Trends, 63*(5), 564–569.

Hunley, S., & Schaller, M. (2006). Assessing learning spaces. In D. Oblinger (Ed.), *Learning spaces* (n.p). Educause. https://www.educause.edu/research-and-publications/books/learning-spaces.

Kim, J. (2017). How Google has not taken over the higher ed classroom. *Inside Higher Ed.* https://www.insidehighered.com/blogs/technology-and-learning/how-google-has-not-taken-over-higher-ed-classroom.

Knight, A. (2013). Reclaiming experience: The aesthetic and multimodal composition. *Computers and Composition, 30*(2), 146–155.

Langstraat, L., & Doe, S. (2014). *Generation vet: Composition, veterans, and the Post 9/11 university*. University of Utah Press.

Lederman, D. (2018, November 7). New data: Online enrollments grown, and share of overall grows faster. *Inside Higher Ed.* https://www.insidehighered.com/digital-learning/article/2018/11/07/new-data-online-enrollments-grow-and-share-overall-enrollment.

Lederman, D. (2020, July 29). Virtual learning will be better this fall. Right? *Inside Higher Ed.* https://www.insidehighered.com/digital-learning/article/2020/07/29/will-virtual-learning-be-better-fall-will-it-be-better-enough.

Lieberman, M. (2019, March 27). Discussion boards: Valuable? Overused? Discuss. *Inside Higher Ed.* https://www.insidehighered.com/digital-learning/article/2019/03/27/new-approaches-discussion-boards-aim-dynamic-online-learning.

Memic-Fisic, I., & Bijedic, N. (2017). Teaching materials for ESL distance learning. *Explorations in English Language and Linguistics, 5*(1), 40–54.

Michinov, N., Brunot, S., Le Bohec, O., Juhel, J., & Delaval, M. (2011). Procrastination, participation, and performance in online learning environments. *Computers & Education, 56*(1), 243–252.

Molina, D., & Morse, A. (2017). Differences between military-connected undergraduates: Implications for institutional research. *New Directions for Institutional Research, 17*, 59–73.

Murphy, E., & Ciszewska-Carr, J. (2007). Instructors' experiences of web based synchronous communication using two-way audio and direct messaging. *Australasian Journal of Educational Technology, 23*(1), 68–86.

Oblinger, D. (2006). Space as a change agent. In D. Oblinger (Ed.), *Learning spaces* (Ch. 1). Educause. https://www.educause.edu/research-and-publications/books/learning-spaces.

Office of the Deputy Assistant Secretary of Defense for Military Community and Family Policy. (2017). *2017 Demographics: Profile of the military community.* https://download.militaryonesource.mil/12038/MOS/Reports/2017-demographics-report.pdf.

Ott, L., Morgan, J. K., & Akroyd, H. D. (2018). Impact of military lifestyle on military spouses' educational and career goals. *Journal of Research in Education, 28*(1), 30–60.

Pandza, H., & Masic, I. (2013). Distance learning perspectives. *Acta Informatica Medica, 18*(4), 229–232. https://www.ncbi.nlm.nih.gov/pmc/articles/PMC3818743/.

Ruffini, M. (2012). Screencasting to engage learning. *Educause Review.* https://er.educause.edu/articles/2012/11/screencasting-to-engage-learning.

Shaw, D., Tham, Y. S. S., Hogle, J., & Koch, J. (2015). Desire: A key factor for successful online GED adult learners. *Journal of Research and Practice for Adult Literacy, Secondary, and Basic Education, 4*(1), 21–35.

Smucny, D., & Stover, M. (2013). Enhancing teaching and learning for active duty military students. *American Sociological Association, 41*(3), 1, 8.

Snow, S., Altman, H., & Athey, P. (2019, December). 750 soldiers with 82nd Airborne headed for CENTCOM, additional 4,000 troops expected to deploy as Iran tensions mount. *Military Times*. https://www.militarytimes.com/flashpoints/2020/01/01/750-soldiers-with-82nd-airborne-headed-for-centcom-additional-4000-troops-expected-to-deploy-as-iran-tensions-mount/.

Spellman-Cann, S., Luong, E., Hendricks, C., & Roberts, V. (2016). Social learning in online environments. In W. Kilgore (Ed.), *Humanizing online teaching and learning* (Ch. 6). Pressbooks.

Starr-Glass, D. (2011). Military learners: Experience in the design and management of online learning environments. *MERLOT Journal of Online Learning and Teaching, 7*(1), 148–158.

Starr-Glass, D. (2013). Experiences with military online learners: Toward a mindful practice. *MERLOT Journal of Online Learning and Teaching, 9*(3), 353–363.

Steen, S., Mackenzie, L., & Buechner, B. (2018). Incorporating cosmopolitan communication into diverse teaching and training contexts: Considerations for our work with military students and veterans. In J. D. Wallace & D. Wecker (Eds.), *The handbook of communication training: A best practices framework for assessing and developing competence* (pp. 414–422). Routledge.

Student Affairs Administrators in Higher Education. (2013). *Measuring the success of student veterans and active duty military students*. NASPA. https://www.naspa.org/report/measuring-the-success-of-student-veterans-and-active-duty-military-students.

Sumner, J. (2000). Serving the system: A critical history of distance education. *Open Learning, 15*(3), 267–285.

Taleo, W. (2016). Affordances of technology to humanize instruction. In W. Kilgore (Ed.), *Humanizing online teaching and learning* (n.p.). Pressbooks.

Wahlstrom, B. (2004). Extreme pedagogies: Teaching in partnership, teaching at a distance. In T. Bridgeford, K. S. Kitalong, & D. Selfie (Eds.), *Innovative approaches to teaching technical communication* (pp. 304–327). Utah State University Press.

Weheba, D. M., & Kader, M. A. (2007, Spring). The impact of communication in teaching: A two-way communication approach. *Tourismos: An International Multidisciplinary Journal of Tourism, 2*(1), 127–143.

Wilkerson, M. (2020, July 28). 4 top asks students have for colleges. *Inside Higher Ed*. https://www.insidehighered.com/views/2020/07/28/four-key-things-students-say-they-want-college-fall-opinion.

Wilkes, L. (2017). An earned accommodation: Framing a veterans syllabus statement for faculty members. *Journal of Veterans Studies, 2*(1), 112–116.

Yarnell, A. M., Dullea, C., & Grunberg, N. E. (2018). Military communication. In F. O'Connor, E. B. Shoomaker, & D. C. Smith (Eds.), *Fundamentals of military medicine* (pp. 165–178). The Borden Institute.

Chapter 13

"So We Hit It with a Hammer, and It Worked"

How Active-Duty Marines Communicate, Collaborate, and Forge Their Own Learning Communities in Social Media

Bree McGregor

It was a sweltering summer evening in Yuma, Arizona, when I met Benjamin for an interview at Marine Corps Air Station Yuma. Using my laptop, Benjamin guided me through his private, military-connected Facebook group, Skid Kids. Created by its members, current and veteran Marine Corps helicopter mechanics from around the world, the group operates outside of the Marine Corps' official purview. As Benjamin navigated the endless scroll of conversations within the group, which included personal stories, memes about the challenges of military life, photos from combat deployments, and videos of helicopter training missions, he summed up the group's purpose through this example:

> I have seen a guy [active-duty Marine] who's desperate. They're trying to figure out a problem they've never seen, and they throw it up there. One guy will pop in like, "Oh yeah. Back in 1970, we had the same problem with this. All we did was hit it with this hammer, and it all of a sudden worked." And they [active-duty Marine] come back like three hours later, like, "So we hit it with a hammer, and it worked."

Skid Kids' members span multiple generations of service, often embodying different perspectives and experiences, yet they are joined together productively through a shared desire to support one another—in this instance trouble-shooting a helicopter maintenance issue. The scenario that Benjamin described underscores members' self-directed resource-seeking behaviors, illustrates creative cross-generational collaboration, and reflects a culture of trust and support cultivated through the shared purpose and communication practices of the group's members.

The interview I conducted with Benjamin became part of a mixed methods study that set out to investigate how active-duty Marines participate in social media. This chapter presents key findings from the study, with a special focus on Marines' communication and community-building practices within[1] military-connected social media groups[2] (MCGs). Survey and interview results illustrate the self-directed ways in which Marines cultivate knowledge and engage in mentorship activities within these groups. Using the social learning theory *communities of practice* (Lave & Wenger, 1991; Wenger, 1998) as a framework for analysis, this chapter analyzes the ways Marines' MCGs reflect members' flexibility and preparedness to socialize into new learning environments in the same ways they professionalize into different kinds of workspaces. These results align generally with an assets-based approach to supporting military-affiliated learners (MALs) as they transition into higher education, which begins by recognizing the skills and abilities present in MALs, as opposed to what is absent or problematic (Kretzmann & McKnight, 1993). Specifically, participant responses indicate that administrators and faculty may support MALs by creating learning environments that model key characteristics of Marines' MCGs and facilitate opportunities for informal, collaborative learning.

BACKGROUND: TRANSITION, COMMUNITY MEMBERSHIP, AND IDENTITY RECONCILIATION

The Veterans Metrics Initiative (2018) estimates that approximately 200,000 service members are expected to transition to civilian life each year over the next four years. With financial support from the Post-9/11 G.I. Bill, many will enroll in undergraduate and graduate programs across the country. In 2018 alone, a total of 708,069 student veterans, service members and their eligible family members used the Post-9/11 G.I. Bill, resulting in $10,673,744 in educational funding assistance (U.S. Department of Veterans Affairs, 2019).

As this influx of MALs continues to increase, they bring with them experiences that both support and complicate their transition from the military into higher education. In a two-year, grant-funded study of military veterans in college writing classrooms, which included a national survey along with site visits and interviews at more than fifty post-secondary institutions, researchers found that "most faculty report high achievement among veterans, as well as a high sense of initiative, professionalism, and leadership" (Hart & Thompson, 2011,

p. 4). Many of the skills learned through military service reflect preparedness to acculturate into new learning environments. Military training, for example, emphasizes the critical nature of expeditious learning (Simmons, Fisher, & Simmons, 2015), and it provides "opportunities for personnel to work with people from different backgrounds and geographical locations, and exposes them to different cultures across the world" (Sorrells et al., 2015, p. 202).

Yet even as MALs bring diverse professional experiences into higher education, research indicates that the markedly different social environment they encounter presents its own set of challenges for student veterans and service members: "While social cohesion served a valuable purpose in combat, the intensity of it at the time made it difficult later on for student veterans to feel like anyone in the civilian world, including at college, could understand them" (Naphan & Elliot, 2015, p. 43). The missing component of social cohesion, when paired with the unique professional experiences that set them apart from more traditional college students, may compound MALs' sense of alienation. In a study of more than 8,500 military members and dependents, participants described "adjustment to civilian culture" and "applying military-learned skills to civilian life" as two of the most significant transitional challenges they experienced (Zoli et al., 2015). These relational challenges, when sustained, can be detrimental to MALs' retention and success in higher education, even when students' knowledge and commitment to learning are strong (American Council on Education, 2018; Arminio et al., 2015; Coll & Weiss, 2015; DiRamio & Jarvis, 2011; Hamrick & Rumann, 2013; Naphan & Elliott, 2015).

In response to these challenges, colleges and universities have sought to bridge the gap between the military and the university through the creation of resources and support for MALs, to include on-campus veterans' centers, as well as training for faculty and administrators who support this population of students. However, for training and support to be successful, it must account for the diverse complexity of the student veterans' experiences and identities in ways that invite participation and enrich the learning environment for all students (Arminio et al., 2015; Coll & Weiss, 2015; Doe & Langstraat, 2014; DiRamio & Jarvis, 2011; Hamrick & Rumann, 2013). In support of this view, recent studies (Blaauw-Hara, 2016, 2017; Hinton, 2013) have shifted the focus to assets-based approaches that recommend leveraging MALs' prior experiences to support their transition into higher education. Through military training, for example, student veterans and service members have developed a mission-oriented approach to completing tasks by "by explicitly spelling out the ways colleges and courses work" (Blaauw-Hara, 2016, p. 818). When we attend to the strengths that MALs bring to our classrooms and campuses, we begin to construct the bridge between their military and academic communities, easing their transition into higher education.

Communities of Practice: A Framework for Investigating Marines' Group Learning Preferences

A useful framework for research that examines MALs' transition from the military into higher education is the social learning theory *communities of practice* (Lave & Wenger, 1991; Wenger, 1998), because it examines how "building an identity consists of negotiating the meanings of our experience of membership in social communities" (Wenger, 1998, p. 145). In the following communities of practice model, we look for the ways in which informal, group learning invites participation, how participation cultivates membership, and how membership shapes identity. Then, we consider how to create communities of practice that leverage these prior social learning experiences of MALs.

A community of practice comprises three dimensions: "a community of mutual engagement; a negotiated enterprise; and a repertoire of negotiable resources accumulated over time" (Wenger, 1998, p. 127). Put simply, within a community of practice members are joined together through a shared interest or goal around which members participate in meaning-making processes through communication and collaboration activities. Because learning within a community of practice is social by design, communities of practice are protean in nature; members' contributions lead to knowledge construction over time and reflect both evolving practice and evolving members. Through their engagement within the community of practice, members learn "certain expectations about how to interact, how people treat each other, and how to work together . . . As an identity, this translates into a form of individuality defined with respect to the community" (Wenger, 1998, p. 152). Through our memberships in numerous communities of practice, which Wenger defines as *multimembership*, we experience these identity translations in our everyday lives. For example, we may be instructors in a university's communications department, volunteers in a community safety program, and members of a local writing group, and our identities are shaped through our "nexus of multimembership" (Wenger, 1998, p. 149). As a result, our "membership does not determine who we are in any simple way; hence, generalizations and stereotypes miss the lived complexity of identity" (Wenger, 1998, p. 146). In other words, the way we see ourselves and our place in the world is a result of a complex network of relations. Through this negotiation, we are "constantly creating bridges—or at least potential bridges—across landscapes of practice" (Wenger, 1998, p. 161). The way we interact with our students shapes the way we see ourselves as mentors, and we leverage that when we volunteer to lead our neighborhood watch group, or when we offer feedback to the aspiring author in our local writing group. That is not to say that the work of bridging landscapes of practice is always easy; "the work of [identity] reconciliation may be the most significant challenge of learners who move from one community of practice to another" (Wenger, 1998, p. 160). As a research framework, communities of practice offer valuable insights into the work of identity reconciliation for MALs. It provides opportunities for analyzing their relationship- and skill-building progress into new learning communities, and it helps make visible the bridges—or at least potential bridges—across MALs' landscapes of practice. Further, communities of practice demonstrate the ways that we may support transitioning students, invite engagement, privilege prior knowledge and experiences, and balance formal instruction with autonomy in participation. In doing so, this resolves the paradox that "one needs an identity of participation in order to learn, yet needs to learn in order to acquire an identity of participation" (Wenger, 1998, p. 277).

When undertaking the investigation of MALs' communities of practice, social media offers useful space in which to trace and analyze their communication and community-building activities. In the twenty-first century, social media has changed interpersonal engagement for both civilian and military populations, leading scholars to extend communities of practice research into virtual environments (Dubé et al., 2005; Kietzmann et al., 2013). In 2009, The Department of Defense published policies that eased restrictions and lifted a partial ban on social media for service members. In the decade that has followed, service members have increasingly used social media to bridge the military-civilian gap, connecting with military and non-military communities (Hart, 2011; Grohowski, 2015; Rahbek-Clemmensen et al., 2012), sharing their stories with public and private audiences (Branham, 2016; Carter & Williamson, 2010; Penich-Thacker, 2010), and even communicating their experiences directly from combat zones (Bender, 2014; Burden, 2006; Silvestri, 2014).

Although communities of practice have traditionally been studied through in-person engagement, technology advancement has led scholars to expand the study of communities of practice to include virtual communities of practice (Dubé et al., 2005) and the role of mobile technologies, such as smartphones, in communities of practice (Kietzmann et al., 2013). A community of practice is deemed *virtual* when its members primarily locate their practice within new information and communication technologies (ICT). Although members may also meet face-to-face, "several factors, including geographical dispersion and busy schedules, make communicating through ICT much more efficient" (Dubé et al., 2005, p. 148).

Virtual communities of practice offer researchers new ways of understanding how participants organize, communicate, and collaborate within and across organizational boundaries (Kietzmann et al., 2013), providing a site for studying both synchronous and asynchronous knowledge sharing among professionals who are geographically dispersed (Dubé et al., 2005; Liao, 2017; Rodman & Trespalacios, 2018; Wenger et al., 2002). As I will show in the discussion section, Benjamin described his access to a community of Marine helicopter mechanics within Skid Kids who span geographical and generational divides and underscores this potential. Social media sites, which often preserve a traceable history of participant engagement through text, video, and images, provide valuable spaces for studying these types of communication and social learning experiences (Buck, 2012; Davies, 2012; Dubé et al., 2005; Kietzmann, et al., 2013). Initial research indicates that for military organizations, virtual communities of practice may enhance professional development and support by offering "a mechanism for individuals to keep each other current in the developments of a shared discipline . . . by providing multiple and more direct modes of disseminating information and ideas" (Daniels et al., 2006, p. 53). Workplace studies, however, have reported both benefits and potential challenges presented by virtual communities of practice. While they provide convenience and levels of access that traditional, face-to-face environments may not always provide, computer-mediated engagement may limit trust-building and identification processes for members (Handy, 1995; Hildreth et al., 2000) and limit meaningful engagement (Cramton, 2001). In response to findings that student veterans and service members, particularly those with combat experiences, often feel socially isolated on campus, "schools could use virtual spaces to help create a space where student veterans can find other veterans who go to the same school" (Naphan & Elliot, 2015, p. 45). Given the changes to military-affiliated members' social media access over the last decade, and the potential benefits of information sharing and support within virtual communities of practice, MALs' use of social media warrants further research (Naphan & Elliot, 2015; Sorrells et al., 2015). By using a communities of practice lens to analyze how active-duty Marines connect with service members and veterans in social media prior to their exit from the military, my study offers scholars and service providers insight into the community-building and identity-negotiation abilities this population of learners already possess.

ELUCIDATING COMMUNITY LEARNING PREFERENCES AND PATTERNS THROUGH A MIXED METHODS STUDY

With a goal of "reporting multiple perspectives, identifying the many factors involved in a situation" (Creswell & Creswell, 2018, p. 182), I undertook a mixed methods approach to investigating active-duty Marines' social media engagement. By using both quantitative and qualitative protocols, I sought a more complete understanding than I could obtain from one form of data collection alone (Creswell & Creswell, 2018; Tashakkori & Creswell, 2007; Tashakkori & Teddlie, 2010). The approach to research with social media participants in this study is informed by the methodological choices pioneered by Davies (2012) and Buck (2012) who traced digital literacy practices of research participants in social media through semi-structured interviews and guided social media tours. The design is further influenced by the work of Branham (2016) and Silvestri (2015) who studied the rhetoric and communication of service members and veterans in social media through participant interviews and digital artifacts collected through Facebook. Through the use of surveys, interviews, and screen-recorded social media tours, I sought to provide a more holistic account of Marines' engagement in social media by "converging several sources of data or perspectives from participants" (Creswell & Creswell, 2018, p. 200), and "generally sketching the larger picture that emerges" (p. 182). These methods also supported my reflexivity in response to my positionality as both a researcher and military spouse. Throughout this study, I remained cognizant of potential biases resulting from my membership and experiences within the Marine Corps community; furthermore, when designing the study and conducting research, I sought to limit these biases through both informal and formal mechanisms for feedback from members of the research participant community, by discussing my research goals and data collection procedures with active-duty Marines in my personal social network, and by completing the Marine Corps' formal research review process.

In addition to completing university institutional review board requirements, I also obtained permission to conduct research with active-duty Marine participants through the Marine Corps' Human Research Protection Program (HRPP). At the conclusion of the HRPP review, I received written approval from Headquarters Marine Corps which, in turn, facilitated my access to the participant community and enhanced the participant recruitment processes. Research took place over four months, and a description of research methods is illustrated in table 13.1.

Designing and Conducting the Survey

Survey questions were organized within three categories: general habits and attitudes regarding social media; communication and community-building practices in social media; and participant demographic information. The digital survey, hosted in Qualtrics, consisted of thirty-four closed-ended questions and two open-ended questions asking for further thoughts on social

Table 13.1 Implementation of a Mixed Methods Study

Research Phase I: April & May 2018	
Data collection procedure:	Digital survey
Participant recruitment sites:	MCAS Yuma; Facebook; Reddit
Products collected:	Survey results
Research Phase II: June & July 2018	
Data collection procedures:	Semi-structured interviews Screen-recorded social media tours
Participant recruitment site:	Digital survey
Products collected:	Interview transcripts Social media screenshots

media and feedback on the survey instrument. A final question asked Marines if they would be willing to participate in a follow-up interview. Those who selected "yes" were asked to provide a personal email address or phone number for follow-up contact.

In order to recruit a participant group that would be large and diverse enough to support generalizable survey results, over a two-month period I used several recruitment approaches on base at the Marine Corps Air Station (MCAS) in Yuma, Arizona, and within social media sites Facebook and Reddit. With permission from the Station Commander, I visited the Chow Hall (cafeteria) of MCAS Yuma at midday, where I distributed recruitment flyers and spoke with uniformed Marines while they waited in food lines that often stretched out into the parking lot. With permission from the Director of MCAS Yuma's Marine Air Group University (Avendano, 2014), I pitched my study to small cohorts of Marines attending professional development courses in makeshift classrooms scattered around the base. Within Facebook, I used the private messaging feature to leverage my personal network of active-duty Marines. I asked them to complete the survey and then share the survey link with other active-duty Marines in their digital social networks. Within Reddit, I published a recruitment pitch and survey link to the public r/USMC page. When the recruitment period concluded, 295 Marines started the survey with 200 completing, for a 68 percent completion rate. This study limits the findings to completed surveys. The Marines who participated in this study represent all demographic characteristics of the current active-duty population. For reference, a snapshot of survey participant demographics in comparison to current U.S. Marine Corps demographics (Headquarters Marine Corps, 2017) is presented in table 13.2.

Designing and Conducting Interviews

Once the first phase of research was completed, I used the survey results to identify and recruit interview participants. Survey participants were

Table 13.2 Demographics Comparison: Survey Participants and U.S. Marine Corps Personnel[1]

	Survey Participants	*U.S. Marine Corps*
Age		
25 or Younger	47.5%	66%
25 or Older	52.5%	44%
Gender		
Male	68%	91.7%
Female	31%	8.3%
Prefer not to answer	1%	0.0%
Race & Ethnicity		
American Indian or Alaska Native	0.5%	1.3%
Asian	1.5%	3.7%
Black or African American	3%	10.3%
Hispanic, Latino, or Spanish Origin	12%	17.8%
Native Hawaiian or Other Pacific Islander	0.5%	0.8%
Other[2]	4%	1.4%
Prefer Not to Answer	9%	1.3%
White	69.5%	63.6%

[1]Although socio-demographic data was analyzed using descriptive statistics, no significant patterns of differences emerged in this study.
[2]Two Marines identified as "American" and four identified as multiethnic.

designated for follow-up interviews when they indicated interest and provided personal contact information at the end of the survey. The resulting list was filtered to include only those who completed the survey and indicated that they participate in MCGs. Although fifty survey participants met the filtered criteria, only four Marines accepted the follow-up interview invitation. As a result, findings from the interviews are illustrative rather than generalizable, a common approach in mixed methods design when the goal is "to locate and obtain information from a small sample [to build on] extensive information from [the larger] sample" (Creswell & Creswell, 2018, p. 219). The interviews were loosely organized around sixteen semi-structured questions asking participants to describe the design and operation of their MCGs, their motives for joining, and their engagement within the group, while also providing a screen-recorded, guided tour of the group.

All interviews took place outside of Marines' scheduled work hours, and three interviews took place in person. Two participants elected to meet on base (outside of their worksites), in private conference rooms at the MCAS Yuma Education Center. One participant elected to meet off base, at a local coffee shop. A fourth interview was conducted remotely because of the participant's geographically distant duty station. I used my personal laptop to screen- and audio-record all interviews and social media tours, using QuickTime for

Table 13.3 Description of Interview Participants and Their Military-Connected Social Media Groups

Benjamin	Benjamin's MCG: Skid Kids
White male	Platform: Facebook
Mid-career enlisted Marine	Status: Unofficial; Closed
Helicopter mechanic	Members: >7,000 Marine helicopter mechanics
Maria	Maria's MCG: USMC Career Planners
Hispanic female	Platform: Facebook
Early-career enlisted Marine	Status: Unofficial; Closed
Career adviser	Members: 417 career adviser Marines
Everette	Everette's MCG: Team House
White male	Platform: Facebook
Mid-career enlisted Marine	Status: Unofficial; Secret
MOS[1] undisclosed	Members: 737 undisclosed Marines
Alan	Alan's MCG: Squadron DH
White male	Platform: WhatsApp
Late-career Marine officer	Status: Unofficial; Secret
Attack pilot	Members: 8 senior Marines officers

[1]A MOS (military occupational specialty) refers to the specific job a Marine performs within the Marine Corps.

in-person interviews and Zoom for the remote interview. Screenshots were taken from the screen-recorded interviews, and screenshots only depict content within the MCGs that participants elected to share and discuss. Pixelmator, a photo editing tool, was used to obscure personally identifiable information of interview participants and group members. Pseudonyms were assigned to participants to ensure anonymity, and a description of each interview participant and their MCG is listed in table 13.3.

Each of the interview participants elected to share and discuss a group formed around their military occupational specialty (MOS). Interview participants' groups controlled access through privacy settings and member-screening processes, and the groups operated as unofficial, grassroots communities with no official military oversight. Benjamin shared a group created for current and veteran Marines who perform different types of maintenance work on H-1 (Huey) helicopters. Maria shared a group created for current and veteran Marines whose job is to provide active-duty Marines with formal career guidance through all stages of their careers. Because the active-duty and veteran members of Everette's group share a MOS that performs classified operations, he elected not to disclose a specific description of his group and its members. Alan, an attack pilot, and squadron commander shared a group he created for active-duty, senior Marine officers within his squadron as a means of privately communicating timely information and making operational decisions.

UNDERTAKING A QUALITATIVE APPROACH TO DATA ANALYSIS

While survey results provided quantitative data that indicated trends related to how and why Marines join and participate in MCGs, interview participants' comments and screenshots from their social media groups offered a more nuanced understanding of their motives and experiences. I used a qualitative approach to data analysis because I wanted to understand participants' perspectives and experiences in relation to one another (Bazeley, 2013; Miles & Huberman, 1994; Saldaña, 2015).

I used the elemental coding method, concept coding, to capture *processes* or convey *ideas*, rather than organize data into content topics, which could be used in abstract and generalizable contexts (Mihas, 2014; Saldaña, 2015, 2016). Concept coding uses words or phrases that "symbolically represents a suggested meaning broader than a single item or action a 'bigger picture' beyond the tangible and apparent" (Saldaña, 2016, p. 119). Because survey data indicated, more broadly, ways in which Marines' MCGs offered participants' valuable sites for communication and community-building activities, I created rich coding concepts that "employ creative phrases rather than static nouns or simple gerunds" (Saldaña, 2016, p. 122), that would capture observable instances of engagement that indicated participants' sense of value through their membership and participation. A brief description of each coding concept is provided in table 13.4. Using these four coding concepts, I analyzed interview transcripts and screen-recorded social media tours for instances of language and images that indicated the presence of these themes.

Table 13.4 Concept Codes Used for Qualitative Data Analysis

Concept Code	*Definition*	*Description*
1. Personal-Professional Relationships	Cultivation of mutual respect, trust, and friendship	Communication that productively blends personal and professional topics in response to member needs and interests.
2. Collaborative Culture	Rank-flattened, informal, dynamic environment	Negotiating of information, disagreeing or arguing, through non-hierarchical engagement.
3. Participatory Empowerment	Confidence cultivated through participation	Ongoing engagement in information-sharing and meaning-making communication activities
4. Membership Identity	Sense of belonging	How they see themselves in relation to members; how valuable they perceive their contributions to be within the group

RESEARCH RESULTS: CONNECTING, COMMUNICATING, AND COLLABORATING IN SOCIAL MEDIA

Survey results from this study, illustrated with examples from interviews, define ways in which Marines' MCGs blend social and professional engagement within collaborative learning environments, how opportunities for sustained engagement in these groups enhance members' participation, and how participation shapes individuals' identities with respect to community membership. As Marines simultaneously participate in the formal learning environments of their workplaces and the informal learning environments of their MCGs, they demonstrate flexibility and adaptability across sharply juxtaposed landscapes of practice. Through these communication and community-building experiences, they cultivate self-efficacy that supports a military-to-higher-education learning trajectory.

Marines Blend Social and Professional Engagement

Because the practice both reflects and shapes the purpose of a community, Marines' motives for joining and participating in MCGs are important to track. Among 200 surveyed Marines, 73 percent reported that they participate in MCGs. In response to the survey question, "What are you seeking from military-connected groups in social media? Select all that apply" (including a write-in option), Marines' indicated interest in both social and professional engagement. Table 13.5 lists Marines' six most frequently selected responses organized by highest response rate.

Within a community of practice, the interplay of professional and social engagement fosters cohesion among its members. Marines call this cohesion camaraderie: mutual trust and friendship, and its presence in MCGs not only reflects Marines' interest in developing a social identity, but also their interest in learning and improving within a profession, which supports the development of a professional identity as well.

Table 13.5 Types of Engagement Marines Seek from Military-Connected Social Media Groups

Engagement or general entertainment	69%
Camaraderie (mutual trust and friendship)	68%
Stories about others' military experiences	61%
Advice or support	52%
Opportunities to mentor others	35%
Privacy that military-connected group provides	23%

While all four interview participants discussed instances of social and professional interactions within their groups, Everette's description of his MCG, Team House, highlights these dual modes of engagement in particularly deliberate ways. The small cohort of Marines in Team House are scattered around the world, and they perform secret and highly specialized operations. Everette's motives for participating in Team House reflect ways in which "our identities are not something we turn on and off, our various forms of participation are not merely sequences in time" (Wenger, 1998, p. 159). Everette, who at the time of our interview was assigned to a temporary administrative job outside of his MOS community, described how he relied on Team House to maintain professional identity as well as social connectedness:

> What's going on with the community? What are people feeling? I'm very removed from the majority of us, doing what I do now. I'm very alone and unafraid [laughs]. So, it's good to hear what's going on . . . because I'm not there anymore . . . I don't know all the good gossip.

In Everette's case, participating in Team House enabled him to continue the work, while geographically isolated from his peers, of "building an identity [which] consists of negotiating the meanings of our experience of membership in social communities" (Wenger, 1998, p. 145). Without access to this virtual community, these identity-building processes would be considerably more challenging for Everette to accomplish.

While Team House helps members stay connected and exchange information, it also provides space for levity in response to the high-stress nature of their work. The group has established a relaxed, informal communication environment, which Everette illustrated with Team House's banner image. The banner depicts Chuck Norris, a famous martial artist, doing a roundhouse kick while wearing Daisy Dukes and black cowboy boots. The static photo stretches across the top of Team House's homepage, remaining visible even as we scrolled through the group's conversation history. Everette explained how the banner image lightened the mood within the group: "So we have a very serious and sober job . . . And then one of the moderators was just like, 'Yeah, we're going to have a little fun here.' So this is Chuck Norris in some Daisy Dukes."

Reflecting, Everette added, "Well, we do [take ourselves seriously] when it counts but yeah. . . this is a fun, very free, open environment, and that [banner image] kind of sets the tone." Everette explained how Team House enables social and professional engagement to blend in critical ways: "I think the purpose of the group is primarily camaraderie, but I think with that camaraderie comes the mentorship, comes the advising, comes the 'I'm going to help you out of a tough spot'." Everette's description underscores ways in which the community of practice creates opportunities for social engagement, which enables professional engagement to occur in more impactful ways, and participation "suggests both action and connection" (Wenger, 1998, p. 55). Similarly, as we support MALs' transition into higher education, it becomes important to consider how we provide opportunities for both socializing these students into college and professionalizing them into a different kind of workspace.

Rank-Flattened Learning Environments Facilitate Collaboration

Because collaboration is central to both workplace and academic learning environments, it is important to examine the operational characteristics of MCGs, such as the group design and norms for participation that support opportunities for collaboration. Although the Marine Corps maintains official accounts across several social media platforms (United States Marine Corps, n.d.), only 4 percent of Marines surveyed in this study reported belonging to an official MCG. Survey results indicate that Marines' MCGs are predominately unofficial[3] and provided privacy settings that, at a minimum, screened membership and prevented non-members from viewing members' communication activities within the groups (table 13.6). These unofficial, rank-flatting communication environments indicate Marines' interest in dynamic, informal learning communities that operate differently from their more traditional, hierarchical professional environments.

While all four interview participants provided examples of collaboration within their MCGs, Alan's Squadron DH group offered compelling insights into a rank-flattened, collaborative learning environment that directly impacted a squadron's operations. In an effort to work around communication challenges in the workplace, Alan created Squadron DH, a secret group hosted in WhatsApp (offering end-to-end encryption), and invited seven of his senior officers into the group. Alan explained that the group enabled "each member to share information based on real time operations, but in a collaborative or a common way, so we all see what's happening, and we can all kind of back each other up."

As they prepared for a deployment, Squadron DH members shared time-sensitive information and made critical operational decisions. Alan

Table 13.6 Characteristics of Marines' Military-Connected Social Media Groups

Group Status:	86% Unofficial	4% Official	10% I don't know
Privacy Settings:	74% Closed or Secret	15% Public	11% I don't know
Social Media Platform[1]:	67% Facebook	6% Reddit	27% Other

[1]Facebook and Reddit were used by the researcher as sites for participant recruitment. Survey results include YouTube (6%), Instagram (6%), and additional social media sites reported by fewer than 3% of Marines.

communicated daily in Squadron DH and said his reliance on it became "almost a safety net for communication." Communication in Squadron DH took place privately, outside of formal meetings, official email exchanges, and even hallway conversations. Alan explained, "I think what I like about this is it's unfiltered, and I don't have to be so choosy with my words." For Alan, a commanding officer who represented the highest point in the squadron's chain of command, there was immediate value in the interrupted hierarchy, especially when it facilitated swift and direct conversations about the safety of Marines and aircrafts. Within Squadron DH, Alan aimed to cultivate members' leadership skills and empower them to take active roles in decision-making processes. He explained:

> I need them, as planners, to go figure this out: Can we do training? Can we still fix the airplanes? Can we get the flight hours we need? . . . And what's great about this is, since it's somewhat informal, I think they chime in with their opinions and stuff very freely, which is also good because we can have kind of a round table discussion a little easier sometimes.

As Alan scrolled through jargon- and acronym-laced threads of conversation in Squadron DH during our interview, he paused to share an example. As a conversation about operational needs unfolded within the group, instead of disseminating orders Alan's only comment was, "Put me wherever you need me," which enabled three members to exchange information, make decisions, and effectively manage an immediate operational concern.

Within communities of practice, "learning—whatever form it takes—changes who we are by changing our ability to participate, to belong, to negotiate meaning" (Wenger, 1998, p. 226). In the instance of Squadron DH, we see how participation in a virtual community of practice enables types of mutual engagement in collaborative, meaning-making activities among Marines that would not be possible within a formal, hierarchical environment. Among the surveyed Marines who participated in MCGs, 86 percent stated their groups were unofficial, indicating how even as they participate successfully in a hierarchical organization like the military when given opportunities, Marines also seek informal learning communities. In higher education, learning environments that balance formal instruction with frequent informal student engagement creates opportunities for a similar "kind of rhythm and . . . allow learning and teaching to inform each other," (Wenger, 1998, p. 266). When students have opportunities to take active roles in learning experiences, these collaborative activities enable learners to "invest themselves in communities of practice in the process of approaching the subject matter" (Wenger, 1998, p. 271).

Frequent Engagement Enhances Member Participation

As we consider how to support MALs' transition into new learning environments, it becomes important to examine how new members become participants in their MCGs. Communities of practice support paths to participation for new members when they provide opportunities for frequent and ongoing engagement, and when the practice directly supports new members' needs and interests. For these reasons, Marines' access to MCGs, and their interest in participating in these groups become important to investigate. In this study, surveyed Marines described the ease of access to social media through one or more personal handheld, portable devices: 100 percent used smartphones to access social media, and 46 percent also used tablets. Further, when asked, "How valuable is social media to you?" only 7 percent of surveyed Marines indicated that social media held no value for them. When the ease with which Marines can access their MCGs and the value they ascribe to social media are considered in unison, there is a clear potential for the active, sustained engagement among members within these groups.

During our interview, Maria exemplified how a newcomer's participation can change over time when there is a clear path to participation. Members of USMC Career Planners, who are stationed around the world, use the group to share information that helps them perform their jobs successfully, which was critical for Maria, a relative newcomer to her MOS who provided career guidance to more than 200 Marines in her unit. When she joined the group, she set an alert on her smartphone to receive notifications whenever a member communicated in the group, which in turn increased the amount of time she visited and read content within the group. Although the group is unofficial and military rank is not formally recognized within the group, Maria initially felt self-conscious about her junior professional status. Through continuous visits in the group, however, Marina learned how to become an active participant in the practice of the community. She explained:

> At first, I was afraid to post anything on there, because I didn't really know anybody in the MOS. And I knew there was higher leadership in there, so I wasn't sure how a simple question, how somebody would respond to it. Obviously, observing the page a lot more made me understand that it's more open and free to ask questions.

Maria discovered a professional community that allowed newcomers to voice questions and opinions without repercussions, along with experienced members who offered balanced perspectives. She explained:

> The Marine Corps is constantly changing. Literally every year, we have to try and catch up to their changes. So, when somebody posts something like, "We're

doing this change, I don't agree with it," then a retired Marine can go in there and say, "Well actually this is way better, because back then, or during this time, we had to do it this way. So if you look at it from that perspective, actually this is a good idea." And then, it'll open up, "Well that makes sense. No wonder the Marine Corps is doing it that way."

Over time, Maria gained the confidence to join these conversations. She explained, "If I read some of the comments and I realize that nobody has really mentioned my thought process on it, then I'll usually participate." As Maria's participation continued to evolve, it reflected not only her understanding of professional subject matter but also of the day-to-day challenges related to the work they perform. In one example she discussed, Maria had created a memehumorously depicting a difficult scenario career planners often deal with: Marines' disappointment with their careers. The U.S. president had recently directed the Pentagon to begin planning the Space Force, a sixth independent military service branch. Career planners stationed around the world quickly jumped into the online group, asking questions, and sharing information and humor. In the meme she shared, Maria compared Marine Corps base 29 Palms to the idea of being stationed on Mars, showcasing a soldier in a space helmet standing in the front of a red, endless abyss. The meme is captioned "well this is fucking bullshit.". Located in a remote desert area of southern California with extreme summer temperatures, 29 Palms is a duty station well-known and often reviled by Marines.

To an outsider, Maria's use of sarcasm and profanity may have seemed unprofessional. However, because she had spent time studying communication in the group, and increasingly participated over time, she had learned the nuanced communication practices of the group and how they differed from her workplace environment. Her meme received twenty-seven "reactions" from members indicating they liked it or found it humorous. Maria conceded that she would not have shared the meme in her workplace because she was the only career planner in her unit, and her unit colleagues did not understand the day-to-day challenges of the work she performed. Through her sustained engagement, Maria had developed a self-awareness and flexibility that reflected her ability to simultaneously participate in formal and informal professional communities, in accordance with the norms for participation unique to each community.

Maria's participation in USMC Career Planners helped her fill knowledge gaps and enhanced the support she provided to Marines within her own unit. Her ongoing, self-directed engagement in USMC Career Planners reflected trust in the knowledge-sharing practices of the community that provides "relations of mutual accountability that becomes an integral part of the practice" (Wenger, 1998, p. 78), which were unavailable to Maria within her workplace. Through her sustained engagement, by first reading and then contributing to the practice of the community, Maria became a part of the meaning-making activities that define the community. Maria's evolving engagement reminds us that MALs are reflexive learners who may be actively adapting to new learning environments even as their outward engagement initially appears passive, and it is important that our learning communities support new members' trajectories toward active participation.

Participation Cultivates Member Identity

Because participation in the community's practice creates opportunities for members to develop a sense of belonging and identity with respect to the community, it is important to investigate how Marines experience this within their MCGs. Through Marines' descriptions of their own engagement, we can better understand how they contribute to the meaning-making practices of these learning communities, and in turn, how it reaffirms their identities as members.

Survey results indicate Marines' engagement in MCGs supports social learning through mentorship activities that leverage members' prior experiences and encourage support-seeking behaviors among peers. In response to the survey question, "How would you describe your participation in military-connected groups in social media? Select all that apply" (including a write-in option), table 13.7 lists Marines' six most frequently selected responses organized by response rate.

Benjamin described Skid Kids, composed of active-duty and veteran Marines, as a place where members come to share their experiences. He explained, "It's cool because you read stories and you compare what you do. It's a contest. It's, 'You did this really cool thing, but let me tell you what I did'." There is an undercurrent of competition within Skid Kids that reflects members' interest in developing and sharing professional expertise. Benjamin explained, "You bicker with each other a lot about who's right, who's wrong about something. You'd be surprised how often our actual work comes into play in groups like this."

Table 13.7 Marines Describe Their Participation in Military-Connected Social Media Groups

Read about the experiences of others	87%
Offer advice or information	52%
Use humor	52%
Share my experiences	38%
Ask questions or receive advice	38%
Encourage or inspire others	30%

Benjamin's own participation reflected his role as a conscientious, experienced member who sought to level-up the knowledge-sharing practices within the community. To illustrate this, he explained, "A guy in here tried to tell me that this one way was the way that you're supposed to do the [helicopter] maintenance action. I just happened to be at work on my phone, on lunch, and I was like, 'No, dude. I got the maintenance publication right here'." Through his participation, Benjamin aimed to uphold high standards of accountability, which he said was common practice among participating members. When a debate unfolded about helicopter maintenance that might have jeopardized the safety of Marines, Benjamin asserted not only his knowledge but also his authority, reaching outside of the social media group to contact the relevant unit directly about the issue. He explained, "I didn't know them [the Skid Kids' Marine], but I knew their unit and I know somebody at their unit. It was kind of like, 'Hey, dude. I just talked to your NUGGET[4] and he's an idiot and he's on Facebook and he's arguing with me'." Benjamin's participation reflects how the community's joint enterprise "is not just a stated goal, but creates among participants relations of mutual accountability that become an integral part of the practice" (Wenger, 1998, p. 78). By holding all members to high professional standards and ensuring Marines' safety, Benjamin's contributions to the community reinforced his own sense of collective responsibility within the community.

The sense of individual value with respect to collective identity is further reinforced through the content members share: in particular, photos and videos of helicopter training operations. The work Skid Kids members perform is one component of larger squadron operations, and the significance of their work is realized when helicopters take to the skies. As Benjamin scrolled through Skid Kids' conversation history, he often paused to share and describe videos and photographs of helicopters in flight. Benjamin said he also shared this type of content in the group, and he explained how it affected his sense of pride and belonging within his professional community:

> As a maintainer, I see these birds sitting in pieces more than I do flying. Once they're up, and I bring it up and I send it on her way, I don't know what she's out doing. It could be two hours till I see it again. It's just nice sometimes to watch videos like this. Also, it's kind of bragging rights once again . . . That's a Whiskey. That's my baby, right there.

Benjamin's experiences in Skid Kids, along with surveyed Marines' descriptions of their participation in MCGs, indicate how "a shared practice thus connects participants to each other in ways that are diverse and complex . . . [and] reflect the full complexity of doing things together" (Wenger, 1998,

p. 77). Rather than building up MALs' sense of individual college identity and belonging in order for them to learn, we should consider the reverse: by creating participatory, project-based learning opportunities, we enable these students to contribute to the practice of the greater learning community and in doing so raise their sense of belonging.

RECOMMENDATIONS: FORGING CAMPUS AND CLASSROOM COMMUNITIES OF PRACTICE

Marines' simultaneous professionalization into workplace and social media groups illustrate participants' abilities to participate flexibly across diverse learning environments. By providing access to virtual communities that complement on-campus veterans' centers, and through facilitating peer-learning activities in classrooms that complement formal instruction, administrators and faculty can create informal, social learning environments that offer MALs additional points of entry into campus communities. Furthermore, these communities can help socialize MALs into higher education.

Create Virtual Communities to Enhance On-campus Programs

In addition to being a low-overhead option for supporting service members and veterans, creating military-affiliated groups in social media may extend the reach of existing campus programs by connecting with students who are unable to use on-campus resources. Research indicates that student veterans and service members are more likely than traditional college students to have full-time jobs, to support families, and to have limited time to complete their degrees due to their continued active service or use of the G.I. Bill (Hawthorn et al., 2013). These factors may limit their time spent on campus and their access to on-campus resources. Alternatively, military-affiliated groups hosted on social media platforms would offer students instant and continuous access to timely information, resources, and people. Compared to more traditional modes of communication employed by veterans' centers—for example, monthly newsletters, emails, or occasional webpage updates—information within virtual social media groups can be shared in more immediate, timely ways. Participants from this study depict a digitally-connected population of learners: 100 percent of Marines described accessing social media through one or more handheld, portable devices. Through social media, on-campus veterans' programs have the potential to access broader, digitally-connected audiences.

However, it is important to note that participation across different communities of practice presents its own set of challenges: "Different practices can make competing demands that are difficult to combine into an experience that corresponds to a single identity" (Wegner, 1998, p. 159). The college campus culture is an example of a place in which multimembership disperses our attention and can make finding a community more difficult for military-affiliated students. The more attention we can give to creating spaces that combine aspects of multimembership, particularly for non-traditional student populations like MALs, the more we enable students to be their whole selves. Creating online military-affiliated groups offer students inclusivity and serves multiple purposes within a single space.

Online military-affiliated groups may also better attend to MALs' preferences for privacy and autonomy over their military identities. Research indicates that student veterans and service members may prefer not to disclose their military-affiliated status (DiRamio & Jarvis, 2011; Thompson, 2014; Weiss, 2015). The desire to keep their military status private, further complicated by the public nature of veterans' centers (often centrally located on campus), may deter some from utilizing veterans' centers on campus. Some minority members within the military-affiliated community may be further deterred from visiting the on-campus veterans' center due to prior experiences within the military's traditionally "aggressive, male-dominated ethos" (Diramio & Jarvis, 2011, p. 53) that does not reflect the true diversity of its members (Iverson & Anderson, 2006; Pelts, Rolbiecki, & Albright, 2015). Using a social media platform as a gathering space, and undertaking group design and administration to include privacy settings and member-screening processes, military-affiliated virtual communities have the potential to serve as welcoming environments for a diverse demographic of MALs. Moreover, digital meeting spaces empower members through the autonomy of self-presentation as opposed to on-campus veterans' centers which provide less ability to control self-presentation. Findings from this study reflect the ways in which these modes of participation in the community's practice increases participants' own sense of identity and belonging in positive ways.

Rather than creating one large virtual community to serve an institution's entire military-affiliated student population, it may be more useful to form multiple smaller online groups to better attend to MALs' specific needs and interests. Similar to the Marines' MCGs analyzed in this study, social media groups could be formed by characteristics deemed useful to each institution's local context: for example, a military-affiliated group for women, or a group formed around an academic major, or a group for MALs who are highly concentrated at university's remote satellite campus.

When creating military-affiliated social media groups, institutions should privilege community frameworks that cultivate informal, social learning environments. Veterans centers may want to enlist military-affiliated members through a grassroots approach to help create, recruit, screen, and administrate the social media group. One way this could be accomplished is by establishing a digital task force composed of the center's military-affiliated student employees. These employees will be best positioned to understand the needs and experiences of students that the group will serve. Furthermore, they may already have a virtual network of military-affiliated academic peers to tap into as the social media group is developed and begins to grow. Because these groups should ideally be formed around the characteristics that are specific to each institution's military-affiliated population, the centers' military-affiliated student employees are best positioned to offer their insights and play a valuable role in the group's development.

Finally, the design and operation should work in concert to share reliable resources and information, as well as to cultivate camaraderie, enabling mutual trust and respect to evolve through participation. Results from this study indicate that participants are interested in building knowledge, sharing their experiences, and participating in mentorship. Social media groups should offer members opportunities to support one another in response to the day-to-day challenges of military-affiliated student life. Additionally, social media groups should create privacy settings and processes for screening new members. Rather than adding an authority figure to the group to moderate participation, which could negatively impact the potential for a dynamic peer-learning environment, rules or expectations for participation within the group should be published to the group and along with the understanding that participants have a collective responsibility to abide by these rules or expectations. As Marines unofficial MCGs formed around professional development demonstrated, even without official oversight and formal hierarchy, their participation embodied the enduring core values of military service, which include integrity, honor, and respect.

Facilitate Peer Engagement to Build Classroom Learning Communities

Transition support does not end with campus organizations; there should be continuity of support that flows into classrooms as well. By creating classroom learning environments that adopt the characteristics of Marines' virtual communities—blending social and professional engagement, flattening classroom hierarchy, encouraging collaboration, and inviting participant engagement—we can support MALs by leveraging their prior communication and collaboration experiences in military-affiliated groups and have the added benefit of "leveling up" the learning experiences of all students in our classes.

Within classroom learning communities, we can help strengthen participant engagement and membership in two ways. First, we must enable members' prior experiences to contribute to the community's practice. By doing so, we recognize the value it adds to the learning environment. This facilitates the second objective, which is to provide MALs with clear paths to participation. One way we can support MALs' paths to participation into our classrooms is through the use of small learning groups, or *student squads*, which is a strategy I have used with success for more than a decade to cultivate communities of practice that support traditional and non-traditional students in both on-campus and online undergraduate writing classes.

In the military context, a squad is a small and specialized group of Marines within a battalion who are trained and ready to be deployed. Together they have endured hardships, overcome adversity, and built camaraderie (Ricks, 1997). At the start of the semester, I disseminate a digital survey to identify students' prior professional and academic experiences as well as their learning and communication preferences, and I use the survey results to form the squads. This does not mean placing all MALs into one group; instead, I look for characteristics that will bridge the military-civilian gap. In my first-year composition courses, which are required for all undergraduate students, I often take academic majors into account when forming groups. I have found these groups quickly established a sense of shared identity and mutual understanding formed around the unique rigors of their disciplines and their similar professional goals. As an added benefit, it has afforded students opportunities to share subject-matter expertise when researching and writing about topics specific to their disciplines and professional fields. To help members quickly establish this common ground, I facilitate the work built into MCGs through their design and operation: I provide regular opportunities for groups to collaborate in low-stakes activities to accomplish learning objectives, and upon their first meeting, I provide a one-page written report to each group that describes survey results they share in common as well as a few key differences that will strengthen them collectively.

Regular informal engagement—even if brief—is important to the cohesion of student squads. While MALs' prior experiences in hierarchical military environments may reflect a more traditional, formal model of learning in which the expert imparts knowledge on the novice, Marines' participation in MCGs indicates experience with informal, social learning as well. As a result, when we move away from the lecture model of instruction and decenter the classroom instructor as the expert, we invest MALs (and indeed, all students) in the activity of learning, and as a result, "taking charge of learning becomes the enterprise of a community" (Wenger, 1998, p. 272). When students participate, they become invested in the activity of learning, which enables them to cultivate identities of membership within our classroom communities. As a counterpoint to research that indicates "the student veteran often needs to have a clear impression of who is in charge in the classroom and who to go to with questions or concerns" (Coll & Weiss, 2015, p. 107), or as students who may struggle with a decentered model of learning (Weiss & Gross, 2013), findings from this study indicate that MALs are already successfully engaged in informal, decentered learning communities, and these environments bridge knowledge gaps and provide a valuable complement to their structured learning environments. When the classroom environment becomes a space of learning activity, "instruction does not cause learning; it creates a context in which learning takes place" (Weiss & Gross, 2013, p. 266), and students have the opportunity to become participants in the meaning-making activities, it results in an increased sense of membership within the learning community.

Because practice is central to the community, it is important to facilitate regular engagement, providing students with opportunities to draw on prior knowledge and experiences and participate in meaning-making activities, which are the "challenges and responsibilities that call upon the knowledgeability of students yet encourage them to explore new territories" (Wenger, 1998, p. 272). In my classroom, these activities occur weekly as low-stakes communication and community-building assignments at each stage of students' researching, writing, and revising processes. Through their ongoing support of one another, students contribute experiences and perspectives that enrich the learning dynamic within their student squads, and they become part of a practice that binds them together.

SUCCESSES AND LIMITATIONS

This study, while providing wide-ranging data and valuable insights, was not without its challenges and limitations. Some of the strategies I implemented, such as eliciting formative feedback from active-duty Marines as I designed the study, helped offset these challenges and limitations, and in the process helped me cultivate trust with research participants. Feedback from Marines helped me limit my own research bias and design research protocol that resonated with participants and better accounted for their diverse backgrounds and experiences. Additionally, obtaining research approval from the Marine Corps HRPP not only granted me formal access to my research community, but also helped establish credibility with key gatekeepers on base. Finally, I cannot discount the fact that my preexisting community membership also supported my data collection efforts. I had an innate understanding of Marine Corps culture and protocols, and I felt at ease with conducting research with Marines and on a Marine base.

However, even with all of these advantages, I still found recruiting active-duty Marines to be a challenging experience. While I felt the recruitment methods I employed were successful, I would have liked to extend the recruitment period, as 500 survey participants may have elucidated stronger statistical patterns that could not be identified from 200 participants alone. I also encountered unexpected difficulties recruiting interview participants, and if I were to expand this study, I would offer the option for participants to complete an interview questionnaire through email or Facebook messenger, and include the option to decline the screen-recorded social media tour portion of the interview. Several Marines who initially indicated interest in interviews eventually declined because they were wary of maintaining their anonymity. Even without the screen-recorded social media profile tours, interview questionnaires would have provided additional, useful insights that could not be gained through survey data alone.

Finally, while the broad scope of this study helped me illustrate a new landscape, it limited the depth of my findings and analysis as a result. This study is a first attempt to make Marines' social learning within social media more visible, but attention to the finer details of their communication and community-building experiences is still needed. Findings from this study confirms that there are many opportunities for further research.

CONCLUSIONS

MALs are already participating in informal, collaborative learning environments where they engage in meaning-making and mentorship activities. As administrators and faculty, we should acknowledge the opportunities to leverage these prior experiences to support MALs' transition onto our campuses and into our classrooms. In addition, we should encourage the use of social media technologies to facilitate the development of informal learning communities, while still attending to the ways face-to-face learning allows students to receive direct support. As technology continues to change the way service members and veterans connect and communicate with military and civilian audiences, further research will provide a clearer understanding of how they leverage this technology to support their time in service as well as help them transition from the military into new civilian communities.

This study forges new territory into MALs' communication and community-building experiences prior to their transition into higher education. This study represents one aspect of social media use by a subsect of service members within a larger military-affiliated learning population, and in doing so, it provides a useful starting point for analyzing the self-directed, social learning experiences that MALs bring to higher education. The small data set presented within this chapter is illustrative, rather than generalizable, but it offers a new perspective on the strengths and abilities that MALs bring to higher education, and it indicates that social media offers a rich resource for further research. Scholars who wish to take up this line of inquiry could replicate a social media study with other branches of the military, recruiting Army, Air Force, Navy, or Coast Guard participants. Scholars could also replicate this study with a focus on subcommunities within the military, such as female service members, for example, to better understand how traditionally underrepresented military-affiliated members build support networks. Likewise, a longitudinal study of participants' engagement in military-connected groups would offer deeper insights into the ways in which community membership and identity evolve through sustained engagement, and how the collective knowledge and purpose of the community continues to evolve in response.

NOTES

1. Pseudonyms were assigned to all interview participants.
2. This study defines military-connected social media groups (MCGs) as subcommunities for current or former service members or about military topics.
3. Within the survey, the following language was used to describe survey terminology: *Official*: "created or managed by the U.S. government or military branch or unit." *Unofficial*: "created by an individual or group without government/military oversight." *Public*: "anyone can view and participate." *Closed*: "searchable; permission from administrator to join." *Secret*: "not searchable; administrator or member must invite you to join."
4. An acronym used unofficially among Marines that stands for "new useless guy getting excellent training."

REFERENCES

American Council on Education. (2018). *Toolkit for veteran friendly institutions*. https://www.acenet.edu/Documents/Veterans-Toolkit-2018.pdf.

Arminio, J., Grabosky, T. K., & Lang, J. (2015). *Student veterans and service members in higher education*. Routledge.

Avendano, U. (2014, January 16). Making the grade at MAG-13: How MAG-University will better equip Marines into the future. *Defense Visual Information Distribution Service.* https://www.dvidshub.net/news/119278/making-grade-mag-13-mag-university-will-better-equip-marines-into-future.

Bazeley, P. (2013). *Qualitative data analysis: Practical strategies*. Sage.

Blaauw-Hara, M. (2016). "The military taught me how to study, how to work hard": Helping student veterans transition by building on their strengths. *The Community College Journal of Research and Practice, 40*(10), 809–823.

Blaauw-Hara, M. (2017). "Learning shock" and student veterans: Bridging the learning environments of the military and the academy. *Composition Forum, 35*, 12.

Bender, A. (2014). Exploring student-veteran expectations about composing: Motivations, purposes, and the influence of trauma on composing practices. In S.

Doe & L. Langstraat (Eds.), *Generation vet: Composition, student veterans, and the Post-9/11 university*. University Press of Colorado.

Branham, C. (2016). *Wounded warrior or war hero? Or maybe, neither?: Resisting common tropes of the veteran and developing digital literacy practices via narrative building and identity presentation in social networking spaces*. Doctoral Dissertation, University of Central Florida.

Buck, A. (2012). Examining digital literacy practices on social network sites. *Research in the Teaching of English, 47*(1), 9–38.

Burden, M. C. (2006). *The blog of war*. Simon and Schuster.

Carter, G., & Williamson, B. (2010). Diogenes, dogfaced soldiers, and deployment music videos. *dot mil: Rhetoric, Technology, and the Military. Special issue of Kairos, 14*(3). http://kairos.technorhetoric.net/14.3/topoi/carter-williamson/index.htm.

Coll, J. E., & Weiss, E. L. (2015). *Supporting veterans in higher education: A primer for administrators, faculty, and advisors*. Lyceum Books.

Cramton, C. D. (2001). The mutual knowledge problem and its consequences for disperse collaboration. *Organization Science, 12*(3), 346–371.

Creswell, J. W., & Creswell, J. D. (2018). *Research design: Qualitative, quantitative, and mixed methods approaches* (5th ed.). Sage.

Daniels, C., Grove, D., & Mundt, E. (2006). Command and communities of practice. *Air & Space Power Journal, 20*(4), 52–62.

Davies, J. (2012). Facework on Facebook as a new literacy practice. *Computers and Education, 59*(1), 19–29.

DiRamio, D., & Jarvis, J. (2011). *Veterans in higher education: When Johnny and Jane come marching to campus*. John Wiley & Sons, Inc.

Doe, S., & Doe, W. W. (2013). Residence time and military workplace literacies. *Composition Forum, 28*(Fall), 151–183.

Doe, S., & Langstraat, L. (Eds.) (2014). *Generation vet: Composition, student veterans, and the Post-9/11 university*. University Press of Colorado.

Dubé, L., Bourhis, A., & Jacob, R. (2005). The impact of structuring characteristics on the launching of virtual communities of practice. *Journal of Organizational Change Management, 18*(2), 145–166.

Handy, C. (1995). Trust and the virtual organization. *Harvard Business Review, 73*(3), 40–50.

Hart, D. A., & Thompson, R. (2013). 'An ethical obligation': Promising practices for student veterans in college writing classrooms. *Results of a 2011 CCCC Research Grant.*

Hamrick, F. A., & Rumann, C. B. (Eds.) (2013). *Called to serve: A handbook on student veterans and higher education*. Jossey-Bass.

Hawthorne, B. A., Bauman, M. C., & Ross, L. E. (2013). Student veterans organizations and student self-advocacy. In F. A. Hamrick & C. B. Rumann (Eds.), *Called to serve: A handbook on student veterans and higher education* (pp. 221–252). Jossey-Bass.

Headquarters Marine Corps Marine & Family Programs. (2017). *The Marine Corps demographics update*. http://www.usmc-mccs.org/mccs/assets/File/2017%20Demographics%20Booklet%20Jun(1).pdf.

Hildreth, P., Kimble, C., & Wright, P. (2000). Communities of practice in the distributed international environment. *Journal of Knowledge Management, 4*(1), 27–38.

Hinton, C. (2013, Fall). 'The military taught me something about writing': How student veterans complicate the novice-to-expert continuum in first-year composition. *Composition Forum, 28*(Fall), 1–16.

Iverson, S. V., & Anderson, R. (2013). The complexity of veteran identity: Understanding the role of gender, race and sexuality. In F. A. Hamrick & C. B. Rumann (Eds.), *Called to serve: A handbook on student veterans and higher education* (pp. 89–113). Jossey-Bass.

Kretzmann, J. P., & McKnight, J. L. (1993). *Building communities from the inside out: A path toward finding and mobilizing a community's assets.* ACTA Publications.

Kietzmann, J., Plangger, K., Eaton, B., Heilgenberg, K., Pitt, L., & Berthon, P. (2013). Mobility at work: A typology of mobile communities of practice and contextual ambidexterity. *Journal of Strategic Information Systems, 3*(4), 282–297.

Lave, J., & Wenger, E. (1991). *Situated learning: Legitimate peripheral participation.* Cambridge University Press.

Liao, T. H. (2017). Developing an antecedent model of knowledge sharing intention in virtual communities. *Universal Access in the Information Society: International Journal, 16*(1), 215–224. https://doi-org.mutex.gmu.edu/10.1007/s10209-016-0452-5.

Mihas, P. (2014, May 22). *Load-bearing codes: Coding the connections* (Paper Presentation). International Congress of Qualitative Inquiry, Urbana, IL, United States.

Miles, M. B., & Huberman, A. M. (1994). *Qualitative data analysis: An expanded sourcebook.* Sage.

Naphan, D. E., & Elliott, M. (2015). Role exit from the military: Student veterans' perceptions of transitioning from the U.S. military to higher education. *The Qualitative Report, 20*(2), 36–48. http://www.nova.edu/ssss/QR/QR20/2/naphan3.pdf.

Pelts, M. D., Rolbiecki, A. J., & Albright, D. L. (2015). Wounded bonds: A review of the social work literature on gay, lesbian and bisexual military service members and veterans. *Journal of Social Work, 15*(2), 207–220.

Penich-Thacker, D. (2010). Don't ask, don't tell. *dot mil: Rhetoric, Technology, and the Military. Special Issue of Kairos, 14*(3). http://kairos.technorhetoric.net/14.3/disputatio/penich-thacker/index.html.

Rahbek-Clemmensen, J., Archer, E. M., Barr, J., Belkin, A., Guerrero, M., Hall, C., & Swain, K. E. (2012). Conceptualizing the civil–military gap: A research note. *Armed Forces & Society, 38*(4), 669–678.

Ricks, T. E. (1997). *Making the Corps.* Scribner.

Rodman, L., & Trespalacios, J. (2018). Knowledge-sharing and potential virtual communities of practice in the U.S. Coast Guard's Afloat Community: A qualitative pilot study. *TechTrends, 62*, 602–611.

Saldaña, J. (2015). *Thinking qualitatively: Methods of mind.* Sage.

Saldaña, J. (2016). *The coding manual for qualitative researchers.* Sage.

Silvestri, L. E. (2014). *Friended from the front: Social media and 21st century war.* Doctoral Dissertation, Iowa Research Online. doi:10.17077/etd.pmbeen3g.

Simmons, C., Fisher, A. K., & Simmons, B. (2015). Cognitive complexity and student veterans. In J. E. Coll & E. L. Weiss (Eds.), *Supporting Veterans in Higher Education: A Primer for Administrators, Faculty, and Advisors* (pp. 30–54). Lyceum Books.

Sorrells, A. M., Wills, S., Reagins-Lilly, S., Cate, C., & Moe, J. (2015). Veterans integration to academic leadership: A collaborative university-U.S. Department of Veterans Affairs outreach program. In J. E. Coll & E. L. Weiss (Eds.), *Supporting veterans in higher education: A primer for administrators, faculty, and advisors* (pp. 199–215). Lyceum Books.

Tashakkori, A., & Creswell, J. W. (2007). Editorial: Exploring the nature of research questions in mixed methods research. *Journal of Mixed Methods Research, 1*(3), 207–211.

Tashakkori, A., & Teddlie, C. (1998). *Mixed methodology: Combining qualitative and quantitative approaches*. Sage.

Tashakkori, A., & Teddlie, C. (Eds.) (2010). *SAGE handbook of mixed methods in social & behavioral research* (2nd ed.). Sage.

United States Marine Corps. (n.d.). Social media. *Marines: The Official Website of the United States Marine Corps*. https://www.marines.mil/News/Social-Media/.

U.S. Department of Veterans Affairs. (2019). *Annual benefits report*. https://www.benefits.va.gov/REPORTS/abr/docs/2018-abr.pdf.

Weiss, L., & Gross, G. (2013). Postmodernism and education. In J. Ainsworth & G. Golson (Eds.), *Sociology of education* (pp. 594–596). Sage.

Wenger, E. (1998). *Communities of practice: Learning, meaning, and identity*. Cambridge University Press.

Wenger, E., McDermott, R. A., & Snyder, W. (2002). *Cultivating communities of practice: A guide to managing knowledge*. Harvard Business School Press.

Chapter 14

Collaborative Apprenticeship Approaches to Writing Instruction in the Military

Ashley Ludewig

Though the number of military-affiliated learners (MALs) enrolled in colleges and universities has grown steadily since the passage of the *Post-9/11 Veterans Educational Assistance Act of 2008* (henceforward the *Post-9/11 GI Bill*), research about MALs has been somewhat limited (and limit*ing*) in scope, relying primarily on teachers' anecdotes about MALs in their classes. Currently, the literature narrowly focuses on resistance, conflict, and underpreparedness. This literature often describes MALs who were reluctant to participate in personal writing assignments, who struggled with the transition to higher education, and who were in need of some kind of intervention (Doe & Langstraat, 2016; Hart & Thompson, 2013). A similar reliance on the deficit model can be seen in research on MALs more broadly in higher education, which is often devoted to outlining disadvantages MALs face (Bonar & Domenici, 2011; Elliott et al., 2011; Madaus et al., 2009). The early work on MALs seemed to focus on issues of administrative and staff support, such as advising, student services/affairs, and disability resources rather than on strategies for academic support, especially in the specific context of writing and communication. This research often makes recommendations about what universities can do to raise and then sustain MAL retention rates but considers little else about how MALs learn or write.[1]

A closer examination of the experiences of MALs in the context of writing and communication is necessary—at least in part—because so much of academic success is grounded in a student's ability to participate in academic discourse, whether through formal assignments or by navigating administrative structures on campus. Moreover, the argument can be made that writing-intensive courses can have considerable influence on MALs' educational experiences due to their relatively small class sizes and the personal, potentially emotional nature of the course content. If MALs take these courses early in their college careers, as in the case of first-year Composition courses, then these experiences may have an even more significant effect on persistence and degree completion. Thus, our efforts to better understand and serve MALs within the first-year Composition classroom may have a tremendous impact on overall success.

As the more specific educational experiences of MALs have started garnering more focused attention, it is worth noting that the themes of trauma, emotional strain, and conversion consistently run through many major publications and special collections.[2] Thus, in spite of good intentions and an increasing focus on writing classrooms as a crucial site of learning for MALs, few of the MALs' early appearances in Rhetoric and Composition research did much to advance our understanding of their literacy practices. Up to this point, as Mallory and Downs (2014) later argued, "the majority of research on veterans in college actually *overlooks the very point:* student learning" (p. 53; emphasis in original).

There have been a few major publications focused on MALs in Rhetoric and Composition since the 2009 special issue of *Teaching English in the Two-Year College,* including a 2013 special issue of *Composition Forum*, a 2014 collection titled *Generation Vet: Composition, Student Veterans, and the Post-9/11 University*, and a 2016 special issue of *Reflections: A Journal of Public Rhetoric, Civic Writing, and Service Learning.* Generally speaking, much of the published research on MALs in Rhetoric and Composition has focused on the following themes:

- some discussion of creative, therapeutic, or community-based writing with MALs (Mapes & Hartley, 2016; Paquette et al., 2016; Schell, 2013; Schell & Klein, 2014; Springsteen, 2014, 2016; Whitworth, 2016);
- analyses of MALs' military and academic literacies (Cleary & Wozniak, 2013; Doe & Doe, 2013; Hadlock & Doe, 2014; Handley, 2016; Hinton 2013, 2014; McGregor & Fernandez, 2016; Mallory & Downs, 2014; Morrow & Hart, 2014; Thompson, 2014); and
- examples of programmatic approaches to supporting MALs (Doe & Langstraat, 2016; Hart & Thompson, 2016; Keast; 2013; Shivers-McNair, 2014; Selting, 2014; Sura, 2016).

Redirecting the conversation about MALs away from a deficit model or a focus on their affective experiences in college classes and with instructors or classmates means engaging in more thorough examinations of the issues represented by the second category; namely, their experiences with literacy.

After all, even though they may be "first-year students" in our classes, military students and MALs (like many of our students) are not necessarily "novices." To consider them as such is, according to Hinton (2013), "asking a lot from first-year MALs writers who bring a pre-existing expertise into the composition classroom that complicates embracing a novice position: a wealth of educational and professional experiences, ingrained cultural values, and a strong collectivist identity" (para. 28).

Although closer analyses of the literacy practices of MALs are beginning to take shape, the cumulative body of research remains rather small; what we do have suggests that we do not know as much as we could, or should, about the literacies of MALs. This chapter contributes to this conversation by answering the following research questions: (a) What are the literacy practices that characterize military discourses? (b) What are the resulting points of tension and connection that can shape MALs' experiences as they learn to write in academic settings? The emphasis of this chapter is not necessarily to point out the "gaps" that writing and communication instruction could or should fill, but rather to highlight the fact that MALs bring a host of literacies with them that can connect with and transfer into contexts they will encounter in higher education. Furthermore, MALs are employing complex literacy practices as they navigate higher education that are not inherently oppositional to traditional understandings of academic literacy practices (Hadlock & Doe, 2014; Mallory & Downs, 2014; Morrow & Hart, 2014). While this assertion has long been made in regards to adult learners more broadly (Beaufort, 1999; Gere, 2001; Lytle, 1991; Michaud, 2011), "student-veterans' former military workplace is generally less well understood by faculty and traditional students alike," which makes it harder for faculty to see the connections between military and academic experiences (Doe & Langstraat, 2014, p. 13). As an examination of military instruction practices, these findings can help us push back on deficit-model narratives about the skills MALs supposedly "lack." I contend that a more nuanced representation of military writing and its associated instructional practices can help educators understand the experiences of MALs and help MALs apply existing communication skills across new and varied contexts.

DEFINING KEY TERMS: LITERACY PRACTICES AND LITERACY EVENTS

Investigations of local, everyday language-based activities can reveal a great deal about larger webs of discursive practices and cultural values (Goldblatt, 2007). Barton, Hamilton, and Ivanič's *Situated Literacies: Reading and Writing in Context* (2000) defined *literacy practices* as "the general cultural ways of utilizing written language which people draw upon in their lives" and add that:

> practices are not observable units of behavior since they also involve values, attitudes, feelings and social relationships. This includes people's awareness of literacy, constructions of literacy and discourses of literacy, [and] how people talk about and make sense of literacy. (p. 7; also see Street, 2005)

Brandt (2001) took a similar stance, noting that "individual literacy exists only as part of larger material systems, systems that on the one hand enable acts of reading or writing and on the other hand confer their value" (p. 1). Ultimately, literacy practices are larger cultural patterns of discourse that can give us a fuller picture of the social attitudes and relationships among the users of that discourse.

Literacy events, on the other hand, are what Heath (1983) described as "any occasion in which a piece of writing is integral to the nature of the participants' interactions and their interpretative processes" (p. 93). Later, Barton et al. (2000) expanded that definition to all "activities where literacy has a role . . . Events are observable episodes which arise from practices and are shaped by them," effectively including virtually all forms of communication (p. 8). Essentially, literacy events are the singular encounters with language that over time come to form the boundaries of available literacy practices. As an individual engages in new literacy events (e.g., being assigned a new genre of writing or being tasked with an unfamiliar project in the workplace), they have the opportunity to apply their existing literacy practices, which may come more easily, and to acquire or create new literacy practices that are appropriate in the novel context, which may be more challenging.

I argue that by carefully examining the specific literacy events that MALs are able to recall, we can better characterize their literacy practices more broadly and resist focusing solely on what *abilities* participants have or lack. Instead, we can map patterns of literacy events over time—including patterns of how specific military literacies are taught—and the values and attitudes about literacy that are embedded in those patterns. Such insights are useful in two ways. The first is fostering a better understanding of military communication practices among faculty and staff in higher education. In doing so, we can be better prepared to recognize the strengths MALs bring into the classroom and to incorporate or be responsive to those experiences in the classroom. The second, closely-related benefit is that we can be more prepared to facilitate explicit conversations with MALs about the differences and similarities they will encounter as they engage with new academic literacies, and offer concrete strategies for productively pivoting from military to academic contexts.

METHODOLOGY

Description of Research Site

This study was conducted at a public, four-year, PhD-granting institution in the South with a total enrollment of approximately 23,000 students. At the time of the study, approximately 930 MALs attended the university.[3] The primary point of contact for MALs on campus is the Office of Military and Veteran Student Services (OMVSS), which is staffed by a VA Certifying Official and an Academic Counselor. Together, they help MALs navigate the process of applying for a variety of education benefits, transferring credits from other institutions, withdrawing from or being readmitted to the university due to deployments or calls to active duty, and making connections with other veteran-related services and student groups on campus.

Recruiting Participants

The OMVSS provided a list of 149 students who had self-identified as MALs and who had been tagged as being in their first semester of study, but only 25 of whom were true first-year students; 122 others transferred in with credits from other institutions and two others had incoming credit designations listed as "unknown." Through a combination of direct emails to individuals on this list and visits to first-year composition classrooms, I was able to recruit nine participants. While this sample size is relatively small, it is important to note that this study is rooted in qualitative methodology scholarship that asserts that having fewer participants does not inherently undermine the credibility of the researcher's findings (Dukes, 1984; Tracy, 2010). In this study, the goal was "not to generalize the information . . . but to elucidate the particular, the specific" (Creswell & Poth, 2018, p. 158), and interview data were gathered to represent the depth of experience in a specific context rather than breadth of experience across many contexts (Tracy, 2010, p. 845).

All nine participants were men; eight were white and one was Hispanic. All nine were enlisted; seven in the Army, one in the Navy, and one served in both the Marines and the Army. Because all of the participants were enlisted, it is important to note that these findings only reflect experiences with writing instruction that happened in relatively informal, "on-the-job" settings; that is, none of these participants recalled having been taught to write for the military in a classroom setting like a MAL in an ROTC program or at the U.S. Military Academy at West Point might experience. Participants' time in the military ranged from four to twenty-four years.

Data Collection and Analysis

The interview process in this study is based on what Seidman (2005) termed "in-depth, phenomenological interviewing" (p. 9), in which the data gathered is used to create profiles of individual participants and to identify themes (Owen, 1984) across a range of participants. This approach is undergirded by a phenomenological philosophy where I sought "to elucidate the meaning, structure, and essence of the lived experience of a phenomenon for a person or group of people" (Patton, 2002, p. 482). In this case, the "phenomenon" is the participants' experience learning to write in and for the military. By conducting interviews, I gathered details about the lived experiences of MALs as they encountered these new literacy events.

Interviews took place over the course of one academic year. Initial interviews and follow-up interviews were one hour long, semi-structured, and based on a series of primarily open-ended questions. Interview questions were a combination of general questions about the participants' experiences with writing[4] and more specific questions targeted at the major themes of the study; specifically, different styles of literacy sponsorship[5] and perceived tensions or connections between military and academic literacy practices.[6] I audio-recorded each interview for transcription purposes and all participants were assigned pseudonyms to protect their anonymity.

In reviewing the interview recordings, I relied primarily on *in vivo* coding (Creswell & Poth, 2018, p. 193) to identify "structural invariants" (Dukes, 1984, p. 201) across the participants' accounts of their experiences with literacy broadly as well as the role that specific sponsors like the military and the academy have played in their learning. Using constant comparison, I was able to identify a clear pattern of "significant statements" (Creswell & Poth, 2018, p. 201) in the recollections participants shared about learning to write in the military and then to reconstruct that essence through direct quotes from multiple participants. What emerged from that composite description was an apprenticeship model of military writing instruction that also fostered collaboration between novices and experts.

RESULTS AND DISCUSSION

In what follows, representative quotes from participants illustrate how their experiences can be characterized as collaborative apprenticeship and how that process reflects and supports both the military's hierarchical structure and its value system. Using theories of collaborative learning and apprenticeship to frame those experiences, I explain (a) how military writers engage in multiple forms of collaboration during the writing process, especially while in the role of an "apprentice," but also while in other roles, and (b) how a collaborative apprenticeship model of military writing instruction is both a means of instilling military values and preparation for sharing in the labor and stakes of military communication.

Defining and Complicating Collaborative Learning

Participants' stories about learning to write in and for the military suggest that military writing culture sometimes follows a model of collaboration described by Kenneth Bruffee in his seminal works "Collaborative Learning and the 'Conversation of mankind" (1984) and *Collaborative Learning: Higher Education, Interdependence, and the Authority of Knowledge* (1999). There are three key tenets of Bruffee's (1984) theory of collaborative learning: (a) that language, conversation, and knowledge are all socially constructed and interdependent; (b) that education is a process of "reacculturation" in which students, through conversation (written, verbal, and internalized) transition into new communities of knowledgeable peers; and (c) that each community of knowledgeable peers has an accepted mode of conversing about, establishing, and maintaining knowledge that is their "normal discourse." Participants' reflections revealed the means by which that "reacculturation" is achieved and why it is so important that the military culture of writing follows these principles.

Calling specifically on the work of Richard Rorty in *Philosophy and the Mirror of Nature* (1979), Bruffee (2009) claimed that "any effort to understand conversation requires us to understand the nature of community life that generates and maintains conversation" and these communities, he explicated, rely on certain epistemological and discursive means for establishing and maintaining knowledge, which Rorty (1979) termed "normal discourse" (p. 550). This "community of knowledgeable peers," Bruffee (2009) continued, "is a group of people who accept, and whose work is guided by, the same paradigms and the same code of values and assumptions" (p. 552). In the case of military writers, the knowledgeable peers are one's superiors and they are responsible for initiating new recruits into the military discourse community.[7] In this context, even mundane communicative tasks become highly collaborative because participants share in the stakes of those endeavors and higher-ranking individuals also stand to benefit in multiple ways from the success of their subordinates.

Military Literacy Sponsorship as Collaborative Apprenticeship

Despite marked differences in their branches, ranks, lengths of enlistment, and ages, the participants of this study shared strikingly similar narratives of their experiences learning to write in and for the military. As all participants were enlisted, the vast majority of the writing instruction they recalled from their time in the military happened in informal workplace settings (i.e., not in a formal classroom space), and through interactions with their direct supervisors. These interactions involved intensive one-on-one instruction and task modeling and focused on correctness and mastery as their primary objectives. In some ways, the writing instruction process participants described appears to be a consistent and structural enactment of fairly standard understandings of literacy sponsorship, which Brandt (1998) defined as "any agents, local or distant, concrete or abstract, who enable, support, teach, model, as well as recruit, regulate, suppress, or withhold literacy-and gain advantage by it in some way" (p. 166). Sponsors are ultimately responsible not only for the ways an individual is introduced to new literacy practices but also how they employ them and the values they ascribe to them. As Brandt (2001) described, the relationship between sponsor and sponsored is a complicated one even in the most straightforward of situations:

> Usually richer, more knowledgeable, and more entrenched than the sponsored, sponsors nevertheless enter a reciprocal relationship with those they underwrite. They lend their resources or credibility to the sponsored but also stand to gain benefits from their success, whether by direct repayment or, indirectly, by credit of association. (p. 19)

Thus, literacy sponsorship is not as simple as a "more educated" individual "teaching" a "less educated" individual something new. The sponsor is inherently in a position of power. Moreover, the sponsor is contributing resources of some kind (time, money, reputation), and the sponsored individual is typically expected to do something with their new skill that will, at the very least, reflect well on the sponsor, or potentially even result in financial gains for the sponsor. Embedded in these relationships are nuanced and evolving distributions of social power and authority which are not always balanced or even advantageous to the sponsored individual.

Superiors initiate new military writers through direct instruction on specific documents,[8] thereby preparing them to write in ways that will be understood by their community and that will be beneficial to their careers. In that sense, there is a clear and pragmatic relationship between the skills being taught and the needs of the institution and the individual, which is not always the case in a college course. As such, the military serves as a powerful sponsor of literacy[9] (Brandt, 2009, p. xiii), and the exchange is straightforward in a way many MALs' appreciate; civilians join the military and receive specialized instruction (including in writing and communication) that may be of use to them after the end of their service, and in return, the military is staffed by highly trained individuals who may go on to bring any number of honors onto the institution itself.

According to participants, the military values precision and concision in language across a wide range of documents, and the stakes are often high as fellow service members' careers or even lives might be on the line. As a result, teaching subordinates to complete mandatory writing tasks accurately is an important part of the higher-ranking individual's role. Almost unanimously, participants described being taught how to write job-specific documents[10] for the military through individual instruction from their direct superior. Ranging from an hour or two to half a day or more, these sessions involved looking at examples of the higher-ranking person's past work, working on documents together, and discussing immediate critical feedback on the subordinate's efforts. Working side-by-side with their superior, the new writer learns not only what the military at large expects from them but also what their direct superior demands of them.

In the participants' reflections, it is apparent that direct mentorship and individual instruction were integral parts of the learning process for new military writers; even if they were able to suss out some of the expectations by looking at examples, they were still expected to practice with their superior before attempting the process on their own. This collaborative process, then, also serves as an apprenticeship process. Ching (2011) put in plain words the benefits of this model in the article "Apprenticeship in the Instructor-Led Peer Conference," in which he argued that student learning is enhanced by having the opportunity to work side-by-side with someone more experienced with the skills required by the task at hand. Ching (2011), falling more or less in line with Bruffee's (2009) theory, wrote that "typically, an apprentice learns how to perform tasks, or modified versions of them, with assistance from (or in close proximity to) a more experienced practitioner" (p. 102). Ching (2011) based their understanding of apprenticeship on Vygotsky's (1978) theory of the zone of proximal development, claimed that in order to grow beyond their current skill level, "learners must participate in collaborative problem solving alongside someone who is 'more 'capable' in some relevant way" (p. 107).

In the context of military writing, less experienced individuals "gain access to the meanings and values teachers ascribe to various features of student writing" and are positioned as "novice readers and writers working in close proximity to their . . . expert instructors" (Ching, 2011, pp. 102–103). In combination with Bruffee's (2009) theory of collaborative learning, an apprenticeship metaphor helped elucidate the ways in which new military writers are expected to inherit or internalize accepted writing practices by observing the "master" while also working *with* the "master" to craft messages and create meaning in appropriate ways. For example, when I asked Jackson[11] how he learned to write in the military, he said:

> You kinda just like, you'd have one of the older guys just sit down with you and like, teach you and then just . . . I mean you don't really get to do stuff like that for a while, so you just watch them. Like, you'll constantly sit down with them and do it. And then eventually they'll let you start doing that kind of stuff once they feel that you're confident and ready.

Jackson's explanation suggested that instruction was direct and individualized and that taking on new tasks independently came only after a significant period of both observing and being observed. John described the way he learned to write performance evaluation reports very similarly:

> [My superior] sat me down one time for like, I think like a half a day. We sat in his office and I went through them with him and he would tell me what he looked for and he would—'cause you know you'd have a regulation too to kind of help you out—and he did that for about a half a day and after that I was like, "Okay, I know what he wants."

Although John only referred to one instructional session, he does make it clear that a significant amount of time was invested in his learning and that it involved both models and a clear articulation of his superior's expectations. Later, John described his own teaching style as "baptism by fire," but when he actually offered details about his approach it seemed to be anything but:

> You know, I mean, and I'd get those ones that'd say "Look, well look, I've never wrote this before' okay?" And I would say "Well come in here, I have examples." and I'd be like "Well this is how it should structure, you know, you're, you'd have your" . . . We used to call it cause and effect. He did this cause, you know, supervised some kind of cause, something . . . and this is what happened. Your result of it. "He watched five people to this, it made him do everything faster," just for a simple example.

Indeed, rather than asking the junior service member to take a leap and attempt a writing task without much guidance the first time, which is suggested by the "baptism by fire" metaphor, John took the time to work through examples of particular documents and discuss specific requirements for language and formatting, much like his superiors had done for him.

Each of these reflections depicts a collaborative learning environment where junior service members look to a "community of knowledgeable peers," e.g., their superiors, and work side-by-side with them until they understand and can emulate the "normal discourse" of their military community (Bruffee, 2009, pp. 551–552). More specifically, through an intensive collaborative apprenticeship process with their direct superiors, new military writers not only learn the ins and outs of military discourse—that is, not only specific forms and genres but also accepted language and modes of communication more broadly—as well as the value systems and cultural norms of their military environment. These findings are in line with other recent research that suggests that tensions between military discourses and academic discourses can include contradictory approaches to concepts like independent inquiry and hierarchies of authority, which ultimately reflect specific differences in the larger institutional contexts of those discourses (Mallory & Downs, 2014).

Collaboration in Document Composition

The instructional process that participants described was not collaborative *only* in the sense that trainees learned how to complete specific tasks by working closely with a knowledgeable superior. The process is also collaborative in the more commonly understood sense, which is that all parties involved work together (to varying degrees) to compose the text and shape its meaning. One way this collaborative process is enacted is when participants have to pass documents up the chain of command for editing and revision. How far up the document has to go in order to be approved depends on the significance of the document; recommendations for a significant award, for instance, need to be signed off on by superiors farther up the chain than a typical end-of-job report, for example. To these participants, passing something up the chain meant more than seeking approval for the document wholesale; rather, each reader physically marked up the document, often with red ink and/or more fully-fledged notes. While this process initially resembles common "peer review" practices in academic contexts, there are some important differences. Based on participants' comments, ownership of texts was perceived as collective rather than individual and feedback was more directive than suggestive.

Participants used words like "check" to describe the reviewer's job and one participant claimed that it was very different from the type of response one might get from a writing professor. Different, that is, because as the use of red ink suggested, review of these documents was sometimes harsh and documents would come back down the chain filled with marks and suggestions;

the worse a writer the author was perceived to be, the more ink he found on return of his draft. One participant explained, "I had ones where it'd look like . . . like you took a red highlighter and just did this [scribbling gesture]," suggesting it was not uncommon to have a document returned with significant changes.

Of course, not all documents needed so much revision and not all reviewers would be so fierce as editors, but participants' descriptions of this process made it clear that there was *always* something to be picked at:

> There's kind of like a . . . I don't know if you'd say a culture or a stigma, that no matter what you send up to like the battalion level staff for awards and evaluations, you're never gonna get perfect. Ya know, it's like, "Ya need to switch this period." And ya know, it's different than having an English teacher critique a paper. [. . .] You typically send it up two or three times to get reviewed before it makes its way through.

In other words, military writers learned to expect that editorial changes were made any time a document was reviewed by their superiors and that the process could turn out to be a lengthy one. Moreover, rather than seeing the writing task as open-ended or feeling like they ultimately had agency over the text as they might in an academic context, the feedback process participants described here reinforced the idea that there was a "right" approach and that they must concede to their superiors' demands.

Overwhelmingly, the concerns participants described finding marks on their work (and marking on others' work) were related to surface-correctness: spelling, punctuation, formatting corrections, syntax, and so on. And yet, the number and kinds of changes suggested both have the potential to significantly alter the document and its message, underscoring how the writing process was almost never an individual effort. As Mark accounted:

> Over time, as you would submit your paragraphs and your summaries, they would make certain changes to it. So you'd have usually one or two people in your chain—from where I was sitting—one or two people above me that would look at 'em and bless off or no and make corrections and add to it.

Mark's description of what happened when a document was passed up the chain of command is significant in that he goes beyond approval/rejection and surface corrections to include additions (and presumably subtractions) from a text. Although there is not much opportunity for creativity in organization or structure, there is certainly still room for personal choices of expression, and it is in that sense that demands for editing and revision open the door for a re-negotiation of meaning with one's superiors. Mark's statement highlighted the collaborative nature of military writing even well beyond the apprenticeship phase and highlights the extent to which the collaborative apprenticeship model of instruction prepares military writers for the review process that will continue throughout their career.

Collaboration in High-Stakes Writing and Collective Responsibility

Collaboration between masters and apprentices in the military is a reflection of both the hierarchical structure of the military and the value of consistency of protocol in that environment. In the following sections, I illustrate how the collaborative apprenticeship model participants described also serves to prepare them for shared responsibilities. Ultimately, in yet another divergence from the relationships and stakes common to university classrooms, the culture participants described also featured a shared investment in the everyday operations of their units and shared responsibility for the outcome of their communicative tasks.

The culture of ongoing collaborative apprenticeship in the military also has a great deal to do with the wide and significant audiences for military texts and the high stakes of the writing some members of the military do. The stakes are more obviously considered "high" as military personnel climb in the ranks or move into more specialized military occupations; however, all participants brought up concerns that suggest a keen sense of both audience awareness and the stakes of their work that frequently goes beyond whether they might be perceived as "good" or "bad" writers. Thus, the collaborative apprenticeship model also acts as a structured way to prepare military writers for the gravity of the communication they will be responsible for in the future.

For example, in the latter part of his career as an Anti-Terrorism Intelligence Specialist, Mark wrote reports in which he prioritized and summarized large amounts of large raw intelligence and made recommendations to Army commanders about possible courses of action. He shared,

> We would take a certain subject, do a lot of research on a certain subject and then write a detailed analysis of a subject . . . an entity . . . that we want to either develop more information on—so we would recommend to our command, and ultimately going up the chain of command, what resources we would like to see put against a certain target or entity to gather more information—or if it . . . it's something we think ultimately should be taken out.

Mark clarified that "[Intelligence Summaries] would then get published in a daily report command-wide up and out. So, anybody can read it if they just log onto the website," making the quality of his work of paramount importance as his recommendations could have significant impacts on commanders' decisions and he was open to constant scrutiny. Long-standing collaborative apprenticeship with his superiors not only prepared Mark to fulfill the significant demands of this kind of writing but also meant a structure was in place for internal review and quality checking before a text was widely circulated.

Chris had a similar experience composing Area Studies for a Civil Affairs Unit in the Army. During his tours in Afghanistan, Chris and his team were responsible for gathering as much knowledge about the local culture as possible in order to guide the American military through interactions with civilians and leaders in Afghani communities. Chris remarked,

> I learned to be really analytical with writing reports for that. [How so?] Umm . . . You tend to leave out a bunch of fluff. So, you get straight to the point, you learn what's important, how to structure a good paragraph or good report. Your reports aren't just being read by some lieutenant, they're being read by colonels and generals and congressmen and etcetera.

Like other participants, Chris' description of his military writing underscores the expectations of concision, directness, and structure and an acute awareness of a wide and powerful audience. The quality of the intelligence Chris and his team gathered and the clarity with which they conveyed it to their fellow soldiers influenced everything from policy decisions to negotiations with local leaders to resolving conflicts with civilians. Because the military relies so heavily on cooperation with local populations, improper or offensive interactions could have dire consequences, making the stakes for Chris' work very high and reinforcing the importance of strong writing apprenticeship as part of his military training.

Even participants who did more "mundane" or "every day" writing in the military remarked that they felt pressure to do that work well: pressure that was accentuated both by the training they received from their superiors and the institutionalized hierarchical review process for their writing. Accordingly, however simple the "form" might appear to be for an annual review document like an NCOER, award recommendation, or end-of-job report, the review process was intensive and demanded conscientious work. Participants described the importance of sending quality work up the chain because it reflected not only on oneself but also on one's immediate superiors and the individual being written about. John offered an example of how attention to detail in end-of-job reports and NCOERs included word choice: "We wouldn't put 'fail.' that's very bad. You know, that's very . . . If you tell someone they failed, you know that'd be a red flag. You'd be like 'continue to work on,' you know, or 'needs improvement'." Indeed, despite specific expectations from superiors and significant constraints on space and organization, participants also found creative ways to convey messages, sometimes with multiple layers of meaning.

These participants' comments are consistent with the results of a study completed by Hadlock and Doe (2014), who echoed some of Mallory and Downs' (2014) claims about MALs and their intensive experiences with particular forms of written communication in the military. That is, these revelations disrupt the assumption that all of that writing was done in lock-step fashion and delineate several ways that military discourse is rhetorically complicated, including its underlying motivations (e.g., advocacy and assertion) and built-in procedures for giving and receiving feedback. These findings suggest that there are far more connections between the literacy instruction and practices in the military and those of academia than one might assume and that MALs may very well be, perhaps with some guidance, well-prepared for the variety of rhetorical situations they face in the classroom.

The participants' understanding of the varied yet significant stakes of their writing not only highlights the role of strong and continuous collaborative writing apprenticeship, but it also points toward another reason *why* such an infrastructure is in place. The hierarchical structure of the military is such that no member is accountable solely to themself; rather, all members are responsible for one another in some way and higher-ranking officials are responsible for the actions of all those working below them. This is not to say that an individual will not face consequences for an error, but that an individual's direct superiors will also share in that responsibility. The same premise holds, then, for the consequences of the writing done in the military. As a result, to ensure that you are represented well by your subordinates and that their writing both has a desirable impact and does not reflect poorly on you as a leader and instructor, you must invest the time into training them well. Through both direct (synchronous, specific instructions about how to complete a writing task) and indirect (asynchronous response to/collaboration on a particular writing task) means, the military writing "experts" instill the principles of military discourse in their subordinates so that inadvertent errors in meaning do not bring unwanted consequences on all the parties involved.

CONCLUSION

Although labeling the military as a "literacy sponsor" is accurate, on its own the term occludes the exact nature of the relationship between new and experienced military writers. Looking specifically at the relationships described by these participants and understanding the concepts of collaboration and apprenticeship helps to illustrate the nuanced nature of those relationships and illuminate the pedagogy being enacted. The collaborative apprenticeship model serves to prepare military writers for the process of creating, reviewing, and revising texts that will continue for their entire military careers. That ongoing collaborative process also reinforces hierarchical structures and ensures that writing tasks are completed accurately, that the messages are rhetorically appropriate, and that undesired outcomes are avoided. This approach to literacy instruction can have a significant impact not only on how sponsored individuals write for the military but also how they will eventually teach others to write for the military *and* how they will expect writing instruction to happen in situations outside of the military. Despite being necessary in a military learning environment, the time invested in providing one-on-one/side-by-side instruction and the rigidity of the expectations these participants described are unlikely to be found in college classrooms, which can lead to frustration for MALs.

Fostering Explicit Connections and Conversations

A more nuanced understanding of military literacy practices can be used as a basis for explicitly addressing some MALs' unique expectations and needs. Additionally, this knowledge can facilitate opportunities for helping MALs reframe their experiences and employ their skills in new academic settings. For instance, passing documents up the chain of command for feedback is not altogether unlike a common peer review activity. While the situations are far from identical, comparing the two types of review activities as a *starting* point could demonstrate to MALs that their military experiences have value in academic contexts and that, in the case of this particular activity, they already have at least a basic understanding of its purpose. Given that the power dynamics between peers are vastly different from the power dynamics between superiors and subordinates in the military, the peer review process may need to be clarified for MALs.[12] With some guidance, MALs could be particularly adept at this kind of activity.

Encouraging Engagement with Support Structures

Additionally, I recommend using these findings to better recognize MALs who would benefit from more one-on-one instruction or direct modeling and encourage them to visit more during office hours or to take advantage of individual tutoring resources, whether they are offered in university writing centers or elsewhere. In this study, few participants were aware that their campus had a writing center with tutors from both undergraduate and graduate levels, meaning there was a significant missed opportunity for them to engage in the kind of one-on-one writing support they were accustomed to. The opaque nature of the university system can occlude the numerous means of support in place not only for MALs but for all students. On one hand, one could argue that it is the student's responsibility to reach out and ask about available resources. On the other hand, in the case of MALs, the act of asking for help can be particularly fraught with concerns about who the appropriate point of contact is and how asking for help will be perceived. And, even if MALs are aware that a resource exists, it may be even more important to them that there is clear communication about what kind of guidance they provide.

Working toward "Real-World" Contexts

While participants readily talked about the importance of rhetorical choices in their military writing (word choice impacting the future of a subordinate's career or an intelligence analysis informing decisions about troop movements, for instance), their reflections on academic writing revealed very little sense of "real life" impact. This is not much of a surprise considering the artificiality of much of the writing students are asked to do. Indeed, Hart and Thompson (2013) found that 71.3 percent of the 205 institutions surveyed still require personal narratives as part of their first-year writing curriculum (p. 9). Of the nine participants in this study, none of them had taken a writing course with a service-learning or community engagement component, suggesting that all of them had probably been asked to compose primarily in genres that could only imply *theoretical* stakes in a *hypothetical* rhetorical situation. But their artificiality is not always lost on the students engaging with them and that can impact not only the seriousness of their response but also the way students perceive their usefulness.

This frustration participants described over the value of imagined writing scenarios extended to genres we may assume are inherently "relevant" to all students. One participant, Alex, was enrolled in Technical and Science Writing when we met. When I asked how the class was going in general, he immediately shrugged and gave me a look that said, "Do you really want to know?" I pressed him to detail what this gesture meant, and he went on to say that although the course was required for his major, he felt like a lot of the writing he was doing was useless to him. His primary example was the résumé assignment. Not only had he completed a résumé during a course he had already taken in the Army, making the "practice" with a class assignment feel unnecessary, he also recognized that his military experience did not translate well into the more "traditional" style of résumé he was being asked to produce.

Alex's experience highlights two key concerns. First, the instructor seems to be making assumptions about the value of their assignment and relying (perhaps too heavily) on an implicit authority, both in their position as an instructor and in the assignment itself. This can be particularly frustrating for MALs who already have significant training and experience; as Leonhardy

(2009) warned, "instructors must first recognize that we have much to learn from veterans, just as we have much more to do for them" (p. 338). In this case, by taking for granted the usefulness of the assignment, the instructor also signals to students like Alex that they know what is most valuable for students without much input from the students themselves. Second, this disconnect signals to students like Alex that there is no "real-world" value to the writing he is producing, diminishing his investment in the task and the course.

Thus, I suggest doing more to honor and value the experiences of MALs. As Hadlock and Doe (2014) concluded, "To make the initial bridge, student-veterans may need to see both the importance and the applicability of their prior experience and the relevance of their current education to the significant civilian challenges that lie ahead" (p. 93). For some, as mentioned above, building this bridge involves allowing or encouraging MALs to write about military and war experiences, sometimes in deeply emotional ways, as part of composition coursework (Braswell, 1999; Corley, 2012; Martin, 2012). For others, this includes learning more about MALs strengths in the classroom (as conscientious small-group leaders, for instance) and allowing them to put those skills to use (Dalton, 2010; Doe & Langstraat, 2016; Mallory & Downs, 2014; Morrow & Hart, 2014).

Reframing Resistance

Another way in which this study contributes to larger discussions of the influence of existing literacies on academic experiences is by exploring the roots of the participants' anxieties about issues like "correctness" and "mastery." Several participants expressed a desire to master some aspect of academic writing (including grammar and APA or MLA style). Similarly, several participants remarked during their interviews that they sometimes wished their instructors would tell them exactly what they wanted in response to assignment prompts. Jackson, for example, said, "I like to just know what the teacher wants so I can just do that." For these students, a successful writing assignment was often one where they "gave the teacher what they wanted" or "told them what they wanted to hear," suggesting that what these students valued most was getting the "right" answer and the "good" grade, not learning to think in new ways or exploring a new topic of interest in an open-ended way. This response is due, in part, to the conditions under which these students wrote in the military. In Braden's words,

> Like they [college instructors] ask . . . Like, most of the writing I've done for my classes here are like, "What do you feel about this? How do you feel about this?" And you don't have to write about anything like that in the military. It doesn't matter at all.

I suggest reconsidering these comments in light of the reality that in the military, mastery over specific forms of writing can represent one's authority in a given situation or even open doors to new forms of power. From this angle, we can see how an emphasis on "correctness" and "mastery" in academic writing reflects the desire for a familiar way of demonstrating authority, asserting power, or meeting certain benchmarks of success. This is a useful perspective not only for understanding why MALs might seem preoccupied with correctness and mastery in academic writing but also why some MALs resist open-ended writing assignments or situations where they are being asked to consider an issue from multiple perspectives.

Casting these behaviors simply as resistance to critical inquiry and staunch adherence to dualistic thinking (e.g., right or wrong, all or nothing) would mean overlooking the sophisticated kinds of thinking and writing these participants described having done in the military. Instead, I argue that we can understand this resistance as the result of a history of situations where reaching the "correct" answer involved serious material stakes—whether in terms of one's social power or in terms of loss of life or costly resources. Again, it is crucial for faculty to engage in explicit conversations with MALs about academic understandings of agency, critical thinking, and correctness to help facilitate these distinctions from military contexts. Hadlock and Doe (2014) also reached this relatively common conclusion, which is that MALs may be best served by "direct explanation of the ideological differences between [the] two worlds" (p. 89; see also Mallory & Downs, 2014; Morrow & Hart 2014). In order to encourage MALs to pivot the skills they acquired in the military toward new academic contexts, there needs to be more open communication about the similarities and differences between military and academic institutional and ideological structures.

Limitations and Future Research

Admittedly, the findings of this study are limited by the number and type of participants that I was able to recruit. Despite the fact that the MAL population at this university was rather large (at least 932) and presumably somewhat diverse in terms of gender and race, I was only able to interview nine enlisted men, eight of whom were white. Therefore, the demographics of my participants are not representative of either the MAL population at this university or the military as a whole. Though broad generalization was not the goal of this study, this reality underscores the need to expand our research with a wider variety of military writers. Moreover, because all of the participants were enlisted, this study is narrowly focused on a very specific type of military literacy sponsorship that occurred only in the professional-yet-informal situations described. There are, of course, benefits to this narrow scope; by focusing on a particular type of literacy sponsorship I am able to give a much richer description of the relationship between sponsor and sponsored, the associated power dynamics, and the teaching strategies espoused. This is particularly true due to the uniformity of my participants' reflections. Nevertheless, the narrow focus of this inquiry also puts constraints on any claims I could make about military literacies and military literacy sponsorship more broadly, particularly those established in formal classroom spaces and manuals, for example.

One avenue for future research in this area is to complete more in-depth investigations of military writing instruction in different contexts. For instance, there would be value in comparing the experiences of these enlisted participants with other MALs whose training takes place in more formal settings like university ROTC programs or officer training institutions like West Point. This work could be especially useful in two ways. The first is fostering a better understanding of the breadth of military writing styles and pedagogies. In doing so, we can be better prepared to recognize and value the diverse experiences of MALs. The second is that we would be more prepared to facilitate explicit conversations with MALs about the differences and similarities they will encounter as they engage with new academic literacies and to offer concrete strategies for productively transferring their skills into new rhetorical contexts.

Future research in Veterans Studies would also benefit from more longitudinal observations of MALs as they navigate their college/university experience. The National Veteran Education Success Tracker report (Cate et al., 2017) is one version of this type of study in that it uses nationwide data to illustrate the educational patterns of *Post-9/11 GI Bill* veterans. The scale of their data makes it possible to track factors like degree completion and persistence, numbers and types of degrees earned, and patterns in the types of institutions MALs enroll in. Their study revealed promising trends more or less across the board, suggesting that MALs are—in some ways—out-pacing and out-performing their traditional counterparts. Their quantitative, data-driven approach helps make a compelling case for continuing *Post-9/11 GI Bill* benefits and against the assumption that MALs struggle in college, but it only reveals one element of the MAL experience.

To complement big-picture approaches to research on MALs, the field needs more in-depth investigations of the nuances of MAL success, including the impacts of existing MAL-focused initiatives on college campuses. In some ways, scholars in Writing Studies and Communication are particularly well-equipped for research in settings like our classrooms and writing centers; however, we can also employ critical frameworks to pursue similar lines of inquiry in a wide range of veteran-serving campus settings in order to investigate the extent to which MALs are being granted *access* to the full benefits of higher education and provided adequate *support* in their educational pursuits.

Finally, there is much more work to be done in training instructors, faculty, and administrators to work alongside MALs. Moreover, increasing support and advocacy for MALs in and out of the classroom, as well as developing best practices for instruction that permeate beyond our own disciplinary boundaries would greatly benefit the MAL community. A looming question for researchers is how we can establish sustainable interdisciplinary relationships that facilitate the support of MALs in other classrooms, departments, and even campus-wide contexts.

NOTES

1. This body of research is also distinct from Veterans Studies writ large, which has a much broader scope and includes research in the social sciences, medicine, history, literature, and cultural studies. This study is positioned at the intersection of the Rhetoric and Composition subset of Veterans Studies *and* the Veterans Studies subfield within Rhetoric and Composition.
2. For example, in a 2009 special issue of *Teaching English in the Two-Year College*, Galen Leonhardy's "Transformations: Working with Veterans in the Composition Classroom," discusses how to help MALs transform their narratives about war and trauma into effective academic compositions and offers suggestions for responding to assignments where veterans disclose some of their troubling experiences. Melanie Burdick's "Grading the War Story" and Sylvia A. Holliday's "Gladly Teach and Gladly Learn" both take up similar concerns about responding to student writing with distressing themes.
3. This number does not include other MALs who are not using the *Post-9/11 GI Bill* to attend college or students in the Air Force and Army ROTC programs.
4. For example, Did you like to read and write when you were younger? How would you describe yourself as a writer?
5. For example, What can you tell me about the kind of work you did in the military? Did you have much experience writing in the military? Do you remember how you learned to do that writing?
6. For example, What kind of criteria are important for you to be successful in a college writing course or on writing assignments in other courses? What skills, attitudes, or behaviors that you learned in the military have you been able to apply in college? Have any of these been obstacles rather than helpers for you?
7. Within the context of this study, it is impossible to know whether the superiors my participants mention received other forms of writing instruction (e.g., formal training at academies or military schools) during their own careers.
8. The two forms of military writing most frequently referenced by participants were (a) the Noncommissioned Officer Evaluation Report (NCOER), which is a yearly performance review completed by one's direct superior, and (b) the

Recommendation for Award form, which is used to recommend service members for honors like Commendation or Meritorious Service Medals. Individual participants also made note of other genres more specific to their own specialties, like maintenance logs, Intelligence Summaries, and Area Studies, which are detailed in more detail later in the chapter.

9. Defined as "agents, local or distant, abstract or concrete, who enable, support, teach, model, recruit, regulate, suppress, or withhold literacy and gain advantage by it in some way" (*Literacy and Learning* xiii).

10. By this, I mean that participants typically described moments when they were asked to take on a new level of responsibility that also required them to learn new genres of communication. For example, after a promotion, a service member's new duties might include writing performance evaluation reports; or, after moving to a position involving equipment maintenance, a service member might have to contribute to maintenance logs for the first time. It was in these kinds of situations that participants described working side-by-side with a more experienced individual to master these new genres. Although a few participants made very brief mentions of more formal sources of instruction (such as manuals), they *all* recalled these intensive sessions with supervisors first and returned to them most often in our conversations.

11. Pseudonyms were assigned to all interview participants.

12. Particularly in light of participants' comments about critiques of their writing in the military being overwhelming in number and often focused on surface-level issues.

REFERENCES

Barton, D., Hamilton, M., & Ivanič, R. (Eds.) (2000). *Situated literacies: Reading and writing in context.* Routledge.

Beaufort, A. (1999). *Writing in the real world: Making the transition from school to work.* Teachers College Press.

Bonar, T., & Domenici, P. (2011). Counseling and connecting with the military undergraduate the intersection of military service and university life. *Journal of College Student Psychotherapy, 25*(3), 204–219. doi:10.1080/87568225.2011.581925.

Brandt, D. (1998). Sponsors of literacy. *College Composition and Communication, 49*(2), 165–185. www.jstor.org/stable/358929.

Brandt, D. (2001). *Literacy in American lives.* Cambridge University Press.

Brandt, D. (2004). Drafting U.S. literacy. *College English, 65*(5), 485–502. https://www.jstor.org/stable/4140731.

Brandt, D. (2009). *Literacy and learning*. Jossey Bass.

Braswell, S. M. (1999). War stories: "Truth" and particulars. *War, Literature & the Arts: An International Journal of the Humanities, 2*(11), 148–156. https://www.wlajournal.com/wlaarchive/11_2/braswell.pdf.

Bruffee, K. (2009). Collaborative learning and the "conversation of mankind." In S. Miller (Ed.), *The Norton book of composition studies* (pp. 545–562). W.W. Norton and Company.

Bruffee, K. (1999). *Collaborative learning: Higher education, interdependence, and the authority of knowledge* (2nd ed.). Johns Hopkins University Press.

Cate, C., Lyon, J., Schmeling, J., & Bogue, B. Y. (2017). *National veteran education success tracker: A report on the academic success of student veterans using the Post 9/11 GI Bill.* nvest.studentveterans.org/wp-content/uploads/2017/03/NVEST-Report_FINAL.pdf.

Ching, K. L. (2011). Apprenticeship in the instructor-led peer conference. *Composition Studies, 39*(2), 101–119. http://www.jstor.org/stable/compstud.39.2.0101.

Cleary, M. N., & Wozniak, K. (2013). Veterans as adult learners in composition courses. *Composition Forum, 28.* compositionforum.com/issue/28/adult-learners.php.

Corley, L. (2012). Reconsiderations: "Brave words": Rehabilitating the veteran-writer. *College English, 74*(4), 351–365. www.jstor.org/stable/23212905.

Creswell, J. W., & Poth, C. N. (2018). *Qualitative inquiry & research design.* Sage Publications, Inc.

Dalton, K. S. (2010). *From combat to composition: Meeting the needs of military veterans through postsecondary writing pedagogy.* [Unpublished Master's Thesis]. Georgetown University.

Doe, S., & Doe, W. W. (2013). Residence time and military workplace literacies. *Veterans and Composition Forum, 28.* compositionforum.com/issue/28/residence-time.php.

Doe, S., & Langstraat, L. (Eds.) (2014). Introduction. *Generation vet: Composition, student-veterans, and the Post-9/11 university* (pp. 1–30). Utah State University Press.

Doe, S., & Langstraat, L. (2016). Faculty development workshops with student-vet participants: Seizing the induction possibilities. *Reflections: A Journal of Public Rhetoric, Civic Writing, and Service Learning, 16*(2), 151–184. http://reflectionsjournal.net.

Dukes, S. (1984). Phenomenological methodology in the human sciences. *Journal of Religion and Health, 23*(3), 197–203. https://www.jstor.org/stable/27505782.

Elliott, M., Gonzalez, C., & Larsen, B. (2011). U.S. military veterans transition to college: Combat, PTSD, and alienation on campus. *Journal of Student Affairs Research and Practice, 48*(3), 279–296. doi:10.2202/1949-6605.6293.

Gere, A. R. (2009). Kitchen tables and tented rooms: The extracurriculum of composition. In S. Miller (Ed.), *The Norton book of composition studies* (pp. 1081–1096). W.W. Norton & Company.

Goldblatt, E. (2007). *Because we live here: Sponsoring literacy beyond the college curriculum.* Hampton Press.

Hadlock, E., & Doe, S. (2014). Not just 'Yes sir, no sir': How genre and agency interact in student-veteran writing. In S. Doe & L. Langstraat (Eds.), *Generation vet: Composition, student-veterans, and the Post-9/11 university* (pp. 73–94). Utah State University Press.

Handley, D. G. (2016). Steal veterans and citizenship pedagogy in the first year writing classroom. *Reflections: A Journal of Public Rhetoric, Civic Writing, and Service Learning, 16*(2), 106–127. http://reflectionsjournal.net.

Hart, A., & Thompson, R. (2013). *An ethical obligation: Promising practices for student veterans in college writing classrooms*. National Council of Teachers of English. www.ncte.org/library/NCTEFiles/Groups/CCCC/AnEthicalObligation.pdf.

Hart, A., & Thompson, R. (2016). Veterans in the writing classroom: Three programmatic approaches to facilitate the transition from military to higher education. *College Composition and Communication, 68*(2), 345–371. https://www.jstor.org/stable/44783565.

Heath, S. B. (1983). *Ways with words: Language, life, and work in communities and classrooms*. Cambridge University Press.

Hinton, C. (2013). "The military taught me something about writing": How student veterans complicate the novice-to-expert continuum in first-year composition. *Veterans and Composition Forum*, 28. compositionforum.com/issue/28/novice-to-expert.php.

Hinton, C. (2014). Front and center: Marine student-veterans, collaboration, and the writing center. In S. Doe & L. Langstraat (Eds.), *Generation vet: Composition, student-veterans, and the Post-9/11 university* (pp. 257–281). Utah State University Press.

Holliday, S. A. (2009). Gladly teach and gladly learn. *Teaching English in the Two-Year College, 36*(4), 368–378. https://library.ncte.org/journals/tetyc/issues/v36-4.

Keast, D. (2013). A class for vets, not by a vet: Developing a veteran-friendly composition course at city college of San Francisco. *Composition Forum, 28*. http://compositionforum.com/issue/28/class-for-vets.php.

Leonhardy, G. (2009). Transformations: Working with veterans in the composition classroom. *Teaching English in the Two-Year College, 36*(4), 339–352. https://library.ncte.org/journals/tetyc/issues/v36-4.

Lytle, S. L. (1991). Living literacy: Rethinking development in adulthood. *Linguistics and Education, 3*(2), 109–138. https://doi-org.spot.lib.auburn.edu/10.1016/0898-5898(91)90002-Z.

Madaus, J. W., Miller, W. K., & Vance, M. L. (2009). Veterans with disabilities in postsecondary education. *Journal of Postsecondary Education & Disability, 22*(1), 10–17. https://www.ahead.org/professional-resources/publications/jped/archived-jped/jped-volume-22.

Mallory, A., & Downs, D. (2014). Uniform meets rhetoric: Excellence through interaction. In S. Doe & L. Langstraat (Eds.), *Generation vet: Composition, student-veterans, and the Post-9/11 university* (pp. 51–72). Utah State University Press.

Mapes, A. C., & Hartley, M. T. (2016). Re-authoring narratives: Reflective writing with veterans with spinal cord injury. *Reflections: A Journal of Public Rhetoric, Civic Writing, and Service Learning, 16*(2), 61–82. http://reflectionsjournal.net.

Martin, T. L. (2012). Combat in the classroom: A writing and healing approach to teaching student veterans. *Writing on the Edge, 22*(2), 27–35. www.jstor.org/stable/43157456.

McGregor, B., & Fernandez, L. (2016). Writing faculty on the marine corps base: Building strong classroom communities through engagement and advocacy.

Reflections: A Journal of Public Rhetoric, Civic Writing, and Service Learning, 16(2), 129–150. http://reflectionsjournal.net.

Michaud, M. (2011). The reverse commute: Adult students and the transition from professional to academic literacy. *Teaching English in the Two-Year College, 38*(3), 244–257. https://library.ncte.org/journals/tetyc/issues/v38-3.

Morrow, S., & Hart, A. D. (2014). Veterans in college writing classes: Understanding and embracing the mutual benefit. In S. Doe & L. Langstraat (Eds.), *Generation vet: Composition, student-veterans, and the Post-9/11 university* (pp. 31–50). Utah State University Press.

Owen, W. F. (1984). Interpretive themes in relational communication. *Quarterly Journal of Speech, 70*(3), 274–287. doi:10.1080/00335638409383697.

Paquette, P., Anderson, A., Burnette, L. T., Canady, J., Carr, B., Coakley, N., & Teamer, Y. (2016). A story worth telling: Sharing stories and impacting lives in the veterans' book group project at Fort Benning. *Reflections: A Journal of Public Rhetoric, Civic Writing, and Service Learning, 16*(2), 83–105. http://reflectionsjournal.net.

Patton, M. Q. (2002). *Qualitative research & evaluation methods* (3rd ed.). Sage Publications.

Schell, E. (2013). Writing with veterans in a community writing group. *Composition Forum, 28*. compositionforum.com/issue/28/writing-with-veterans.php.

Schell, E., & Kleinbart, I. (2014). "I have to speak out": Writing with veterans in a community writing group. In S. Doe & L. Langstraat (Eds.), *Generation vet: Composition, student-veterans, and the Post-9/11 university* (pp. 119–139). Utah State University Press.

Seidman, I. (2005). *Interviewing as qualitative research: A guide for researchers in education and the social sciences.* Teachers College Press.

Selting, B. The value of service learning for student-veterans: Transitioning to academic cultures through writing and experiential learning. In S. Doe & L. Langstraat (Eds.), *Generation vet: Composition, student-veterans, and the Post-9/11 university* (pp. 240–256). Utah State University Press.

Shivers-McNair, A. (2014). A new mission: Veteran-led learning communities in the basic writing classroom. In S. Doe & L. Langstraat (Eds.), *Generation vet: Composition, student-veterans, and the Post-9/11 university* (pp. 216–239). Utah State University Press.

Springsteen, K. (2014). Closer to home: Veterans' workshops and the materiality of writing. In S. Doe & L. Langstraat (Eds.), *Generation vet: Composition, student-veterans, and the Post-9/11 university* (pp. 140–155). Utah State University Press.

Springsteen, K. (2016). Veterans writing and a rhetoric of witnessing. *Reflections: A Journal of Public Rhetoric, Civic Writing, and Service Learning, 16*(2), 20–34. http://reflectionsjournal.net.

Street, B. V. (2005). At last: Recent applications of new literacy studies in educational contexts. *Research in the Teaching of English, 39*(4), 417–423. www.jstor.org/stable/40171646.

Sura, T. (2016). Articulating veteran-friendly: Preparing first-year writing instructors to work with veterans. *Reflections: A Journal of Public Rhetoric, Civic Writing, and Service Learning, 16*(2), 187–205. http://reflectionsjournal.net.

Thompson, R. (2014). Recognizing silence: Composition, writing, and the ethical space for war. In S. Doe & L. Langstraat (Eds.), *Generation vet: Composition, student-veterans, and the Post-9/11 university* (pp. 199–215). Utah State University Press.

Tracy, S. J. (2010). Qualitative quality: Eight "big-tent" criteria for excellent qualitative research. *Qualitative Inquiry, 16*(10), 837–851. doi:10.1177/1077800410383121.

Vygotsky, L. S. (1978). *Mind in society: The development of higher psychological processes*. Harvard University Press.

Whitworth, M. (2016). Writing to bear witness. *Reflections: A Journal of Public Rhetoric, Civic Writing, and Service Learning, 16*(2), 35–60. reflectionsjournal.net.

Chapter 15

Adapt and Overcome

Military-Affiliated Learners and Transfer of Learning

Catherine St. Pierre

In 2013, I taught Argument and Research Writing at the education center on a large Marine Corps base in the southeastern United States. In the same semester, I observed three instances of transfer, an occasion when "learning in one context or with one set of materials impacts on performance in another context or with other related materials" (Perkins & Salomon, 1988, p. 3). The three students' experiences showed reuse of class content in a military setting and reuse of military knowledge in an academic setting.

"Justin"[1] enrolled in the class while in the process of medically separating from the Marine Corps. Justin was struggling to compose a document about his injuries and needs for a Department of Veterans' Affairs (VA) board to review and determine his disability rating. He described struggling with this document, but he told me that our class discussion of *pisteis* and using appropriate appeals for different audiences helped him think differently about the task. "Since we talked about *pathos*, I get what the adviser meant about it being too dry, so I'm going to use lots of *pathos* appeals in my next draft." Justin reported transferring our class concepts to achieve his own purpose in a difficult composing task; what he learned in an academic context improved his writing in a military context.

Later that semester, "Walter," a veteran who retired after twenty-three years of service, told me that many concepts about persuasion were familiar from his training to become a Marine recruiter in the 1990s. He expressed surprise that his military experience was not only applicable but also helpful in academic writing. Walter transferred concepts from his military service to improve his work in college.

In contrast to these two examples of positive transfer, "Robert," a senior staff non-commissioned officer, experienced interference, or negative transfer, during an analysis project. Robert explained that in his Military Occupational Specialty (MOS), analysts are trained not to judge or make an argument because that might lead the analyst to ignore facts that do not align with the claim. Because of his eighteen years of analyzing without arguing, Robert reported it was difficult for him to analyze and make an argument. He wanted me to know that other students might experience similar challenges with the assignment.

These interactions demonstrated two important things: as an educator, I lacked knowledge of military discourse and literacy practices, and I needed better understanding of the ways practices in the military and in college affect one another. The semester highlighted to me the need to research transfer. For military-affiliated learners (MALs), military service and military discourses[2] represent important sources of existing or prior knowledge about communicating, speaking, and composing texts. Until now, the concept of *transition*[3] has received the most scholarly attention in post-9/11 veterans' studies (Blaauw-Hara, 2015; Corley, 2012; DiRamio, Ackerman, & Mitchell, 2008; Elliott, Gonzalez, & Larsen, 2011; Hinton, 2012; Rumann & Hamrick, 2009; Rumann & Hamrick, 2010; Sargent, 2009; Selting, 2014; Valentino, 2012; Voigt, 2016). However, for MALs studying and working in the post-9/11 university, *transfer* is also an issue worthy of study and consideration. Yancey et al.'s (2014) study of transfer treats prior knowledge as important, as it can positively support writers in new tasks and contexts, like Walter's use of persuasive techniques from recruiting duty. However, prior knowledge can also mislead learners if they misunderstand a new communication situation or misapply knowledge, like Robert's struggle with analysis and argument.

I begin by adopting an asset frame, viewing MALs through the lens of strength and contributions, instead of deficits or skill gaps (Cleary & Wozniak, 2013). The asset frame steps outside the binary portrayal of military-affiliated citizens as either heroes or ticking time bombs to inquire instead about students' capacities (Lieberman & Stewart, 2014). This correction is needed in higher education, because "many college approaches to student veterans oversimplify their experience and ignore their strengths as learners and individuals, approaching them instead from a deficit mindset that is focused on lack" (Blaauw-Hara, 2015, p. 112). In order to reject deficit models and adopt an asset frame (Lieberman & Stewart, 2014; Hart & Thompson, 2016) for MALs, teachers and scholars who want to holistically connect students' prior knowledge and skills with new content will be better positioned to recognize and develop students' contributions if they understand the assets that MALs claim to bring to campus. MALs in this qualitative study reported repurposing literacy practices, skills, and dispositions[4] sponsored by the military to academic and civilian contexts. MALs represented military professionalism, supported by time management, leadership, and being mission oriented as a source of their academic success. However, other military habits, such as using profanity and insults, clashed with academic discourse, exhibiting negative transfer. MALs negotiate the expectations of both discourse communities and institutions to achieve personal goals.

LITERATURE REVIEW

Transfer Overview

Perkins and Salomon (1988) proposed the dominant contemporary theory of transfer of learning, which "occurs when learning in one context or with one set of materials impacts performance in another context or with other related materials" (p. 3). Transfer can be positive if it improves learning, yet it can also be negative by interfering with new learning. Transfer is metaphorically "near" or "low road" when the conditions are similar enough to trigger automatic behaviors (e.g., driving a truck if you already know how to drive a car) or "far" or "high road" if the learner must abstract a principle, such as using a chess strategy in a military context (Salomon & Perkins, 1989).

Transfer is a foundational premise for communication and composition classes; stakeholders like administrators, faculty, and students expect instruction in communication and critical thinking to benefit students in other university classes, in workplaces, in civic spheres, and in many other contexts (Duffy, 2012; Wardle, 2007). Researchers have suggested that transfer is affected by the curriculum (Smit, 2007; Wardle, 2009; Yancey et al., 2014), the task (Thorndike, 1921; Woodworth & Thorndike, 1901), the context (Beach, 2003; Tuomi-Gröhn & Engeström, 2003; Wardle, 2007), and the individual's role and dispositions (Driscoll & Wells, 2012). This study investigates the role of the individual. Transfer is not only possible but can be enhanced by instruction, particularly by cultivating metacognition and reflection (Jarratt, Mack, Sartor, & Watson, 2007; Nelms & Dively, 2007; Reiff & Bawarshi, 2011; Rounsaville et al., 2008; Wardle, 2012; Yancey et al., 2014).

Beyond positive and negative transfer, however, is the problem of missed transfer, which was also called frustrated integration (Nowacek, 2011). Research suggests that untransferred knowledge is a problem (McCarthy, 1987; Smit, 2007). Students sometimes struggle to see the connections between classes; for example, McCarthy (1987) presented a case study demonstrating that a student, "Dave," was asked to summarize and analyze in three different classes, but he saw each task as unrelated and new. The problem of untransferred knowledge was revisited by Bergmann and Zepernick (2007) and Nelms and Dively (2007). Bergmann and Zepernick (2007) demonstrated that students who place low value on composition courses do not seek opportunities to transfer knowledge from the class; they placed higher value on writing done in the disciplines and higher transfer from the courses they valued more. Nelms and Dively (2007) observed a disconnect between concepts and themes composition faculty reported teaching and skills other faculty observed students transferring. Together, these studies suggest that the role of the individual is significant in transfer.

Rounsaville et al. (2011) show that the first-year students they surveyed have significant experience writing and composing in school, moderate experience with writing for work, and significant experience composing outside school and work. But when writing for first-year composition (FYC), students drew upon a small fraction of their experiences with writing from work or from outside genres. After affirming that FYC serves an important role in transitioning to writing in the university, Rounsaville et al. (2008) call for more research into prior knowledge and its uses:

> Future research into students' prior discursive resources might ask questions such as: To what extent, and in what diverse ways, do students bring these habits and strategies with them to FYW courses? What previous experiences and resources do they draw on? (p. 109)

Their findings on the importance of prior knowledge resonate with my own. By studying the ways military-affiliated writers report using prior knowledge from the military, I address some of the questions Rounsaville et al. (2008) raised. My research considers "prior knowledge" as a variable that influences transfer. Acknowledging that military experience is diverse and varied based on time in service, rank, MOS, and many other factors, MALs still share a common institutional literacy sponsor (Brandt, 1998). Although I do not claim my findings are universally generalizable, I articulate trends in how MALs characterized their experience of transferring and repurposing military skills and dispositions to academic contexts: namely, transferring professionalism, which is associated with time management, leadership, and being mission oriented.

MALs and Transfer

Many Americans want to find value and applications for military experiences in civilian contexts, which we see in programs like "Troops to Teachers," which operate on the premise that military service will carry benefits to civilian contexts. In *Progress in Educating Veterans in the 21st Century*, four veterans with established academic careers threw down a gauntlet for American educators and administrators: "By ignoring what veterans have learned in the military, our society essentially throws away the time and money invested into military training and experience that could be applied to . . . the civilian world" (Herrmann et al., 2011, p. 68). Herrmann et al. (2011) served in Vietnam, and saw their service demeaned or undervalued; I interpret this challenge as an appeal from a previous generation to do better by MALs of this century. In part, that means recognizing and valuing contributions made by MALs to civilian and academic environments.

The desire to recognize, repurpose, or transfer military training manifests in the success of media that argue that leadership and decision-making skills from the military can improve civilian leadership.[5] The desire to apply military training to benefit civilian society is also communicated in government partnerships and programs,[6] mainstream journalism (Klein, 2011; Beauchamp, 2015; Nicholson, 2019), and (of greatest relevance for our discussion) in composition and communication scholarship (Bender Smith, 2014; Blaauw-Hara, 2017; Carroll, 2008; Hart & Thompson, 2013, 2016; Hinton, 2013; Schell & Kleinbart, 2014). The wider cultural trend is to consider the value of leadership, teamwork, or discipline from the military in civilian society, yet I focus on communication and literacy-related transfer. Student veterans cultivate valuable skills, habits, and dispositions, such as professionalism, that transfer to academic contexts, providing prior knowledge they may leverage to succeed in higher education (Blaauw-Hara, 2015; Cleary & Wozniak, 2013; Hadlock & Doe, 2014; Hinton, 2013; Mallory & Downs, 2014; Selting, 2014). However, this prior knowledge is often remixed and transformed through contact with academic disciplines and contexts (Mallory & Downs, 2014; Yancey et al., 2014).

In 2009, education scholars DiRamio et al. (2009) completed one of the earliest peer-reviewed studies on post-9/11 veterans' transitions into higher education. They quote a participant whose remark they characterize as typical:

> I think it [combat/military training] helped me out a lot and it has given me a lot of self-discipline, establishing goals, time management, and everything. There are so many things you can get from the military to help you out as a college student. (Regular Army Service Person) (p. 7)

The traits named by this anonymous participant can be seen as dispositions that facilitate positive transfer between military and college situations. Discipline, setting goals, and managing time are represented as transferable and positive skills developed in the military. Time management and goal setting (under the label "mission oriented") were also identified by participants in this investigation as critical to their learning experience.

MAL Transfer and University Discourse

Some scholars in veterans' studies and composition are working to articulate potential connections between military habits and literacy practices with strength in composition. For example, Hinton's (2013, 2014, 2015) study of Marine student veterans and college writing argues that Marine student-veterans "possess a good amount of knowledge useful in college composition contexts; however, they were generally unaware of these transferable skills or knowledge until asked to articulate their experiences with writing in the Marine Corps" (Hinton, 2014, p. 267). Through the process of reflecting on their military service, MALs were supported in generating new knowledge.[7] Asking MALs to articulate and reflect on their experiences with communication and learning may promote metacognition, a key element of high road transfer.

Four studies on composition transfer for student veterans are especially relevant to research on transfer and MALs: Cleary and Wozniak (2013), Hinton (2013), Hadlock and Doe (2014), and Blaauw-Hara (2015). Cleary and Wozniak (2013) proposed an asset-based heuristic to consider student veterans' strengths. Hinton (2013) argues that participants connected their prior knowledge about learning and literacy from military service to college composition in four areas: topic selection, writer identity, individual writing style, and shared stylistic features. Hinton (2013, 2014) shows that Marine student veterans expect educational procedures from the military, such as consistent focus on leadership and transparency regarding expectations and evaluation criteria, to be part of civilian education. When it is not, student veterans expressed frustration. Hadlock and Doe (2014) argue that service members write in military genres and exert agency in more ways than the student veterans or the faculty who teach them realize. By presenting evidence from surveys, document analysis, and interviews with one dozen veterans, they challenge the stereotype that service members' discourse and actions are uniform, particularly in writing and thinking: "In working closely with and living alongside military service . . . members, we have witnessed their aptitude, agency, and critical thinking, all of which situate them to thrive in higher-education settings" (p. 74). Instead of assuming a deficit, Hadlock and Doe (2014) illustrate ways that MALs asserted agency through writing, such as advocating for others in evaluations, that serve as an asset in college. Finally, Blaauw-Hara (2015) extends Hinton's (2013) work on military teaching strategies and articulates three strengths service members bring to college: drive to complete the mission, enriched understanding of the world and its citizens, and a connection to the larger community (pp. 113–118). Blaauw-Hara (2015) draws from interviews with student veterans to argue that persistence, perspective, and community connections, all inculcated by the military, are strengths that facilitate student success in community colleges.

Collectively, these studies demonstrate MALs bring with them existing knowledge that not only provides a foundation for strong academic writing but overall academic achievement. Missing from these investigations and the collective literature, however, is an understanding of how a MAL's role and dispositions impact transference (Driscoll & Wells, 2012). To better understand how MALs define their role and inherent qualities in transfer, this research project is guided by the following research question:

> RQ: What can be learned about transfer of learning and prior knowledge by studying the experiences of student veterans?

Using existing theories of transfer to explain the instances of transfer research participants identified, the findings add to a growing body of work on post-9/11 veterans and MALs studying and working in the university by discussing transfer and the qualities to which student veterans attributed their success.

METHODS

To answer the research question (RQ: What can be learned about transfer of learning and prior knowledge by studying the experiences of student veterans?*)* and learn more about the literacy practices in the Marine Corps, college, and individual experiences with both institutions, I collected data using online surveys and semi-structured interviews. To be included in this research project, participants had to be currently enrolled in college and either active duty, reservists, or veterans of the USMC. Content analysis was used to analyze the data and identify common themes.

Recruitment

Survey Recruitment

I distributed the survey using personal and professional connections. I posted the survey on my social media accounts, emailed the link to acquaintances, and asked for help spreading the word from members of professional networks like colleagues at the institutions where I was a graduate student and an adjunct faculty member. Members of the Conference on College Composition and Communication Special Interest Group "Writing with Current, Former and Future Members of the Military" shared the link with students on their campuses. The survey was distributed via emails from eleven college and university VA offices (either the academic institutions were represented at the base education center where I taught or I had an affiliation with the institution as a student or faculty member). There were announcements in base bulletins for two weeks, and business cards with a QR code link and invitation to take the survey were available in the education centers and base libraries of two Marine bases in the southeast for three months.

Over a six-month data collection period, 138 people started the survey. One hundred and twelve Marine-affiliated learners completed the survey, equating to an 81 percent completion rate. Participants were active duty, reservists, and veterans. Some participants had multiple military affiliations as they were military spouses and children of service members in addition to having served in the military. The instrument did not specifically invite these multiple levels of affiliation, but I acknowledge them.

Interview Recruitment

Surveys provided a wider corpus of data and a starting point for understanding individual experiences for MALs; however, more information was needed to fully explicate their lived experience. Following the methods of Brandt (1998) and Selfe and Hawisher (2004), to learn more, survey participants were invited to participate in semi-structured interviews. I sent email interview invitations to survey respondents who completed every survey question (including discursive questions), served after 9/11/01,[8] had been stationed in eastern North Carolina,[9] and volunteered to speak to a researcher. I interviewed fifteen student veterans: three active-duty Marines, eleven veterans, and one reservist to gather more detailed information. The shortest interview was completed in forty-four minutes, thirty-eight seconds; the longest lasted eighty-three minutes and forty-seven seconds. Each interviewee either selected a pseudonym or was assigned one.

Instrument Design

In this section, I describe the survey and interview questions that provided information about military-affiliated writers' perceptions of transfer.

Survey

The survey began with a question to establish informed consent; this was the only question that a respondent could not skip. The survey was divided into five sections: informed consent, Marine Corps background (such as years of service, MOS, and B-billets) and memories of literacy practices in the Marine Corps (open ended), college/academic background (such as major, highest degree obtained, and years attended) and perceptions of college literacy practices, comparison and contrast, and follow up information. On the survey, I used open-ended questions (which used "if-then" display logic) to ask about transfer. For instance, the survey asked about thinking, speaking, reading, and writing in the Marine Corps and in college and provided opportunities for participants to reflect on these literacy practices. Participants selected the genres in which they read and wrote as Marines and students and reflected on their training and formal schooling and effects on their literacy practices.

On the advice of three student veterans who tested the survey and provided feedback on the user experience of the instrument, I did not collect demographic data on race, sex, gender, class, marital status, or military rank. The volunteers claimed that they are more likely to lie and are significantly less likely to complete a voluntary survey when these questions are asked, as they assume they and their views will be stereotyped, misrepresented, or distorted because it is viewed through demographic lenses. I considered asking the questions after the categories named, but ultimately respected the members of the community and their vehement recommendations not to collect that data.

However, survey participants answered questions about their careers, so facts are available about the participants. The oldest participant enlisted in 1962 and the youngest in 2014, and open-ended responses specifically mentioned engagements in Vietnam, Panama, Bosnia, Mogadishu, the Persian Gulf, post-9/11 conflicts in Iraq and Afghanistan, as well as time stateside on bases completing more routine training and maintenance. Survey respondents included thirty-eight active-duty Marines, sixty-eight veterans, and six reservists. They served in thirty-three Military Occupational Specialties. Participants responded from twenty-five states in the USA and three regions (North America, Japan/the Pacific, and a ship between Ghana and Nigeria).

Interview

In the semi-structured interviews, I introduced five general topics with each participant: background, military experiences (focusing on literacy practices and learning), college and postsecondary academic experiences (particularly literacy practices and learning), comparison and contrast of military and academic practices, and then concluding questions. The comparison and contrast section of each interview elicited information germane to transfer. The fifteen interviews were audio-recorded, transcribed, member-checked, and coded in several rounds. Member-checking is a process of asking participants to confirm the accuracy of data and of increasing reliability of findings (Holloway & Wheeler, 2002, p. 72). I sent transcripts to each interview participant; some edited the written documents to more clearly reflect their meaning and others approved the initial transcript.

Coding

I coded the survey responses and interview transcript data in several rounds: open, axial, and selective (Strauss & Corbin, 1994). In an early round of open coding, I coded *in vivo* and labeled data with words or phrases that occurred in the text. "Professional" emerged early as an important theme; the word "professional" was the fifth most frequently used content word in responses to the survey, after "emails," "letters," "content," and "Marine." In axial coding, I compared codes to each other, looking for patterns in early *in vivo* codes and themes. For example, I began to see that the code "time—managing and using well" was frequently linked to "professional." The codes "peer leader" and "good example" had been frequently used together, so I wrote a memo to define each and relate them to the larger categories. In a late round of coding, I coded categories from the literature review. I applied the code "prior knowledge" when prior knowledge influenced performance later. I coded direction of transfer, like college to military, military to college, and bidirectional.[10] It is also worth remembering that it can be difficult to pinpoint the source of knowledge. As an anonymous participant shared on the survey, "I was both a marine and a student simultaneously so it is difficult to distinguish what skills came from where." Survey and interview data offer participants' memories, perceptions, and representations of their literacy practices. Although interview participants do not retrieve objective truths from their memories, the act of narrating shapes and makes knowledge and connections (Hinton, 2013; Jarratt et al., 2009).

Table 15.1 summarizes results of the data analysis and explicates with data:

RESULTS

When analyzing the data about transfer, prior knowledge, and MALs, I noticed emphasis on habits of mind and dispositions over specific skills or genres. I did not see "I learned to write [x] in the military and that helped/hurt me in college." Instead, participants identified particular habits like professionalism, which was connected to three subthemes: time management, leadership, and being mission oriented. Then, when coding transfer from military

Table 15.1 Data Analysis Results

Code	*Definition or Description of Use*	*Example from the Data*	*# of uses*
Professionalism	Used to refer to a flexible collection of behaviors and attitudes which participants used to succeed in college.	"I also believe that the professional correspondence I've written and read in the service has shaped the way I think and view the service and life in general."	390—autocode136—professionalism & transfer
Time Management	Used when participants described punctuality, meeting deadlines, reasons why time is important, and talked about strategies to control time efficiently in connection with professionalism.	"I am punctual both to class and with my assignments."	34
Leadership	Used when respondents mentioned learning leadership skills in connection to their success and professionalism.	Q. Did reading or writing affect your military experience in a positive way? A. "It has helped shape my leadership ability professionally."	51
Being Mission Oriented	Used when participants used phrases like "mission oriented," "accomplish the mission," or "complete the mission."	"The biggest influence the Marines had on my college success is the work value that was instilled in me. The willingness and discipline to put the mission (graduation) first."	9 mission 49 with definition from the U.S. Dept of Veterans' Affairs (listed in Results)

into academic settings, profanity and code switching from military speech to civilian norms emerged as a theme.

Professionalism

The United States Marine Corps praises professionalism. The USMC communicates this value through induction training and reinforces professionalism through its service culture. Thus, it was not surprising that "professionalism" was a recurring theme in the data. Participants used the words "professional," "professionally" and "professionalism" 276 times on the survey and 114 times in the interviews. Fourteen of fifteen interviewees used a word related to "professional" to explain the source of success in college before I introduced the word. The frequency of "professional" helps show its importance to participants. But what does it mean to be professional in the Marine Corps and to transfer that to college?

The USMC (1998) publication *Marine Corps Values: A User's Guide for Discussion Leaders* chapter on "Professionalism and Ethics" begins with this paragraph:

> Core Values: Honor, Courage, and Commitment! Conduct beyond reproach! Doing what is right! Integrity! *Consummate professional*! These are basic Marine Corps values which have earned the special trust and confidence of America in her Marines. Any breach in this special trust can have devastating effects on our unit, our Corps and our Country. (4-1/p. 62, emphasis mine)

The publication makes it clear that not only professionalism but *consummate*—complete, ideal, and absolute—professionalism is expected. The stakes are high for each Marine reader's unit, the Corps, and the nation.

Although professionalism is valued by Marines, individuals may define or embody the disposition differently.[11] Acknowledging this nuance, instructors/scholars might choose to pay attention to the disposition of professionalism because the Marine student veterans in this study perceive professionalism as a strength they bring to the communication classroom. Professionalism is a way of using military experiences and scripts to achieve excellence in the classroom (Mallory & Downs, 2014). Further, understanding student veterans' expectations regarding professionalism may be valuable for instructors so they can predict or understand conflicts or make informed choices regarding the ways they construct their ethos and teaching personae.

As the following analysis demonstrates, those participants who spoke about transferring the disposition of professionalism from the Marine Corps to college situations often spoke about distinguishing themselves from peers through time management, leadership, and being mission oriented. Participants applied professionalism from the Marine Corps to college both mindfully (high-road transfer) and reflexively (low-road transfer). They generally articulated the results of the transfer as positive, with exceptions for frustration with others' lack of professionalism. Notably, there was occasional transfer of one unprofessional habit: "speaking Marine" and lapses in their code switching from profanity or insults to academic discourse.

Time Management

MALs represented time as valuable. Their ability to manage time in college through their actions, such as meeting deadlines and completing reading and other homework, was represented as part of professionalism. Twenty-one survey respondents linked time management to professionalism as a source of academic success. For example, when comparing college and the military, one anonymous survey respondent wrote:

> Time management is paramount, as is the ability to compartmentalize the competing interests in my life so as to better focus my energies on the task at hand. That is, I can set aside my feelings about irrelevant events from the day/week and apply my efforts to homework.

This survey respondent manages time and focus to succeed in schoolwork. In addition to the survey results, thirteen interview participants expressed belief that time management strategies from the Marine Corps helped them in college. (The other two indirectly referenced time management because both spoke about not wasting time, implying that they value time as a resource to be used carefully.) For example, Jon claimed that "time efficiency has definitely been a major, major factor in helping me out in college. Professionalism, too." Jon continued to claim that managing time efficiently and professionalism helped him build his credibility and ethos, especially in public speaking.

Ilario expressed pride with his planning, preparation and time management. Ilario explained that he has a Traumatic Brain Injury (TBI) that affects his concentration and memory, particularly with subjects he does not enjoy. Therefore, he reported waking up at 5 a.m. each day to read and retain the content from particularly challenging textbooks. Ilario manages his time as a reflection of his commitment to his studies, dedication to mission success, and his professionalism overall.

In addition to the discipline of working early in the day to limit distractions, Ilario explained that he transferred a specific time management strategy from the Marine Corps, which he called "Reverse Planning." When reverse planning, one works backward from a goal to the present in order to decide what steps to take, when to complete them, and how to divide a large task into smaller steps. Before starting school, Ilario used reverse planning to design a schedule to take the classes needed to earn his associate's and bachelor's degrees. The college Ilario attended required all students take an academic skills class, in which the capstone project was creating a degree plan similar to the one Ilario completed independently before enrolling in college. Ilario explained how he applied his military training in reverse planning to new academic contexts:

> Pretty much all the things I'm learning in that ACA [Academic Skills] 122, I did it before I even signed up for the class. And that's where I blame those kids, I was like, "It's your fault." [Laugh] We had to do a project to where we're going to go for our four-year degree and I was like, I already did that. I already went and talked to them [university admissions counselors and . . . registrar].

Ilario learned reverse planning as a Marine and repurposed it for his academic endeavors. He reported that he transferred a Marine Corps technique for time management and that it improved his efficiency in earning his college degree.

Like Jon and Ilario, Dan claimed that the Marine Corps helped him learn to manage time. "Being in the Marine Corps and college put a lot of things into perspective for me, so in terms of, like, what is actually busy and what's not really busy [Dan laughs]." However, because of his success with time management, Dan reported that he found his peers' lack of this skill to be frustrating. Like the veterans in Mallory and Down's (2014) study, Dan reported arriving fifteen minutes early to academic classes, meetings and events. He explained that his habitual earliness contrasts with peers' tardiness.

> I show up to everything in college 15 minutes early at least. And sometimes, like, 20 or 30 minutes because I wanted to be 15 minutes early and I was afraid I'd be less than 15 minutes early, and that was just a big thing. Anyway, so I learned that if you tell somebody in college that you're going to meet and work on a paper at a certain time, then they're going to be there, like, 20 to 30 minutes after that time and that was incredibly frustrating in college because, you know, if I'm there 15 minutes early, then I'm waiting 30 or 45 minutes for you to get there. So, I think adjusting to that, and just adjusting to, like, the standards.

Mallory and Downs (2014) identified student veterans' habit of arriving fifteen minutes early as a script that student veterans follow even when no one else does (p. 61). This is a form of low road transfer that can be positive if it helps a student veteran, but it can also cause frustration, as Dan testified. Dan characterized this as a "silly" example of Marine Corps transfer. However, the pattern recurred in the data and resonated with scholarly findings. I am not suggesting that instructors should arrive fifteen minutes early to class (many of us have less than fifteen minutes in between classes), yet I am suggesting that starting class or a meeting late is likely to be perceived as unprofessional or as wasting time by some MALs.

Like Dan, Sadie described how her expectations for time management from the Marine Corps caused friction during her academic studies. Sadie explained that in the Marine Corps, there is a constant sense of urgency. Sending an email with a read receipt indicates that the recipient should respond within twenty-four hours as a matter of respect for the sender. During graduate school, Sadie sent a draft of her work to a professor who promised to respond the next day. The professor responded ten days later. Sadie expressed her feelings,

> I was like, 'How disrespectful is that?' And that really hurt me. And I really had to learn to grow up and take my Marine Corps hat off and realize that the world doesn't work like the Marine Corps, but that caused a lot of frustrating days for me.

For Sadie, the broken promise and the delay communicated a lack of respect for her work and for her as a student. But as she came to understand the standards of professionalism in academic settings, she stopped applying Marine Corps standards to new situations, although she indicated she experiences frustration when she perceives that her time is being wasted.

Sadie was frustrated by the delayed response to her writing, but the broken promise was more significant. As a leader and an officer, Sadie remembered that when she promised a Marine that she would respond by a certain time, she delivered. Keeping her word was tied both her credibility and to showing her respect and care for junior Marines. One practical implication for teacher scholars is to give students a realistic estimate for when work will be reviewed and to honor promises and send feedback on the expected date. Sadie did learn to recognize that not every email or assignment requires a response in twenty-four hours. Communicating about time frames for responses to email and other assignments may help a student veteran to set their expectations.

Leadership

Sadie's comment about the importance of keeping promises suggests that student veterans connect professionalism to leadership, and that both dispositions transfer to new contexts. On the survey, twelve respondents connected leadership and professionalism explicitly. For example, one participant wrote: "[Reading and writing] has helped shape my leadership ability professionally." Another person wrote "Professional writing plays a huge part in writing Fitness Reports and awards, both of which are an important part of being an officer and leader." In the second response, the Marine links professionalism, writing, and leadership. Professional writing is an element of evaluating junior Marines, which this person sees as a significant part of the job.

Of the twelve respondents who connected professionalism and leadership on the survey, four wrote about transferring their leadership skills to other situations, which is a small proportion of the survey responses. However, during the interviews, eleven participants spoke about transferring some elements of leadership from the Marine Corps to other settings. For example, Harry and Larry both spoke about adapting training methods to teach classes and make oral presentations.

Leadership is an important element of professionalism; the Marine Corps Institute course *Leading Marines* was required for Lance Corporals (E-3) seeking promotion to Corporal (E-4), the first non-commissioned officer (NCO) rank. It was replaced in 2010 by an online interactive course, *Leading Marines*. Non-commissioned officers serve as leaders for junior enlisted Marines. In a 1995 foreword to one edition of *Leading Marines*, Commandant of the Marine Corps, General Mundy, wrote that "The most important responsibility in our Corps is leading Marines. If we expect Marines to lead and if we expect Marines to follow, we must provide the education" (p. 3). Toward the conclusion, the authors argued, "Trust in the Marine Corps and in unit leaders who consistently set the example expected of military professionals is vital to establishing unit cohesion" (p. 59). My study focused explicitly on the Marine Corps, but all branches of service have similar professional military education. Thus, leadership and professionalism are connected in military instruction. The literature review summarized popular nonfiction (e.g., Carrison & Walsh, 2004) to support the argument that leadership transfer from the military is occurring; this section explores what MAL leadership may look like with peers.

Transferring leadership skills and dispositions from the Marine Corps to college has implications for communication teachers. Morrow and Hart (2014) who are both veterans and teacher-scholars, contend that the first element of engaging veterans in composition classes should be to "capitalize on teamwork and leadership training" (p. 43). Further, they argue that "The value of leadership and teamwork is ingrained in veterans from their first day of basic training. Consequently, student-veterans may often take the lead in class discussions and help other students tap into their own experiences to find meaning through the expression of their ideas" (Morrow & Hart, 2014, p. 43). Morrow and Hart represent military experience with leadership as a potential asset in the classroom because student veterans transfer leadership skills.

Having established that the Marine Corps teaches leadership, I now analyze data to support the argument that student veterans reported transferring leadership skills and dispositions developed in the Marine Corps to college, including composition. The participants in my study with the most seniority (Phil, Gunner, Bob, Carlos, Thomas, Martin, and Sadie) reported that they serve as role models. Martin was completing a master's degree in leadership studies and credited his growth as a leader to his experience as a Marine. The SNCOs (Phil, Bob, and Carlos) discussed the value of their leadership skills for younger students.

In one response, an anonymous survey participant represented mentoring and leadership as professional and positive traits that transferred from the military:

> Because of my military experience I have development [sic] many characteristics that have positively influenced my education. I am punctual both to class and with my assignments, I am devoted to my education, I am able to see the larger picture, and I believe that I am a leader and a mentor for my fellow classmates who are younger than I am.

Although the respondent does not use the word "professional" to explain his or her success, the writer highlights positive qualities related to military habits often associated with professionalism: time management (the student reports being punctual), leadership (behaving as a leader and mentor to younger peers) and commitment to education (mission oriented). In this answer, the participant claims to act as a mentor and leader for other peers.

Interview participants also reported transferring Marine Corps professionalism to other contexts. For example, Phil, who served for twenty-two years, reported positive transfer of the Marine disposition of professionalism to his work as a student. After I asked what he brought from the Marine Corps to college, Phil answered, "the professionalism side of the house. We're taught, you know, leadership traits, leadership principles. I bring that." Phil's answer connects professionalism and leadership as dispositions and skills he transferred to college. Phil described his willingness to risk others' judgment by participating and asking questions and explained that this shows maturity. His active participation, attendance, time management, attention to the instructor, and peer leadership modeled ways to transfer professionalism to college. Phil emphasizes turning papers and assignments in promptly, a form of time management. His "love" for school also disposes him to work hard and to look for opportunities to transfer the habits of mind he spent two decades developing.

Mission Oriented

Another Marine Corps disposition related to professionalism emerged as a theme in the data: being "mission oriented." My data, documents from government agencies, and scholarly sources suggest this disposition transfers to non-military contexts, including college.

Documents from the U.S. Department of Veterans Affairs (2013) demonstrate the "mission oriented" disposition is actively cultivated by all branches of military training. In "What It Means to Be Mission Oriented," which is part of the Veterans Employment Toolkit, The U.S. Department of Veterans

Affairs (2012) defines being "mission oriented" through the following list of examples:

- Committing to the mission
- Setting high standards
- Putting the mission of the unit before the needs of the individual
- Completing tasks needed to complete the mission, despite obstacles that may arise
- Not quitting or accepting defeat
- Being disciplined and following rules
- Working together as a team
- Respecting the team leader and his or her decisions
- Leading a team effectively and motivating team members (p. 1).

This definition incorporates elements of persistence, commitment, collaboration, and leadership. However, being mission oriented is also an element of being "professional." The Department of Veterans' Affairs (VA) cites the "Soldier's Creed" from the Army, which includes the declaration "I am an expert and I am a professional" to understand why veterans are mission oriented. They also claim that Marines, who proclaim the motto "Semper Fidelis" (always faithful) are always faithful to accomplishing the mission (p. 1). The VA constructs being mission oriented as an element of being professional.

Further, in the same handout designed for potential employers and managers of veterans, the VA (2012) argues that being "mission oriented" carries over to other settings after the military, which I interpret as transfer. In addition to claiming that transfer of this mindset occurs, the VA frames it as an advantage and benefit to civilian employers and managers, the intended audience of the presentation. Arguably, this disposition also transfers to academic contexts.

Scholars make similar claims about being mission oriented. In *Generation Vet,* Morrow and Hart (2014) write that "Student-veterans, who are coming from a highly mission-oriented culture, will likely also expect instructors to provide them with clear end goals and rationales for assignments as well as detailed grading criteria" (p. 45). Hadlock and Doe (2014) and Mallory and Downs (2014) also call attention to student veterans' mission orientation and its transfer to academic settings. Blaauw-Hara (2015) articulated "a drive to complete the mission" as a strength of student veterans, citing examples from his data of MALs who persist through obstacles to complete assignments, courses, and degrees—each of which they characterize as a mission (pp. 113–116).

The theme of being mission oriented emerged through my coding process as a property of professionalism. The phrase "mission oriented," "accomplish the mission," and "complete the mission" occurred five times on the survey and in four interviews (Andrew, Harry, Sadie, and Thomas). Yet participants in all interviews discussed some or all of the elements that the VA listed in the definition of "mission oriented." They linked their professionalism and success to military training in committing to achieving goals, adhering to high standards, completing tasks, persisting in the face of obstacles, using self-discipline, working with others, respecting professors (leaders), and leading peers. Ultimately, I coded "mission oriented" in every interview.

Scholars and the Department of Veterans Affairs encourage employers and leaders to be clear with veterans about the parameters of the new mission. However, some student veterans frame a new academic mission for themselves and strive to achieve it. One student veteran reported that he read the mission statements of each public university in his home state and chose to apply to the university whose mission he felt best aligned with his own academic and professional goals. This student veteran envisioned enrolling in an academic institution as participating in furthering the university's mission. When he was rejected by his top choice university, he appealed the decision and composed a petition, which included the sentence "If I am accepted into this University, not only will you have a student that has already committed to the goals of your mission statement, you will gain a student leader that younger students can emulate." In this appeal, leadership and being mission oriented are portrayed as assets the student veteran can contribute to the campus community. The student was proud to report that university admissions counselors reversed the rejection to offer him admission.

On the survey, one MAL articulated graduation as the mission: "The biggest influence the Marines had on my college success is the work value that was instilled in me. The willingness and discipline to put the mission (graduation) first." Another MAL explained that being mission oriented affected the way they read academic texts and study in college:

> I believe that I tend to seek out the main point of the issue, argument, or text immediately. I'm not a fan of dancing around the main points. That I believe, is a direct reflection of my attitude as a Marine. Mission accomplishment is what drives us, and that means getting to the solution as quickly and efficiently as possible.

This participant values "mission accomplishment." I found it interesting that the preference for directness and efficiency is applied to texts, reading, and writing. In this case, the participant perceives a connection between an idea from the Marine Corps and college literacy practices.

Not all participants related mission orientation to literacy practices, even if they thought it did transfer to other aspects of their studies. Gunner transferred a "mission-oriented" mindset, another element of professionalism. Being mission oriented allowed Gunner to demonstrate commitment to his education. Gunner's reflection on questions about transfer from the Marine Corps illustrates his perception that dispositions and habits from the military were valuable when he became a student at a theological seminary, even when his literacy skills felt limited. He shared the following:

> So, did the Marine Corps prepare me to go into seminary? Yes and no. Yes in the fact that I already knew about responsibility, I already understood authority, I understood time management and being able to drive myself towards a goal and accomplish the goal, but no in the aspect of reading and writing.

In his answer, Gunner articulates elements of professionalism that were important skills or habits from his time as a Marine that prepared him for later academic study—time management, being mission oriented (driving toward and accomplishing a goal) responsibility, respect for others, and interpersonal skills (following and leading are linked to understanding authority), which are all part of his professionalism. Even though skill with time management and accomplishing the mission and the disposition of professionalism positively transferred from being a Marine to being a student, Gunner reported he did not *feel* prepared because the literacy practices were so different than his workplace and technical literacy practices.

Carlos, Ilario, Phil, Gunner, and Sadie shared the perception that professionalism includes commitment to learning (demonstrated by repurposing a mission-oriented mindset) and personal responsibility as well as time management. For example, as a single parent and medically-discharged veteran, Carlos experienced additional challenges to living up to the time management standards he set for himself as a student. He reported showing his commitment to his education by driving himself to complete goals and take responsibility for falling short:

> [Service in the Marine Corps has] only benefited me for like . . . not organization, but completing tasks, being on time, accepting responsibility for when you're not on time, and you're remorseful for that. It's not like, [flippant tone] "Uh, sorry." I'm like, [slow and serious tone] "I'm late, and really, I'm sorry." You know, "I'm sorry. There's a reason for it, but it's my fault, I accept responsibility." Now I've had to do that a couple of times this last semester, and I hate doing that. First I was like—being late on an assignment, I've never been late. Then it was I missed a couple of assignments, and oh my God, I was like I was on those soap operas, like [dramatic tone] "Aaaah" around the house. Kids are laughing, I'm like "I didn't turn in my assignment." And I feel like I let everybody down, and it's like "Ugh." And option—there's not an option of

> not turning in an assignment. There's not an option. It's like you fail something, you're failing yourself. I think the Marine Corps has brought that to me. It's just like taking care of somebody, one of your troops. You're taking care of yourself now.

Carlos demonstrates responsibility for his action by apologizing without making excuses, regardless of reasons for being late. He repeated that he does not view skipping an assignment as an option. I interpreted his comment in two ways. First, I saw his statement as an instance of being "mission oriented," committed to accomplishing the mission (the class) by completing all assignments. This echoed Gunner's claim that driving himself to complete goals was a skill from the Marine Corps that prepared him for academic success. However, Carlos's answer also shows a particular attitude or disposition that he learned as a Marine has transferred into his civilian life. He did not report purposefully calling upon his mission orientation, but he recognized and articulated the presence of the dispositions and positive effects for him as a student. In this way, a disposition or habit of mind from the military shows low-road positive transfer.

In contrast to the examples of positive transfer of mission orientation and commitment, Harry spoke about a way that his mindset from the Marine Corps interfered with his learning. Specifically, being mission oriented discouraged Harry from attending overwhelming classes:

> I had this sort of, this unconscious Marine mentality that had been, you know, just, you know, seared into me at some point where like I have to, you know, you have to be successful. You can't fail. You can't get—like mission fails—you can't ever—there's never a mission failure. That being said, I, you know, the failure was—the idea of failing was just really—I mean it was, like shameful, it was like I couldn't—there's never—I couldn't—it was unacceptable. And so, at least initially, I would, I think, like my just knee jerk reaction would be like to—if I didn't understand a class, I couldn't even go in. I would just skip the class, I wouldn't even go in and try and partake, because I failed. And—which is just a bad idea on all accounts. But then like—and eventually, I would—I sort of got the hang of it. Actually, I think it had to do with something—there was one class—I think there was required attendance from the teacher, actually, as part of your grade. You can't just not show up. And I remember I think I did show up, and coming into the class, I found that my—what I thought was a stupid pathetic attempt at like absolute failure was actually basically on par with my peers. And—and so I didn't—I guess from there, at that point, it kind of gave me confidence. At least I wasn't—I may not be like, "Well, I'm a genius at this," but, I'm—I'm safely within a reasonable range of the median.

Harry emphasized his perspective about the impossibility of mission failure by repeating "can't" twice, "couldn't" twice, and building to "never." He expressed shame and avoidance at the thought of failing. Further, Harry seemed to conflate struggling or floundering with failing. Whereas Carlos saw skipping an assignment or mismanaging time as failing, at the start of his academic studies, Harry saw skipping classes as preferable to not understanding and thus feeling danger of failing in class. He attributed his aversion to a deeply held ("seared in") reflexive ("knee jerk" and "unconscious") habit of mind from the Marine Corps. In this case, the drive to succeed or complete a mission morphed into a fear of failure that inhibited learning, growth, or attempting the tasks; Harry's actions seem like an example of Wardle's (2012) "answer getting" dispositions, which are seen when students are so resistant to failure that they refuse to take risks. The need to succeed was a reaction to seeing an academic class as a mission, making it an instance of low road transfer. This is also an instance of negative transfer because the disposition to avoid failure was so intense that it inhibited Harry's learning. A critical incident occurred when the professor required attendance. What he felt was a "stupid, pathetic attempt" he later reassessed as "normal." Harry observed the (ironic) failure of his failure-avoiding behaviors ("a bad idea") and reported that comparing his efforts to his peers helped him position himself along a continuum that permits a wider spectrum for success, including "genius," "median," and "failure." Respecting the professor's inclusion of attendance in the grading policy and adhering to it allowed Harry to learn how to positively transfer other behaviors from the military to college.

Speaking in Profanity and Insults

Having established the value MALs place on professionalism, it may seem contradictory that study participants also reported that Marines, the soldiers of the sea, swear like proverbial sailors. There is no official doctrine or class teaching salty speech, yet learning to use creative (or at least abundant) profanity and funny, memorable, or pointed insults is a change in speech due to immersion in the Marine culture and discourse community. Eleven interview participants alluded to using or learning "bad" or profane language as Marines. Six participants reported that they used profanity, insults, and politically incorrect language more frequently in speech while serving as Marines. However, past indiscretions were connected to changes. For example, Andrew, a Combat Engineer, claimed that a trend of increasing professionalism within the Marine Corps makes creative insults and profanity less acceptable than it was when he enlisted.

Andrew described the situation, changing standards, and his own efforts to update his speaking habits:

> In the Marine Corps, it's equal to walking on a construction site: profanity here, and short concepts there, and tempers are always flaring, and there's always yelling and screaming across the job site at people, so it's really affected that [my speaking] drastically. I've tried working on it, because now the Marine Corps is shifting into a more professional atmosphere. Like, I got talked to the other day by my boss, and we just got new females in the shop. So it's, "Can't really talk like that anymore, we've got females around." It's been an interesting change.

Andrew linked his boss's changing standards for Marine's language use to both increasing standards of professionalism and to gender integration of his unit. He took responsibility for his language ("I've tried working on it") and expressed awareness that profanity, flaring tempers, and angry tones (yelling and screaming) are not considered professional, even though they were socially acceptable for the culture of the group.

Like Andrew, Gunner, Dan, Phil, and Carlos acknowledged past profanities or rough language, and emphasized the changes in their speech habits now, after ending active service and becoming veterans. For example, Carlos explained his experience and evolution:

> We have our own language in Marine Corps, like anybody else has, you know, you know, any other special—oh, what was it—institution is what I'm thinking. Where they have their acronyms and that kind of stuff, and their lingo. So it was very—I don't want to say wrong, but very graphic the way we talked, the way we thought, but again, the Marine Corps was not a place to be like—it's not a soft place. . . .definitely from then to now, totally different. I actually know words, and know how to spell. I have a thesaurus at home that I try to break myself from using the same words that I always say, definitely in papers, to learn more words, because you're very limited in the Marine Corps.

Carlos identified two differences in his speech from Marine Corps indoctrination—using acronyms or jargon and hard, graphic language with a limited vocabulary. Carlos was reluctant to condemn the speech practices and people who use them, yet he emphasized how different the past was from the present, when he concentrated on expanding vocabulary through using resources and his writing. At another point in the interview, Carlos reflected on listening to another Marine speak in ways he used to speak and feeling embarrassed of his past. He reported being more conscious of his speech because of his time in college.

Code Switching

During the interviews, all participants moderated their language, demonstrating success with code switching. This success and participants' discussion about changing their speech when entering the academy indicates an interesting element of transfer. The example of profanity illustrates that an individual can recognize that elements of practices acceptable from one discourse community are unacceptable in other contexts and can change their practice and theories about what is acceptable in speech (code switch).

Phil, who served on recruiting duty, explained that recruiting helped him curtail foul language, but he was very conscious that Marine Corps norms differ from accepted speech in college. He explained that profanity was normalized in his speech: "Like, yes, the 'f' word, I used in place of 'the'. You know, I mean, it was pretty bad. [Laughs] . . . You don't even know what you're saying. It becomes so a part of your culture." Phil's laughter as he described his profanity and worked to gain awareness of his speech and code switch highlighted something that faculty may want to be aware of. Phil worked to self-censor, but he also sought external feedback on the appropriateness of his speech in context.

Reflection can also highlight when a military habit fails in a civilian environment. Phil, who prided himself on his professionalism, recounted a "critical incident," a failure or notable problem that causes a learner to create new knowledge to deal with the problem (Yancey et al., 2014). In this case, the language skills Phil used in military settings were not appropriate in academic contexts:

> I can't do that [use profanity and violent metaphors] to students, either. I learned that the first day. Jokingly told a kid I was going to stab him in the neck, and that [pause] that did not go over well. [Laughs] I was like, "Wow, am I not supposed to say that? I'm joking, man, I'm just kidding." He's like horrified. I'm like, "Oh boy." I had to apologize. Which I don't ever do. It's kind of funny.

When Phil spoke to a college student in terms he had used as a Marine, the younger student's appalled reaction caused Phil to realize that he would need to change the way he speaks and jokes with college students. This is an instance of failed or negative transfer, where the failure caused Phil to change his speaking habits and to code switch. Reflection after a "critical incident" helped Phil to realize that he needed to moderate his speech in the new setting. Reflection, therefore, can be helpful for learning how to enter a discourse community. To some extent, individuals can choose to adapt or maintain helpful practices sponsored by military service, like structure for a presentation, and choose to drop harmful habits, like profanity, but there may be errors in code switching.

RECOMMENDATIONS FOR TEACHING

To implement this research into practice, first consider adopting an "asset frame" (Cleary & Wozniak, 2013). That is, view students through the lens of seeking strength and contributions, instead of deficits or skill gaps. By encouraging teacher-scholars to adopt an asset framework, I am not attempting to deny or undermine challenges of disabilities such as TBI or Post-Traumatic Stress. However, Hart and Thompson (2013) caution that "fixation on these issues alone facilitates views of veterans that do more to inhibit learning than to foster it" (p. 12).

After adopting an asset frame, teachers can scaffold curriculum and assignments to "teach for transfer" (Yancey et al., 2014). Transfer increases when teachers encourage, model, and teach metacognition and reflection. For example, a reflection on each assignment can help students to articulate what skills they learned, what they might use in the future, and what they reused or reinvented from previous classes, work contexts, or other environments. Reflection helps a student generate knowledge about the process and abstract the skills to use in the future.

Research participants overall saw professionalism as positive, so behavior that fails to meet professional norms and expectations provoked negative reactions. When a professor or peer acted in a way that research participants perceived as unprofessional, like starting or arriving to class late (mismanaging time), making excuses (avoiding responsibility, poor leadership), skipping class and assignments (placing low value on the mission), then military-affiliated writers reported feeling disrespected and frustrated.

Communication teacher-scholars might wish to be aware that "unprofessional" behavior can alienate MALs. Awareness of the value of professionalism to student veterans opens new ways to connect with and communicate with students about academic behaviors. For instance, I state expectations for academic professionalism for both students and myself in the syllabus. This statement is evolving; I continue to revise as I learn more through research and experience. A recent version reads:

> *Participation and Professionalism*: I expect you to demonstrate professional academic conduct, which includes participating in class activities and bringing materials like textbooks or drafts with you. Points may be deducted for tardiness

> or unprofessional conduct, like sleeping in class, disrupting discussion, or choosing not to participate in class activities.
>
> I encourage you to email me, call me, come to office hours, or schedule an appointment to discuss any issues so that you may succeed in the course. Please allow 24 hours during the week for me to respond to emails. I do not generally respond to email sent on holiday weekends (for example Thanksgiving) until the Monday following.

In this policy statement, I use examples to suggest ways to demonstrate professionalism in class and acknowledge in class that this list is not exhaustive. I state my attendance policy separately and have a statement about the military community adapted from a colleague in the CCCC standing group Writing with Current, Former, and Future Members of the Military. That statement reads as follows:

> Military Community: I recognize the complexities of being a member of the military community (active duty, reserve, veteran, spouse, adult child, etc.) and also a student. Training, complications with GI Bill disbursement, deployments, rigidly scheduled VA medical appointments, and other military and veteran related developments can complicate your academic life. If you make me aware of a complication, I will do everything I can to assist you and to put you in contact with people who are able to assist you. You can also contact [university veterans' services office] for resources.

I establish expectations for how long it takes me to answer emails and also when I am away from email, I believe this explicit framing is an example of Universal Design for Learning that benefits all students. No syllabus statement is a panacea for every issue, but I try to communicate expectations while also communicating my desire to help students succeed.

Many MALs in this study represented time as a tool that communicates respect and priorities. They communicate respect for education by completing assignments before a deadline. Understanding this view of time may help a professor to understand why a MAL arrives fifteen minutes early to class. While I doubt that any educator begins class or office hours late out of malice, awareness that our use of time tells students where they rank in the hierarchy of research, service, and teaching is valuable information. I also encourage teacher scholars to provide reasonable and reliable estimates about when we will provide feedback or answer email. Sadie's example of feeling disrespected because feedback arrived later than expected reminds me that culture influences how messages are interpreted.

Many MALs have developed leadership skills that they transfer to other contexts. In my classes, I offer opportunities for leadership especially in discussions and group work. For example, MALs like Phil reported participating and asking questions to show their peers how students can maximize learning opportunities in college. One person asking questions can lead others to feel safe to do the same. I praise good leadership skills when I see them. For example, in online discussions during Spring 2020 when our university unexpectedly switched from face-to-face to online classes during the COVID-19 pandemic, one MAL established a positive tone with rigorous critical thinking in his online discussion posts; I saw others mimicking the tone and I pointed out to him that his example lead to more meaningful discussions.

Finally, teacher scholars can help MALs to choose and frame "the mission." This may work best in office hours or advising. However, knowing that MALs place value on accomplishing the mission means that this skill can be transferred, especially if the MAL actively frames their mission as completing a degree, a certificate, a class, or an assignment. Missions may be large (Ilario had mapped out his academic career from community college to a graduate degree) or very specific (like Harry's example of attending class).

To summarize, this research can be implemented through related actions. First, adopt asset heuristics and frames regarding student capabilities. Teach for transfer by encouraging metacognition and reflection. Recognize that professionalism is an important value for MALs. MALs interpret use of time as communication about priorities and respect for students. Classes may offer leadership opportunities. Teacher-scholars can help MALs to define their "mission" in class, in college, or beyond.

LIMITATIONS AND FUTURE RESEARCH

This research, however, is subject to several limitations. This was a qualitative study. The sample size was small (112 survey respondents; fifteen interviews) and focused on individuals affiliated with the United States Marine Corps, which is only one branch. The small sample size may limit generalizability. In addition, I was the only person who coded the data. I shared process memos and in-process reports with my adviser and committee, but I recognize the limitation of a single coder on the study.

I am a military spouse. I am not a veteran, reservist, or service member. My experiences as a spouse affect my knowledge of the military and perceptions of it. My role as a spouse was part of my identity as a researcher when distributing the survey and part of my positionality in conducting interviews, transcribing, member checking, coding, analyzing data, and presenting findings.

Additionally, Marine-affiliated learners self-selected to participate in both the survey and the interview. It is possible that there was some bias in the sample because people who completed it wanted to talk about literacy practices, the military, and college. I did not offer incentives or rewards to participate.

Participants included active duty, reservists, and veterans of the USMC. I did not have IRB approval to include other MALs, such as military spouses, children, parents, or siblings. Although some participants were both active-duty military and military spouses, for example, the population was limited rather than broadly inclusive.

Other researchers might work to account for race or demographic factors in terms of military and veterans' literacy practices. As I explain in the Methods section, I did not collect demographic data on the survey because Marines advised against it. Research that collects data about race, age, class, gender, sexuality, and/or (dis)ability could also be beneficial. Nor did I ask about PTSD/TBI, yet this might contribute to greater understanding of neurodiversity. Further, data about the experience and outcomes of online and face-to-face composition classes for student veterans could prove significant.

Methodological diversity might make new knowledge and inquiry possible. I conducted a survey and semi-structured interviews, but future projects might learn more by analyzing documents, conducting focus groups, observing composing activities through screen capture technology or a think-aloud protocol, and/or through multiple interviews with the same participants. Longitudinal studies could refine theories about transfer from the military and from composition. Quantitative data might reveal other trends and patterns. We have much to learn from, with, and about MALs, discourse, and transfer.

CONCLUSION

The Marine Corps teaches Marines to perceive professionalism as an important value. When communication teacher-scholars understand that many MALs perceive professionalism to be an asset they contribute to the classroom, they are better able to understand students' contributions. In this study, Marine student veterans who discussed "professionalism" did not limit it to a single action or formula. Instead, "professionalism" refers to a flexible collection of behaviors and attitudes. The research participants who claimed that professionalism inculcated by the military enabled them to succeed in college often referred to time management, leadership, and being mission oriented. Many of these values and attitudes assist through low road transfer, but some are invoked consciously, especially time management. I argued that MALs transfer the disposition of professionalism from the Marine Corps to college in numerous ways and acknowledged the challenge of transferring unprofessional habits.

Mallory and Downs (2014) argued that veterans who consciously apply military scripts do so as a way of achieving excellence; transferring military-based dispositions seems to be similar. Several MALs said that professionalism helped set them apart from their peers positively. One participant, Bob, claimed, "In fact, being completely average in these respects [completing all assignments, meeting deadlines] in college will get you far. If you apply the same expectations from the Marine Corps to college, you get noticed quickly." A MAL who strives to meet all deadlines (manage time), model participation and good behavior for peers (leadership), and complete courses as part of accomplishing a goal (mission oriented) is a student communicating their desire for academic excellence.

NOTES

1. All student and research participant names are pseudonyms.
2. Discourse is "a sort of 'identity kit' which comes with costume and instructions on how to think, act, talk, and write, so as to take on a social role that others will recognize [. . .] saying-doing-being-valuing-believing combinations" (Gee, 1989, p. 7). Discourses accrete rather than replace one another.
3. Schlossberg, Waters, and Goodman (1995) defined transition as "any event, or nonevent, that results in changed relationships, routines, assumptions, and roles" (p. 27).
4. Facione, Facione, and Giancarlo (2000) define dispositions as "a person's consistent internal motivation to act toward, or to respond to, persons, events, or circumstances in habitual, and yet potentially malleable, ways" (p. 6).
5. Take, for example, books like *Corps Business: The 30 Management Principles of the US Marines* (Freedman, 2000), *Semper Fi: Business Leadership the Marine Corps Way* (Carrison & Walsh, 2004), *No Excuse Leadership: Lessons from the U.S. Army's Elite Rangers* (Barber, 2004), *Extreme Ownership: How U.S. Navy SEALs Lead and Win* (Willinik & Babin, 2015), *Tribe: On Homecoming and Belonging* (Junger, 2016), and *Leaders Eat Last: Why Some Teams Pull Together and Others Don't* (Sinek, 2017).
6. Four programs sponsored by agencies of the US government illustrate desire to make use of military training in civilian situations: *Operation Homecoming: Writing the Wartime Experience* from the National Endowment for the Arts and Department of Defense which operated from 2004–2012 (Carroll, 2008), the *Transition Assistance Program* (TAP, 2006) cosponsored by the US Department of Labor and Department of Defense, the *Veterans Employment Toolkit* (US Department of Veterans Affairs, 2013), and also a report for the Secretary of Defense titled "What Veterans Bring to Civilian Workplaces: A Prototype Toolkit for Helping Private-Sector Employers Understand the Nontechnical Skills Taught in the Military" (Hardison et al., 2015).

The report accomplishes similar work as TAP by presenting tables about skills developed through military training that are applicable to civilian industries and jobs. However, the audience is employers, not prospective employees; this outreach is an extension of legislation like "VOW," the Veterans Opportunity to Work (VOW) Act of 2011, which aimed to retrain veterans for civilian employment and to offer incentives to hire veterans.

7. Hinton's claim that her interviews were epistemological exercises, through which student veterans generated knowledge about their literacy practices, echoes the findings of a study by Jarratt, Mack, Sartor, and Watson (2009). They interviewed approximately 100 juniors and seniors at the University of California Irvine about their experiences learning to write in college. As students constructed narratives about their skills, the upperclassmen shaped connections between classes, assignments, and experiences. The interview prompted reflection and metacognition about thinking and learning processes. They argue "the interview generates as much as retrieves knowledge" (Jarratt, Mack, Sartor, & Watson, 2009, p. 49). Some students made connections between first-year composition and later writing during the interviews. "Despite their initial insistence that they had carried nothing forward from First year composition, they later mentioned—and, more importantly, commented upon—their use of specific strategies and genres from composition in their upper-division writing courses" (Jarratt et al., p. 63). As Hinton (2014) and Jarratt et al. (2009) suggest, interviews and reflection offer an opportunity to generate knowledge and recognize the connections between practices and ideas.

8. The terrorist attacks on September 11, 2001 initiated a new era of conflict for the U.S. armed forces. Service members who served after September 11, 2001 were eligible for the post-9/11 G.I. Bill. Regardless of whether they used the Post-9/11 GI Bill or another means to finance education (e.g. Tuition Assistance, Montgomery GI Bill), this date is significant for contemporary veterans and military-affiliated learners. Choosing to interview post-9/11 MALs meant that participants' experiences were recent and relevant to contemporary praxis.

9. The selection criteria that interview participants must have served in Eastern North Carolina was related to readings on the importance of place making (e.g. Kinloch, 2009). At the time, my family was stationed at a Marine Corps base in North Carolina, and I hoped to learn something about MALs experiences with place and place making. Despite this selection criteria, location did not emerge as a significant theme in coding the data.

10. One interviewee attended school on an ROTCarship, commissioned, and completed a master's degree while serving on active duty—for this military-affiliated learner, direction of transfer was contingent and multiple.

11. In a 2014 article in the *Marine Corp Gazette*, Sergeant Major J. Willeford highlighted potential ambiguity in defining and enacting professionalism in the Marine Corps: "It is clear that a type of abstract professionalism exists everywhere in the Marine Corps. When we are impressed and want to praise, we may say a person is acting 'professionally'; to reprove, we may say someone is acting 'unprofessionally'. This is a widely recognized and effective rhetorical device. Ironically, because the elements of professionalism are never specifically defined, the word itself presents us with an unidentifiable standard" (p. 95).

REFERENCES

Barber, B. E. (2004). *No excuse leadership: Lessons from the U.S. Army's elite rangers*. Wiley.

Beach, K. (2003). Consequential transitions: A developmental view of knowledge propagation through social organizations. In T. Tuomi-Gröhn & Y. Engeström (Eds.), *Between school and work: New perspectives on transfer and boundary-crossing* (pp. 39–62). Emerald Group Publishing.

Beauchamp, S. (2015, September 28). The military isn't preparing people for private-sector success. *The Atlantic*. https://www.theatlantic.com/business/archive/2015/09/military-business-success/407422/.

Bender, A. (2014). Exploring student-veteran expectations about composing: Motivations, purposes, and the influence of trauma on composing practices. In S. Doe & L. Langstraat (Eds.), *Generation vet: Composition, student veterans, and the Post-9/11 university* (pp. 174–195). Utah State University Press. http://www.upcolorado.com/book/9418.

Bergmann, L., & Zepernick, J. (2007). Disciplinarity and transfer: Students' perceptions of learning to write. *WPA: Writing Program Administration, 31*(1–2), 124–129. http://docs.lib.purdue.edu/writinglabpubs/6.

Blaauw-Hara, M. E. (2017). "Learning shock" and student veterans. *Composition Forum, 35*. http://compositionforum.com/issue/35/learning-shock.php.

Blaauw-Hara, M. E. (2015). *Community, identity, and transition: Student veterans and academic writing at the two-year college*. Doctoral Dissertation, Old Dominion University. http://gradworks.umi.com/37/20/3720263.html.

Brandt, D. (1998). Sponsors of literacy. *College Composition and Communication, 49*(2), 165–185.

Brent, D. (2011). Transfer, transformation, and rhetorical knowledge: Insights from transfer theory. *Journal of Business and Technical Communication, 25*(4), 396–420.

Carrison, D., & Walsh, R. (2004). *Semper Fi: Business leadership the Marine Corps Way*. AMACOM.

Carroll, A. (Ed.) (2008). *Operation homecoming: Iraq, Afghanistan, and the home front, in the words of U.S. Troops and their families*. University of Chicago Press.

Cleary, M. N., & Wozniak, K. (2013). Veterans as adult learners in composition courses. *Composition Forum, 28*. https://compositionforum.com/issue/28/adult-learners.php.

Corley, L. (2012). "Brave words"—Rehabilitating the veteran-writer. *College English, 74*(4), 351–365.

DiRamio, D., Ackerman, R., & Mitchell, R. L. (2008). From combat to campus: Voices of student-veterans. *NASPA Journal, 45*(1), 73–102.

DiRamio, D., Ackerman, R., & Mitchell, R. L. (2009). Transitions: Combat veterans as college students. *New Directions for Student Services, 2009*(126), 5–14. doi:10.1002/ss.311.

Driscoll, D. L., & Wells, J. (2012). Beyond knowledge and skills: Writing transfer and the role of student dispositions in and beyond the writing classroom.

Composition Forum, 26. http://compositionforum.org/issue/26/beyond-knowledge-skills.php.

Duffy, J. (2012, March 16). Virtuous arguments: Essay on the value of first-year writing courses. *Inside Higher Ed*. http://www.insidehighered.com/views/2012/03/16/essay-value-first-year-writing-courses.

Elliott, M., Gonzalez, C., & Larsen, B. (2011). U.S. military veterans transition to college: Combat, PTSD, and alienation on campus. *Journal of Student Affairs Research and Practice, 48*(3), 279–296. doi:10.2202/1949-6605.6293.

Facione, P. A., Facione, N. C., & Giancarlo, C. A. E (2000). *The California Critical Thinking Disposition Inventory*. California Academic Press.

Freedman, D. H. (2000). *Corps business: The 30 management principles of the U.S. Marines*. Harper Business.

Gee, J. P. (1989). Literacy, discourse, and linguistics: Introduction. *Journal of Education, 171*(1), 5–17. doi:10.1177/002205748917100101.

Hadlock, E., & Doe, S. (2014). Not just "yes sir, no sir": How genre and agency interact in student-veteran writing. In S. Doe & L. Langstraat (Eds.), *Generation vet: Composition, student veterans, and the Post-9/11 university* (pp. 73–94). Utah State University Press.

Hardison, C., Shanley, M. G., Saavedra, A. R., Crowley, J. C., Wong, J. P., & Steinberg, P. S. (2015). *What veterans bring to civilian workplaces*. RAND Corporation. https://www.rand.org/pubs/tools/TL160.html.

Hart, D. A., & Thompson, R. (2013). "An ethical obligation": Promising practices for student veterans in college writing classrooms (Conference paper). *2013 Conference on College Composition and Communication*, The Riviera Hotel, Las Vegas, NV, United States.

Hart, D. A., & Thompson, R. (2016). Veterans in the writing classroom: Three programmatic approaches to facilitate the transition from the military to higher education. *College Composition and Communication, 68*(2), 345–371.

Herrmann, D., Hopkins, C., Wilson, R. B., & Allen, B. (2011). *Progress in educating veterans in the 21st century*. CreateSpace.

Hinton, C. E. (2014). "Front and center": Marine student-veterans, collaboration, and the writing center. In S. Doe & L. Langstraat (Eds.), *Generation vet: Composition, student veterans, and the Post-9/11 university* (pp. 257–277). Utah State University Press.

Hinton, C. E. (2013). "The military taught me something about writing": How student veterans complicate the novice-to-expert continuum in first-year composition. *Composition Forum, 28*. https://compositionforum.com/issue/28/novice-to-expert.php.

Hinton, C. E. (2012). *The experiences of Marine student veterans in undergraduate composition courses: A phenomenological study*. Doctoral Dissertation, St. Louis University. http://gradworks.umi.com/35/53/3553100.html.

Holloway, I., & Wheeler, S. (2002). The nature of qualitative research: Development and perspectives. In I. Holloway & S. Wheeler, *Qualitative Research in Nursing* (pp. 3–25). MPG Books Ltd.

Jarratt, S., Mack, K., Sartor, A., & Watson, S. (2007). *Pedagogical memory and the transferability of writing knowledge: An interview-based study of UCI juniors and seniors*. http://writing.uci.edu/wp-content/uploads/sites/30/2016/04/pedmemory.pdf.

Junger, S. (2016). *Tribe: On homecoming and belonging*. Twelve.

Kinloch, V. (2009). *Harlem on our minds: Place, race and the literacies of urban youth*. Teachers College Press.

Klein, J. (2011, August 18). The next greatest generation. *Time*. http://swampland.time.com/2011/08/18/the-next-greatest-generation/.

Lieberman, D., & Stewart, K. (2014). *Strengthening perceptions of America's Post-9/11 veterans: Survey analysis report*. https://gotyour6.org/wp-content/uploads/2015/10/GY6-Survey-Analysis-Report-2014.pdf.

Mallory, A., & Downs, D. (2014). Uniform meets rhetoric: Excellence through interaction. In S. Doe & L. Langstraat (Eds.), *Generation vet: Composition, student veterans, and the Post-9/11 university* (pp. 51–72). Utah State University Press.

Marine Corps Institute, & Mundy, C. E. (2002). *Leading Marines*. http://www.marines.mil/Portals/59/Publications/MCWP%206-11%20Leading%20Marine.pdf.

Morrow, S., & Hart, D. A. (2014). Veterans in college writing classes: Understanding and embracing the mutual benefit. In S. Doe & L. Langstraat (Eds.), *Generation vet: Composition, student veterans, and the Post-9/11 university* (pp. 31–50). Utah State University Press.

McCarthy, L. P. (1987). A stranger in strange lands: A college student writing across the curriculum. *Research in the Teaching of English, 21*(3), 233–265.

Nelms, G., & Dively, R. L. (2007). Perceived roadblocks to transferring knowledge from first-year composition to writing-intensive major courses: A pilot study. *WPA: Writing Program Administration, 31*(1–2), 214–240.

Nicholson, J. (2019.). Council post: Veterans are the future of America's startups, and nonprofit leaders can help them succeed. *Forbes*. https://www.forbes.com/sites/forbesnonprofitcouncil/2019/11/05/veterans-are-the-future-of-americas-startups-and-nonprofit-leaders-can-help-them-succeed/#537391dc61d4.

Nowacek, R. S. (2011). *Agents of integration: Understanding transfer as a rhetorical act*. Southern Illinois University Press.

Palmeri, J. (2012). *Remixing composition: A history of multimodal writing pedagogy*. Southern Illinois University Press.

Perkins, D. N., & Salomon, G. (1988). Teaching for transfer. *Educational Leadership, 46*(1), 22–32.

Reiff, M. J., & Bawarshi, A. (2011). Tracing discursive resources: How students use prior genre knowledge to negotiate new writing contexts in first-year composition. *Written Communication, 28*(3), 312–337. doi:10.1177/0741088311410183.

Ricks, T. E. (2007). *Making the Corps*. Scribner.

Rounsaville, A., Goldberg, R., & Bawarshi, A. (2008). From incomes to outcomes: FYW students' prior genre knowledge, meta-cognition, and the question of transfer. *WPA: Writing Program Administration, 32*(1), 97–112.

Rumann, C. B., & Hamrick, F. A. (2009). Supporting student veterans in transition. *New Directions for Student Services, 2009*(126), 25–34. doi:10.1002/ss.313.

Rumann, C. B., & Hamrick, F. A. (2010). Student veterans in transition: Re-enrolling after war zone deployments. *Journal of Higher Education, 81*(4), 431–458.

Salomon, G., & Perkins, D. N. (1989). Rocky roads to transfer: Rethinking mechanisms of a neglected phenomenon. *Educational Psychologist, 24*(2), 113–142.

Sargent, W. M. (2009). *Helping veterans transition into academic life through the creation of a university veteran support group: So we can better serve those who served us*. http://www.eric.ed.gov/ERICWebPortal/contentdelivery/servlet/ERICServlet?accno=ED505889.

Schell, E. E., & Kleinbart, I. (2014). "I have to speak out": Writing with veterans in a community writing group. In S. Doe & L. Langstraat (Eds.), *Generation vet: Composition, student veterans, and the Post-9/11 university* (pp. 119–139). Utah State University Press.

Schlossberg, N. K., Waters, E. B., & Goodman, J. (1995). *Counseling adults in transition: Linking practice with theory* (2nd ed.). Springer Publishing Co.

Selfe, C. L., & Hawisher, G. E. (2004). *Literate lives in the information age: Narratives of literacy from the United States*. Mahwah, NJ: Erlbaum.

Selting, B. (2014). The value of service learning for student-veterans: Transitioning to academic cultures through writing and experiential learning. In S. Doe & L. Langstraat (Eds.), *Generation vet: Composition, student veterans, and the Post-9/11 university* (pp. 240–256). Utah State University Press.

Sinek, S. (2017). *Leaders eat last*. Portfolio.

Smit, D. W. (2007). *The end of composition studies*. Southern Illinois University Press.

Strauss, A., & Corbin, J. (1994). Grounded theory methodology: An overview. In N. K. Denzin & Y. S. Lincoln (Eds.), *Handbook of qualitative research* (pp. 273–285). Sage Publications, Inc.

Thorndike, E. L. (1921). *Educational psychology: The psychology of learning*. Teachers College, Columbia University.

Tuomi-Gröhn, T., & Engeström, Y. (2003). *Between school and work: New perspectives on transfer and boundary-crossing*. Emerald.

US Department of Labor, Transition Assistance Program, & U.S. Department of Defense. (2006). *TAP Manual (Workshop)*. Department of Labor. https://www.dol.gov/vets/programs/tap/tapmanualmar06.pdf.

US Department of Veterans Affairs. (2012, April 6). *What it means to be mission oriented*. Department of Veterans Affairs. http://va.gov/vetsinworkplace/docs/em_missionoriented.html.

U.S. Department of Veterans Affairs. (2013). *Veterans employment toolkit home* (Program Homepage). https://www.va.gov/vetsinworkplace/.

U.S. Marine Corps. (1998). *Marine Corps values: A user's guide for discussion leaders*. Marine Corps Combat Development Command. http://www.marines.mil/Portals/59/MCRP%206-11B%20z.pdf.

Valentino, M. J. (2012). Serving those who have served: Preparing for student veterans in our writing programs, classes, and writing centers. *WPA: Writing Program Administration, 36*(1), 164–178.

Voigt, J. (2016). A personal challenge of veteran integration. *Pedagogy, 16*(3), 532–539. doi:10.1215/15314200-3600877.

Wardle, E. (2012). Creative repurposing for expansive learning: Considering “problem-exploring” and “answer-getting” dispositions in individuals and fields. *Composition Forum, 26*. http://compositionforum.com/issue/26/creative-repurposing.php.

Wardle, E. (2009). “Mutt genres” and the goal of FYC: Can we help students write the genres of the university? *College Composition and Communication, 60*(4), 765–789.

Wardle, E. (2007). Understanding ‘transfer’ from FYC: Preliminary results of a longitudinal study. *WPA: Writing Program Administration, 31*(1–2), 65–85.

Willink, J., & Babin, L. (2015). *Extreme ownership: How U.S. Navy SEALs lead and win*. St. Martin’s Press.

Willeford, J. D. (2014). What is a military professional? *Marine Corps Gazette, 98*(9), 95–99.

Woodworth, R. S., & Thorndike, E. L. (1901). The influence of improvement in one mental function upon the efficiency of other functions. *Psychological Review, 8*(3), 247–261. doi:10.1037/h0074898.

Yancey, K., Robertson, L., & Taczak, K. (2014). *Writing across contexts: Transfer, composition, and sites of writing*. Utah State University Press.

Chapter 16

From Application to Graduation

Collective Recommendations for Serving Those Who Serve

Amy R. May, Victoria McDermott, and Leandra Hinojosa Hernández

This volume explicates and humanizes MAL identities as they relate to higher education. The chapters range from MALs' long-standing history in academia to effective strategies for transferring their military skills to the higher education classroom. Our diverse range of authors sought to thoroughly define the MAL identity using both their lived experiences as MALs and their extensive work as educators or support staff supporting the MAL community. Utilizing a communication approach to pedagogy and education, each chapter serves as a tool and resource for communication scholars and administrators who work with military-affiliated students. By identifying the barriers MALs may face while working toward their educational goals, institutions of higher learning (IHLs) are challenged to create programs, reimagine instructional strategies, and provide administrative support to better serve this community of learners. MALs bring with them over 75 billion dollars[1] of funding through the Post-9/11 G.I. Bill (CBO, 2019), and with great funding comes great responsibility to ensure progress toward achieving educational goals.

RECOMMENDATIONS

Critically evaluating the current state of MALs in higher education has identified gaps in the level of support currently available for this population of learners. Thus, as members of the academy, we are challenged to examine our current infrastructure and support practices. Gaps in existing programmatic offerings and infrastructure need to be resolved, while programs that serve

MALs and support their development need to be further strengthened and funded. The U.S. Army provides a roadmap for innovation that may serve IHLs: "challenging and changing cultural norms . . . [supporting] the rapid conversion of new ideas into valued outcomes, and leveraging the diverse talents and capabilities of the Total Force, key partners, suppliers, and other collaborators" (U.S. Army, 2016, p. 3). By challenging and changing cultural norms within the academy to be more inclusive of what we've uncovered and explored in this edited volume about military and academic culture, institutions can work toward bridging the cultural divide. Moreover, colleges and universities are ripe with innovative thinkers that challenge the status quo every day (May & McDermott, 2019); continuing to provide inadequate services for MALs fails to harness the immense capital available in our communities of inquiry. Finally, while this edited volume has identified gaps in existing infrastructure, a "Total Force" of caring, talented educators, staff members, and administrators are steadfast in their support and commitment to serving students, including MALs. Continuing to build our capacity by collaborating across campus, within our local communities, and with other institutions, further enriches opportunities for growth, innovation, and supportive care. As such, we provide readers with final recommendations for supporting MALs in academia with full acknowledgment of their limitations. What works at some institutions may not serve others, and these recommendations are limited within the scope of possibilities. They represent a starting point for continuing the discussion based on the literature and findings presented by the contributors in this edited volume. We have broken down these recommendations into strategies for institutions, policymakers, and educators.

For Institutions

Identifying MALs and Assessing Their Experience

Before MALs can be supported, they must be identified and invited to engage with our campus communities. Furthermore, these numbers matter when it comes to opportunities for federal, state, and private sector funding (American Council on Education, 2018). While federal financial forms and benefit forms identify a portion of these students (i.e., veterans using the G.I. Bill as a funding source), some MALs may choose to privately fund their education, choose not to disclose, or fail to identify with the "veteran" label (American Council on Education, 2018; see Chiara, Chiara & Ashlock, chapter 3). Adding more inclusive questions on application forms would help to more accurately capture these data to guide policy change, resource allocation, and direct future research. Moreover, being transparent as to the "why" you are requesting this information by clearly communicating how the data will be used at your respective institution and beyond to design and deliver support/intervention programs. The American Council on Education (2010) recommends asking "Have you ever served in the United States Armed Forces?" to promote self-identification (p. 6). The editors recommend reframing the question to include a more inclusive invitation to disclose, such as "have you or an immediate family member, specifically a spouse or parent, ever served in the United States Armed Forces?" The reframing allows institutions to identify spouses, children, and parental dependents, an overlooked segment of MALs in need of additional support (see chapter 5 by Rossetto & Owlett in this edited volume).

In addition to identifying MALs, campus communities are challenged to assess the campus climate and evaluate the experiences of these learners. The Center for American Progress (2012) introduced the Environmental Evaluation for Veterans Index (EEVI), an assessment tool designed to identify gaps in support that identify barriers for MALs. The EEVI self-assessment provides IHLs assessment data across three dimensions: personnel and services, institutional structures, and social and cultural support. The results of the self-assessment identify gaps that can be addressed using other tools, such as the American Council on Education (2018) Toolkit for Veteran Friendly Institutions. Several of the strategies provided by the American Council of Education (2012) are included and expanded upon in the editor's list of final recommendations.

Policies

Creating university-level policies is an important step to ensuring *all* MALs are treated equally, supported, and accommodated in higher education. Policies do not traditionally extend to their families; however, spouses and dependents are significantly impacted by deployments (Villagran et al., 2013; Huebner et al., & Grass, 2007; Rossetto & Owlett, chapter 5) and should be offered accommodations under institutional policies. Currently, the majority of institutions base policies off of federal (Higher Education Opportunity Act, 2010) and state mandates requiring them to accommodate active-duty and reservist soldiers on official duty orders. Legally required accommodations focus almost exclusively on readmission post-deployment (Higher Education Opportunity Act, 2010). Readmission covers a minute fraction of needed accommodations to ensure holistic support for MALs. Institutions should consider implementing the following at a minimum to better support service members and their families: tuition-waiver deadlines, credit and tuition reimbursement due to mid-semester deployment, course evaluation and makeup process after a mid-semester deployment, and graduate program extensions due to deployment (Military Family Research Institute, 2015a). These policies reflect an institutional awareness of the realities associated with military service and a respect for the demands placed on MALs during and post deployment. Overall, prior to implementing any policy, MALs should be invited to comment and review these policies to ensure they are helpful to the group they are intended to support.

Establishing MAL-informed policies is a critical first step; however, institutions are encouraged to develop a distribution and communication plan to ensure all institutional members are aware of policies and how they should be applied. All policies must be easy to understand and clearly communicated to students, faculty, staff, and administrators with some type of acknowledgment receipt to ensure accountability. A survey of MALs found many of the policies designed to support them are "difficult to interpret and inconsistently applied" (Military Family Research Institute, 2015a, p. 6). Moreover, a knowledge gap exists between administrators and staff when it comes to MAL-specific policies and available options for MALs facing deployment (Military Family Research Institute, 2015a). A MAL facing deployment should not be told different things depending on who they are talking to or struggle to interpret a policy designed to support them.

Finally, establishing policies at the university level helps to ensure faculty are held to a universal standard when accommodations are necessary to support MALs. For example, some instructors may not agree to be flexible with an active-duty student on temporary leave from the classroom, requiring them to meet deadlines even with notice of their field training and missions. Additionally, at the inter and intra department level, policies can vary from professor to professor, which may create additional confusion and frustration as "instruction and evaluation in the military tends to be more consistent across settings" (Borsari et al., 2017, p. 7). Furthermore, the lack of cohesive support can lead to frustration and sometimes failure among MALs who may not be able to meet inflexible course requirements due to military constraints, such as lack of access to technology during field exercises or unstable internet access barring a MAL from connecting to synchronous sessions. Holding all faculty to a consistent standard through university-level policies may be necessary for assuring achievement and degree completion. The editors recognize the importance of academic freedom and do not suggest bypassing faculty senate or other faculty-directed workgroups on campus. Any policy that infringes on classroom administration should be developed and deployed in collaboration with faculty and their representatives otherwise they are one-sided and slated for failure. While the process of creating such policies may be labor-intensive and subject to debate, research suggests that "when the infrastructure is aligned to support a change initiative and a set of organizational values and priorities, then transformation is more likely to happen and to be sustained" (Kezar, 2019, p. 3).

Funding Support Services

MALs "find the logistic challenges of the transition from military to college overwhelming" (Borsari et al., 2017, p. 6). As such, from "application to graduation" is the approach advocated for by the editors of this volume as it speaks to the continued and sustained support needed to address barriers to degree completion. As a starting point, administrators are encouraged to assess current funding allocation for Veterans Affairs (VA) Centers to identify opportunities for additional staff and program development as they serve as "a bridge for both civilians and soldiers to better understand the dynamics of college and military service" (Kirchner et al., 2014, p. 15). Having enough staff to support all MALs within the university system is critical. Support staff may include Veterans Benefit Administration liaisons, MAL-specific advisers (see chapter 8 by Moullet & Belding in this edited volume), and MAL-disability services liaisons who can speak to the unique needs of this population as it relates to mental and physical health.

Campus and community-partnership specialists may also work for or partner with VA Centers. Campus partnership with Reserve Officer Training Corps (ROTC) programs may help further build a sense of military camaraderie on campus communities; moreover, veterans and active-duty soldiers may be able to serve in mentor roles for future service members, something heavily valued by the military as an outlet for sharing their lived experience with the "next generation" of service members and their families (Center for American Progress, 2012). Community liaisons may serve as strategic partners to identify and build relationships with military-affiliated community agencies, such as local military bases, nonprofits, and military-friendly employers, who may serve to fill in the gaps when funding and/or additional resources are unavailable or not feasible for a respective institution.

The editors also encourage VA Center staff to critically evaluate their spaces to ensure VA Centers are inviting for all members of the MAL community. Incorporating specific niches of MALs is a way to connect and support these learners within the university system (Doenges, 2011). For example, VA Center Directors may consider how military spouses and children are invited to their spaces as the name implies inclusion/exclusion of specific groups. Furthermore, exploring additional programmatic offerings that target spouses, children, and adult dependents may facilitate and support their integration into campus communities. A MAL choosing to pursue their degrees online may have less access to physical VA Centers, requiring centers to establish a digital presence to reach these learners. Building virtual spaces and social media communities may help create connectedness and promote retention and degree completion (Moore, 2019; see chapter 13 by McGregor in this volume). When given an opportunity to weigh in on value-added services, MALs recommended Live Chat as a communication channel, especially when communicating with colleges and universities regarding G.I. Bill funding questions and concerns (Steele, Salcedo, & Coley, 2011; see chapter 12 by Cobos in this edited volume). As McGregor shows us in chapter 13, MALs have experience building supportive communities online for everything from emotional support to sharing job-specific knowledge, and IHLs are encouraged to develop and support these communities at an institutional level.

Training

Faculty, staff, and administrators interact with MALs on a daily basis, playing a critical role in defining the service member's experience in higher education. However, faculty may have little to no experience with the military or military culture. As such, they may not be aware of the challenges associated with being a military-affiliated member or the unique cultural values and norms associated with military service (Steele et al., 2011; Southwell et al., 2018). Partnering with the campus VA Center and other military-affiliated community organizations to design and deliver training to address knowledge gaps may help bridge the cultural divide experienced by some MALs. Additionally, training may include specific strategies to help promote relationship development with MALs especially those struggling with mental and physical challenges that impact their ability to initiate and sustain communication with faculty (Church, 2009; Cate & Albright, 2015). Finally, members of the campus community are encouraged to identify training and development opportunities that allow them to challenge their implicit bias regarding military service members and their families. One common bias against veterans, in particular, is related to PTS and perceived instability (Schreger & Kimble, 2017). When members of a community "hold . . . negative biases towards veterans, whether implicit or explicit, this may have a significant impact on the ability of military personnel to integrate back into society" and into our campus communities (Schreger & Kimble, 2017, p. 12).

Moreover, for institutions looking for campus-wide intervention strategies, consider Green Zone. Originally launched at Virginia Commonwealth University (VCU) in 2010, this program has expanded across the globe (Nichols-Casebolt, 2012). This program is designed to support the development of "a cadre of individuals throughout the institution who were supportive and knowledgeable about issues facing our military students and the various resources available to assist them" (Nichols-Casebolt, 2012, p. 29). Members of the campus community completing the Green Zone training may be more informed about this population and serve as advocates for navigating the system, supporting the "application to graduation" journey.

Credit for Military Service

Service members bring with them rich histories to include military-specific training, skills developed through years of service in military occupations (MOS) and lived experiences that cannot be duplicated in the classroom, such as a deployment to a war zone. Institutions are encouraged to explore the feasibility of Credit for Service programs as part of their offerings to MALs (Ryu, 2013). The Department of Defense partnered with American Council on Education (n.d.) to create Joint Services Transcripts (JSPs), which details a service member's relevant education history and corresponding credit recommendations for institutions. These programs not only validate a service member's lived experience, but they also save a MAL's tuition benefits and reduce the time needed for degree completion (Bergman & Herd, 2016). Time and money are barriers identified for MALs as G.I. Bill funding is limited (VA, 2020), and time between duty stations and/or deployments can impact degree completion (see chapter 12 by Cobos in this edited volume). Credit for military service may reduce time toward degree completion by fifteen credits, equating to a full semester at most colleges and universities, and save a MAL anywhere between $1600-$6000 on tuition costs (DoD, 2019). Moreover, some MALs seek out institutions with credit for service programs (Steele, Salcedo, & Coley, 2011), offering a competitive recruitment advantage for some IHLs.

Pay Attention to Private For-Profits and Build Online Programs

MALs are more likely to attend for-profit institutions and take online classes than their civilian peers (Radford, Bentz, Dekker, & Paslov, 2016), and they are attracted to these institutions for a variety of reasons. Financial incentives, flexibility, and on-demand access to needed classes were cited as primary drivers of their decision to enroll in private for-profit institutions (Steele, 2012). Many private for-profits set tuition rates for MALs at G.I. Bill Benefit levels, negating the need to "comparison shop" more affordable tuition rates (Steele, 2012). Furthermore, many for-profits offer Credit for Service Programs and a revolving selection of online courses starting once a week, once every few weeks or on some other less rigid schedule (Steele, 2012). Most traditional colleges and universities are still tied to quarter or semester-based systems heavily driven by face-to-face course offerings, which offers less flexibility for MALs. Colleges and universities looking to compete with these institutions are challenged to evaluate their programs and reimagine their MAL experience. Currently, many IHLs are exploring flexible delivery options in the wake of COVID-19 (Lederman, 2020). While many colleges and universities may see COVID-19 systems as temporary, evaluating the long-term sustainability of these programs is encouraged by the editors as they address MAL-related flexibility concerns and may increase access to education for not only MALs but all historically underserved populations with limited access to physical campuses.

MAL-Specific Learning Communities

Colleges and universities have positively contributed to student success using a learning community model, a cohort-based approach where students take classes together and build relationships through shared campus experiences. A century of research into learning communities affirms their impact, including "enhanced academic performance, integration of academic and social experiences, gains in multiple areas of skill, competence, and knowledge, and overall satisfaction with the college experience" (Zhao & Kuh, 2004, pp. 130–131). Administrators are encouraged to explore the feasibility of MAL-specific learning communities as they frequently cite challenges relating to their peers, describing "a sense of difference and frustration with what was seen as younger students' immaturity and 'sense of entitlement'" (Steele et al., 2011, p. 38; see chapter 3 by Chiara, Chiara, & Ashlock, as well as chapter 2 by May & McDermott in this edited volume). Additionally, research indicates MALs are important sources of support for other MALs; however, they are not always visible on campus, further underscoring the need to develop some sort of self-identified cohort model (Steele et al., 2011). Collectively, cohorting MALs may address some of the concerns related to "differentness;" moreover, the shared lived experience may help build communities and perceived "connectedness" with campus communities (Center for American Progress, 2012). Furthermore, a hidden curriculum, a set of unspoken rules covering rules of engagement, exists on college campuses, especially for traditionally underserved populations, such as first-generation students (Smith, 2013). The most current data available from 2014, identify 64 percent of veterans as first-generation (Postsecondary National Policy Institute, 2018), underscoring the need to explore what MALs do not know about the college experience in addition to the military-civilian-academia cultural divide that may prevent them from fully integrating and thriving socially and academically. All in all, reserving sections or portions of sections for high demand classes of MALs may help address the issue of classes filling up prior to G.I. Bill reimbursement (Steele et al., 2011).

The editors acknowledge cohorting MALs as potentially problematic due to scheduling and outside responsibilities; however, these programs may be designed in a way to blend both face-to-face and online delivery options. Likewise, the cost in terms of time and money for adding an additional MAL-specific class section or adding a specific class to a MAL degree path are legitimate concerns. One strategy for reducing the financial burden of such a program on institutions is to offer faculty service credit for teaching cohorted MALs as part of the tenure and promotion review process. Additionally, the editors suggest these programs should be highly encouraged and supported by VA Centers; however, they should not be required for MAL graduation. In the end, these courses may help MALs "fit in . . . and the camaraderie experienced by those who choose veteran-only classes will likely continue and translate into positive experiences throughout their college careers" (American Council on Education, 2018, p. 9).

For institutions committed to recruiting, retaining, and supporting MALs through to graduation, these recommendations offer a starting point. To optimize support, institutions need to clearly communicate their commitment to MALs through policy and action. Overall, if MALs are not feeling supported in higher education, sharing their experiences can negatively impact the desire or willingness of other MALs to embark on a college degree. However, institutions are not the only group that can support MALs access and success in higher education. The subsequent section explores the strategies and support policymakers can provide for MALs at the state and federal levels.

FOR POLICYMAKERS

MALs may feel like "dollar signs in uniforms" as more institutions vie for G.I. Bill funding (Arnold, 2020). Policymakers are encouraged to listen to the MALs in their communities and advocate for legislation that helps them more easily access promised funding but also ensures institutions are held accountable for ensuring high-quality teaching and learning outcomes. Advocacy groups offer a starting point for listening and learning more. For example, the Iraq and Afghanistan Vets of America (IAVA) organize "Storm the Hill," where members advocate for vets on the U.S. Capital and share the stories of 4.1 million post-9/11 veterans (IAVA, n.d.). Of particular concern for MALs are VA Benefits, opportunities for spouses and dependents, institutional accountability, and transfer/licensure.

VA Benefits

In 2017, Trump signed the Harry W. Colmery Veterans Educational Assistance Act (2020). Known as the "Forever GI Bill," this legislation offers additional benefits and funding for some MALs and specific degree programs, such as science, technology, engineering and math (STEM) (Harry W. Colmery Veterans Educational Assistance Act, 2020); however, it did little to address the complexities associated with accessing the funds. MALs may not know the nuances of the G.I. Bill, such as how to use, transfer, and get the most out of their benefits (Steele et al., 2011). Moreover, late payments continue to plague the reimbursement system, causing undue financial burden on MALs and may result in dropped classes, further delaying progression toward degree attainment (Steele et al., 2011). In addition to benefit access being unclear, service members often report receiving benefits with no explanation. Multiple reimbursements may result in charges of overpayment, which can create further financial burden on service members and their families (Steele et al., 2011). Even with increased funding under the Forever G.I. Bill, reimbursement for degree programs outside of STEM still remains at thirty-five months (Harry W. Colmery Veterans Educational Assistance Act, 2020). For MALs juggling work with other family-related responsibilities, the allotted time for degree attainment may not be achievable. There are also other challenges, such as state-to-state variations, lack of funding for Active Guard reservists, and institutional barriers to certification that delays funding for MALs and complicate an already confusing system (Steele et al., 2011). All of these limitations suggest more reform is needed to improve timely and accurate access to funding for both individual MALs and the institutions that serve them.

Military Spouses

Policymakers are encouraged to consider the experience of military spouses and dependents as part of larger social programs designed to fund service member education and support services. IHLs play a critical role in preparing military spouses and dependents to be competitive in local duty station job markets; however, spouses often struggle to find employment and integrate into local communities (Alwine, 2020; Hernández, 2019; Williams, Mariani, Routh, Keyal, & Hill, 2020). To support spouses and dependents, institutions need funding support to develop job-centric upskilling, expedited training programs designed to train people for in-demand jobs in the local communities they serve while stationed there (Williams et al., 2020). Additionally, the editors encourage policymakers to support the Military Spouse Career Education Act (2020). Among other things, this bill covers the cost of exams granting college credit, licensure exams, continuing education credits, and continued funding for My Career Advancement Account (MyCAA), a program designed to cover costs for spouses in certain degree and certificate programs (Wagnon, 2020). Funding this bill supports military spouses and IHLs that provide access to these educational opportunities. Ultimately, the level of spousal support has a direct impact on service member retention, making this a defense issue of critical importance for policymakers (Wagnon, 2020).

Institutional Accountability

Policymakers are challenged to hold institutions receiving G.I. Bill funding accountable and address the "90/10 loophole." In 1998, Congress expanded existing legislation designed to prevent fraud, waste, and abuse in federal financial programs by enacting the 90/10 rule (Lee & Looney, 2019). The 90/10 limits the shares of revenue an institution can receive from federal financial aid to 90 percent; 10 percent must come from non-federal sources (Lee & Looney, 2019). However, the 90 percent cap does not include military and veteran student benefits, creating a loophole, which has been exploited by some for-profit institutions (Lee & Looney, 2019). Private-for-profit institutions receive over 90 percent of their funding from government grants, contracts, and appropriation, including the G.I. Bill, violating the 90/10 rule (National Center for Education Statistics, 2020, figure 1). In comparison, private colleges and universities and nonprofits receive 20 percent and 31 percent of their funding, respectively, from the same federal programs (National Center for Education Statistics, 2020, figure 1). This is of significant concern for policymakers as some for-profit institutions have been deemed "predatory," failing to deliver quality education and saddling students with significant debt and questionable credentials (Committee on Veterans Affairs, 2019; Lee & Looney, 2019). As previously noted, MALs are more likely to attend for-profit institutions (Radford et al., 2016), and service member advocacy groups, such as the Student Veterans of America, argue "some for-profit institutions are using the loophole to aggressively and deceptively recruit veterans" (SVA, n.d., para. 1).

The editors recognize the complexities associated with the 90/10 loophole and validate a MAL's right to choose an institution that best meets their needs. However, policymakers are encouraged to consider supporting the Protect Veterans' Education and Training Spending (Protect VETS) Act of 2019. This legislation outlines a tiered system of penalties designed to hold *all* institutions receiving G.I. benefits accountable. Penalties for violations escalate over a three-year time period, starting with no new Veterans Affairs (VA) and Department of Defense (DoD) student enrollments in year one and ending with a loss of access to federal funding for at least two years (Protect Veterans' Education and Training Spending [Protect VETS] Act of 2019).

License Portability and Transfer Credit

Of significant concern to MALs are licensure and transfer of credit restrictions as upward of one-third of military spouses work in occupations requiring licensure (Wagnon, 2020). In 2020, the DoD released a report defining the problem and making recommendations for inter-state partnership to lessen the burden on military spouses. IHLs are directly impacted by the licensure challenges as MALs may quit degree programs that require licensure as a direct result of frequent relocation and the unpredictable nature of military service (DoD, 2019). While the DoD has encouraged states to enact reciprocity for military spouses and pass laws for expedited review, "these methods proved insufficient to address the underlying concerns of military spouses" (DoD, 2019, p. 2). Governors and state policymakers are encouraged to take action to create both short and long-term solutions, such as reciprocity through compacts, multi-state agreements between institutions that ease the transfer of credits.

One compact of particular interest as the COVID-19 pandemic rages on is the Multi-State Collaborative on Military Credit (MCMC) (2016), which includes an agreement between thirteen states (Illinois, Indiana, Iowa, Kansas, Kentucky, Michigan, Minnesota, Missouri, Nebraska, North Dakota, Ohio, South Dakota, and Wisconsin) to address the challenges MALs face when pursuing a career in healthcare and social support services (social work, for example). Participating institutions in these states offer accelerated or bridge programs for MALs with credit transferability between institutions to help address relocation concerns within the region (DoD, 2019). Many of these programs, however, target veterans exclusively, offering them credit for military service, underscoring the need for comparable programs supporting spouses.

Policymakers at the federal and state level have a responsibility to advocate for their constituents and work to build stronger, more inclusive communities through policy and legislative action. When the Protect Veterans' Education and Training Spending (Protect VETS) Act of 2019 was announced, Senator Tester, ranking member of the Senate Veteran's Affair Committee proclaimed, "Congress has a responsibility to put our men and women in uniform—and their benefits—first," further reaffirming action must be action to ensure "veterans get the education they deserve" (Committee on Veterans Affairs, 2019, para. 3). While recent legislation, such as the Harry W. Colmery Veterans Educational Assistance Act (2020)[2] and the Protect Veterans' Education and Training Spending (Protect VETS) Act of 2019, support Senator Tester's commitment to service members, more legislative action is needed to support military spouses and dependents to ensure they get the education and opportunities they deserve as vital members of the Armed Forces community.

FOR EDUCATORS AND SUPPORT STAFF

Administrators play a critical role in providing direction and structure to IHLs. Likewise, policymakers ensure access to resources and protections afforded by legislation and state/federal program initiatives. However, when it comes to the day-to-day experience of MALs, no one plays a greater role than educators (professors, instructors, TAs, and so on) and support staff in how well service members adjust and thrive in IHLs. MALs "often [note] that the style of teaching and classroom management used by professors or instructors can make an important difference in [the] adjustment process" (Sportsman & Thomas, 2015, p. 46). Faculty and staff deserve recognition for their continued effort to support all students and their willingness to learn more about MALs and their unique needs through their engagement with this edited volume. As IHLs work to support degree completion for MALs, educators and support staff are encouraged to share their military affiliation, seek out additional educational opportunities to develop military cultural competence, and utilize pedagogical approaches that validate the lived experiences of MALs and their contributions to communities of inquiry.

Identify MAL Affiliation

As with any other student, MALs are looking for opportunities to connect and identify with faculty, staff, and peers. For MALs, having access to information about a faculty or staff member, specifically their connections to military service or community, may help establish credibility through shared lived experience, build relationships based on shared group identity, and affirm MALs lived experiences will be valued in the campus community. Faculty and student support staff are encouraged to consider situating themselves in relation to the military if appropriate and if it feels comfortable sharing this aspect of their personhood (see the chapter by Velez, McDermott, & Marlin in chapter 10 of this edited volume). Along with preferred pronouns and diversity and inclusion statements on syllabi, for example, consider adding any military affiliation, military cultural training (such as Green Zone, for example), research related to the military, or a personal statement regarding your commitment to serving MALs, even if no direct connection to the military exists. Nonverbal displays of military identification/affiliation, such as photos, memorabilia, or insignia, in personal workspaces where MALs are likely to congregate, are also a way to communicate this information in a more informal, yet powerful way. Communicating this information directly to students prior to the start of the semester and continuing throughout the duration of the course may help reduce perceived differentness between a MAL and a faculty/staff member.

While educators and staff members may be encouraged to share a military-connected identity, MALs struggle with whether or not to disclose their military affiliation. As such, MALs may not want to be asked about their military affiliation or experience even if attending class in uniform. MALs are concerned about being politically, socially, or ethically stereotyped by generalizations that do not define their identity (see Howe, chapter 4 in this edited volume). Research indicates MALs receive inappropriate questions regarding their wartime service, suggesting it is best to avoid this topic unless specifically invited to engage by a service member (Steele et al., 2011). Provide a safe space to ensure MALs feel supported disclosing their military affiliation and lived experience, but avoid direct questioning or making assumptions regarding what it must mean to serve unless invited to do by the MAL. MALs may be more comfortable defining the rules for engagement regarding when, where, why, and how they choose to identify and reciprocity (as it relates to disclosure of MAL affiliation) should be encouraged but not expected or required.

Seeking Out Educational Opportunities

VA coordinators assert "educating faculty and staff about the unique needs and contributions of military students is essential for providing relevant, effective support" (Military Family Research Institute, 2015, p. 2). This edited volume provides a starting point for learning more about MAL culture, the unique needs of this population of learners, and the damaging effects of stereotypes and misconceptions about MALs in hopes to contribute to building military competence. However, additional resources may be available on your campus, within your community, and within the larger academic networks to build and promote your overall military competence. On your local campus, partner with the VA Center to identify training opportunities. Seek out community agencies that work with or advocate for veterans and their families. Consider joining and/or attending conference proceedings sponsored by the National Communication Association's (NCA) Communication and Military Division, a group that "promotes scholarly, pedagogical, and community-engaged work that highlights and critiques communication, messages, and discourses within and about the military" (NCA, n.d., Divisions, para. 1). Additionally, the Center for Deployment Psychology at the Uniformed Services University (n.d.) offers a free *Military Culture: Enhancing Clinical Competence* course. Although the target audience is mental health providers, the course offers general level information on military culture, including its history, values, and general operations. Furthermore, educators and staff members tend to serve in "mental health advisory" roles, listening to students disclose mental health concerns and often directing them to campus mental health centers, making the content of the *Military Culture: Enhancing Clinical Competence* course more salient.

The educational opportunities highlighted above identify more formal ways to gain knowledge about this community, but informal interactions may provide additional information on a much more intimate and personal level. For example, one university hosted a faculty and staff luncheon where they were served Meals, Ready-to-Eat (MREs) (Military Family Research Institute, 2015b). The result was a collective sharing of personal and professional experience, including from the service members in attendance. Faculty also extended the learning by inviting service members in attendance into their classrooms, connecting MALs with their peers, and promoting classroom engagement (Military Family Research Institute, 2015b). Educators are also encouraged to physically or virtually visit the VA Center or other resource centers devoted to serving MALs to connect with their support staff. Overall, this allows faculty and staff to communicate their commitment to serving this population of learners through their willingness to meet MALs where they are, physically and mentally/emotionally/cognitively as it relates to supporting their academic achievement and connectedness to their learning communities (Sportsman & Thomas, 2015).

Classroom Policies

Classroom policies are established to clearly define expectations and ensure a standard set of guidelines by which all students will be held accountable. They are necessary and provide needed structure to guide learning communities. While the editors are not advocating for special treatment for one group of students over another, we argue working with MALs may require additional flexibility. MALs should be encouraged to reach out to faculty to discuss specific circumstances that may negatively impact their perceived performance, that is, scheduled field training, deployment, or potential permanent change of station during the semester. A syllabus statement indicating a willingness to discuss strategies to adapt to and overcome perceived barriers is an important first step (American Council on Education, 2018).

Many educators, for example, establish attendance policies to ensure students are physically present as much as possible within the context of the learning community. Absences can negatively impact grades and in some cases, too many absences can result in an automatic failure. Thus, all students are encouraged to schedule appointments, even healthcare appointments, outside of the scheduled class time. Service members may have limited options when it comes to scheduling appointments with the VA, specifically. Some veterans wait up to seventy days for appointments and last-minute scheduling is common (Draper, 2019). Furthermore, an active-duty service member may have a month-long training in the field come up with little to no notice. With careful planning, a dedicated MAL can complete any missed coursework before or after leaving to complete the course. However, not all instructors have provided this level of flexibility for MALs in the past. A service member or veteran who misses class due to VA appointments or training does not equate to a MAL intentionally skipping class due to lack of engagement or motivation (American Council on Education, 2018; see May & McDermott chapter 2 in this edited volume), warranting additional consideration by educators.

In addition to attendance policies, faculty may require class participation as part of a student's overall final grade. These policies are designed to promote active learning and engagement across multiple dimensions. MALs, especially veterans, "may be initially hesitant to actively participate in class discussions" (American Council on Education, 2018, p. 8). The reasons are varied but may include being politically, socially, or ethically stereotyped (see chapter 4 by Howe in this edited volume), perceived differentness, and combat-related disabilities that may impact cognitive processing and heighten communication-related anxiety (Pearrow & Cosgrove, 2009). Educators are encouraged to reimagine equitable classroom participation. Providing multiple channels for class participation may be a strategy for engaging MALs and promoting diversity and inclusion. For example, providing a separate online discussion forum may reduce performance-related stress and provide extra time to formulate a response.

Pedagogical Approaches

Driven by the need to achieve teaching and learning outcomes, educators are united in their efforts to build inclusive and responsive communities of inquiry. However, educators define their role and deploy their teaching practice in many different ways. As such, in this section, the editors offer strategy-driven reflection as a starting point for challenging personal practice. This approach acknowledges there are a plethora of value-added approaches and underscores the importance of academic freedom. The editors encourage readers to review Hernández and Belding chapter 11 in this edited volume for a more comprehensive view of pedagogy and how specific strategies may be employed to better support MALs. The editors acknowledge this section is limited in scope and underscore the intention to provide the opportunity for reflection on a more personal level as we move forward and identify strategies to better support and validate the lived experience of MALs in the classroom. Each reflection opportunity offers a discussion prompt followed by "food for thought" intended to start a conversation while acknowledging the editors have more questions than answers when it comes to meeting the needs of this population.

Reflection #1. How can educators create more inclusive learning communities through language and topic selection? Educators are acutely aware of the impact of diverse and inclusive language on classroom climate. The editors extend this to include the use of inclusive topics that may guide class discussion or content development. As Sciullo (chapter 9) demonstrated, war is a common metaphor used with little regard for people who have experienced it or have been impacted by it, including service members and their families. Moreover, the American Council on Education (2018) recommends educators should not share personal experiences with war or viewpoints about war as it may lead to feelings of uncomfortableness for MALs not ready to confront their own experiences or create further alienation if their lived experiences do not match up to those of the instructors. Likewise, previous research has identified that MALs may feel isolated and disrespected if instructors expressed disapproval of war and/or troops in the classroom (Elliot, Gonzalez, & Larsen, 2011; Syverud, 2019). Taking into account the diversity of experiences and beliefs in the classroom and being cognizant of the language used is vital for including all students within the learning environment.

Reflection #2. How can we validate and recognize the diversity of MALs and the challenges they may face building and sustaining their student identity? For some MALs, entering the classroom and adopting the student identity may be a complete paradigm shift, "a restructuring of their world view requiring [them] to entirely give up their previous assumptions" (Sportsman & Thomas, 2015, p. 45). The paradigm shift may be a slow, tedious process that occurs over many semesters. As faculty work with MALs, especially in entry-level courses, supporting the paradigm shift becomes critically important. To support this shift, educators are challenged to identify opportunities to validate and recognize their lived experiences while encouraging new ways of knowing. MALs bring extensive lived experiences, cultural knowledge, and diverse leadership background that should be utilized within the college classroom (Caddick, 2018). As seen in a variety of chapters from Chiara, Chiara, and Ashlock (chapter 3) to Hernández, De Los Santos Upton, and Aguirre (chapter 7), MALs have a diverse range of potentially competing identities (see chapter 2 by May & McDermott) and helping them connect these differing identities/challenge these identities/reframe these identities within the context of the classroom becomes critically important.

While previous research has identified MALs want faculty members to better acknowledge and understand their viewpoints and perspectives (Barry et al., 2014; Steele et al., 2011; see May & McDermott, chapter 2), female veterans may have different experiences integrating into civilian life and the classroom (Amer & Guoewi, 2018; Tam-Seto, 2019). Often characterized as "Invisible Veterans" female veterans may feel as though their service is not valued or recognized by the public. Moreover, the identity of "veteran" may evoke specific personality and outward appearance characteristics that discount those who do not identify with the masculine. Thus, assuming to know of a MALs previous experiences, viewpoints, and beliefs can negate their voice. It is vital for instructors to engage in cultural awareness, sensitivity, and knowledge when communicating with all MALs, and specifically female veterans (Tam-Seto, 2019). However, the experiences of female and non-gender conforming veterans remains significantly underdeveloped in the literature.

Challenge #3. How can educators connect military learning practices with academic communities of inquiry? As defined by Chiara, Chiara, and Ashlock (chapter 3 in this edited volume) service members are taught to learn in very specific ways, including standardized curriculum and assessments, guided development through mentorship, and strict adherence to military regulations. While most college classrooms are based on an individual educator's relationship to their discipline with considerable academic freedom regarding how content is presented and assessed, there are shared andragogical approaches that may provide a scaffold and opportunities for transference (see chapter 15 by St. Pierre in this edited volume). For example, collaborative activities may be a way to connect MALs with their military-related and service experiences. The military, as an institution, values teamwork, and providing opportunities for MALs to collaborate with their peers to produce a final deliverable may help them connect to learning in a way that is familiar while also challenging them to consider different perspectives (Sportsman & Thomas, 2015). In chapter 14, Ludewig highlights the importance of collaboration during the writing process as an example of how these strategies can be employed within the higher ed classroom.

Reflection #4. How can educators build genuine and informed relationships with MALs? Power is an important consideration within the context of any learning environment. While MALs tend to be older and more experienced (Postsecondary National Policy Institute, 2018), many acknowledge the "rank" system (see May & McDermott, chapter 2) with the educator as the helm of the classroom. Educators are challenged to initiate conversations with MALs to lessen the perceived power distance. Asking questions and listening are fundamental communication skills that serve educators well when interacting with students in general, including MALs. Moreover, check-in with MALs to learn more about their transition to the classroom; this may be an opportunity to connect them with additional support. As noted by McDermott and May (see chapter 2 in this edited volume), MALs may not feel comfortable reaching out for help or initiating conversations themselves (see chapter 3 by Chiara, Chiara, and Ashlock, as well as chapter 4 by Howe in this edited volume). However, be genuine and informed. Research indicates MALs may perceive questions such as "are you ok?" as "meaningless pleasantry" and often respond in the affirmative even if they are struggling (Sportsman & Thomas, 2015, p. 49). As an alternative, validate a MALs contribution by providing positive feedback regarding their academic performance or level of engagement with some aspect of college life (Sportsman & Thomas, 2015).

Reflection #5. How can educators frame instructions to provide meaningful roadmaps for MALs to follow while providing opportunities for higher-order thinking? MALs, especially active-duty service members and veterans, often navigate assignments using detailed checklists and clearly defined instructions. As noted by Sportsman & Thomas (2015), "detailed steps are nearly always outlined for every military task, and strict compliance is expected" (p. 46). As MALs transition into the classroom, they are likely to encounter educators who take a more "open" approach to allow freedom of thought and expression. This can be challenging for MALs. For example, when completing assignments or in-class work, MALs may experience frustration by what they perceive as no, vague, or unclear instructions. As such, the "college way" of doing things may seem lax, unclear, or lacking in specificity" (Sportsman & Thomas, 2015, p. 46). Educators are challenged to reimagine assignment descriptions in ways that meet a MALs' need for structure while guiding them to take on more cognitive "risk."

Collectively, these discussion prompts offer opportunities for educators to be more active in the process of designing intervention strategies to better meet the needs of MALs in *their* classroom based on preferred pedagogical practices and style of teaching. Moreover, open-endedness of this approach gives the collective academic community an opportunity to answer the earlier call for innovation by challenging norms and deploying a Total Force to bring about meaningful and substantive change (U.S. Army, 2016).

FINAL THOUGHTS

MALs encounter many of the same issues and concerns as their civilian counterparts with added challenges from military-specific experiences that impact their transition to college (Drummet, Coleman, & Cable, 2003). For MALs, the transition out of military culture and into the classroom presents a challenging and uncertain crossroads. Administrators, policymakers, educators, and support staff have the ability and power to remove some of the barriers MALs are currently facing on their journey to meet their educational goals. Throughout this edited volume, the editors, along with our contributors, sought to explicate these barriers and provide readers with tangible, actionable solutions for improving MALs experiences and retention in higher education, but there is an incredible amount left to explore when it comes to meeting the needs of this diverse population of learners. Notably missing from the list of recommendations are the following:

- MALs. Collectively, more research is needed from the MAL perspective to better understand how they sensemake their experiences along with the specific strategies they use to promote success, however they define it, in the classroom and across the campus community.
- Female and non-gender conforming MALs. Specifically, research indicates female veterans may have different experiences integrating into civilian life and the classroom (Amer & Guoewi, 2018; Tam-Seto, 2019). These differences, however, are not fully defined nor recognized in IHLs. Even less is known about the experiences of non-gender conforming MALs in higher education.
- Male and non-gender conforming spouses. The bulk of what we know about the experiences of military spouses comes from research driven by female spouses of male service members. How male and non-gender conforming spouses navigate higher education and the realities of military service (deployment, frequent relocations, and so on) is not fully explicated in the literature.
- Students with no connection to the military. A body of literature suggests MALs feel disconnected from peers in the classroom (Steele et al., 2011; see chapter 2 by May & McDermott in this edited volume, as well as chapter 3 by Chiara, Chiara, & Ashlock). Other than anecdotal stories from MALs, researchers know little about how students with no military affiliation perceive MALs and sense make their experiences with MALs in the classroom.
- MALs of Diverse Backgrounds. While Hernández, De Los Santos Upton, & Aguirre (chapter 7) provided a micro-history of MAL diversity, and Pérez (chapter 6) provided an overview of Mexican and Mexican Americans' history serving in the Armed Forces, the literature on the experiences of MALs of color remains insufficient. With the rise in awareness and national support for the Black Lives Matter movement and increasing demand to combat violence against women service members, service members of color, and LGBTQ+ service members, the voices of diverse MALs and their experiences in the higher education classroom demand investigation.
- VA Center Directors. Directors play a critical role in serving this population; however, their voice is currently underrepresented in the literature.

As we look toward an uncertain future for higher education and what it means to be a member of the academy, we are challenged more than ever to advocate for our students and challenge institutional barriers that prevent access and achievement. Our military-affiliated students have the experience, the resilience, the passion, and the gumption to succeed in and enhance IHLs. Thus, we encourage educators around the world to rethink what it means to create and foster more inclusive spaces for MALs of all diverse backgrounds and identities. This includes creating course learning outcomes and assignments with MAL experiences and contextual factors in mind, decolonizing the communication classroom through the introduction of counterhegemonic narratives, and cultivating more supportive spaces to encourage MAL and non-MAL students to engage in identity exploration and collaboration. It is our job as members of the academy to advocate for MALs. Maintaining the status quo of MAL treatment will result in a failure to support a community of learners that so graciously sought to serve us.

NOTES

1. Data reflect funding between 2009 and 2018 for 2.3 million beneficiaries (CBO, 2019).
2. This legislation was signed by Trump in 2017; however, amendments take effect August 1, 2020.

REFERENCES

Alwine, R. (2020, July 23). *Is milspouse employment a national issue with a local solution?* https://www.military.com/spousebuzz/2020/07/23/milspouse-employment-national-issue-local-solution.html.

Amer, S., & Guowei, J. (2018). It's a man's world: A qualitative study on females in the military. *Pennsylvania Communication Annual, 74*, 48–74.

American Council on Education. (2015, Spring). *By the numbers: Undergraduate student veterans.* https://www.acenet.edu/the-presidency/columns-and-features/Pages/By-the-Numbers-Undergraduate-Student-Veterans.aspx.

American Council on Education. (n.d.). *Joint services transcript.* https://www.acenet.edu/Documents/Joint-Services-Transcript-Brochure.pdf.

American Council on Education. (2018). *Toolkit for veteran friendly institutions.* https://www.acenet.edu/Documents/Veterans-Toolkit-2018.pdf.

Arnold, L. (2020). *Dollar signs in uniform': How for-profit education continues to target GI Bill money* [Audio Podcast]. https://www.arnoldventures.org/podcasts/podcast-dollar-signs-in-uniform-how-for-profit-education-targets-gi-bill-money.

Barry, A. E., Whiteman, S. D., & MacDermid Wadsworth, S. (2014). Student servicemembers/veterans in higher education: A systematic review. *Journal of Student Affairs Research and Practice, 51*(1), 30–42. doi:10.1515/jsarp-2014-0003.

Bergman, M., & Herd, A. (2016). Proven leadership = college credit: Enhancing employability of transitioning military members through prior learning assessment. *Advances in Developing Human Resources, 19*(1), 78–87. doi:10.1177/1523422316682949.

Borsari, B., Yurasek, A., Miller, M. B., Murphy, J. G., McDevitt-Murphy, M. E., Martens, M. P., Darcy, M. G., & Carey, K. B. (2017). Student service members/veterans on campus: Challenges for reintegration. *The American Journal of Orthopsychiatry, 87*(2), 166–175. doi:10.1037/ort0000199.

Caddick, N. (2018). Life, embodiment, and (post-) war stories: Studying narrative in critical military studies. *Critical Military Studies*. doi:10.1080/23337486.2018.1554942.

Cate, C. A., & Albright, G. (2015). Supporting student veterans: Utilizing game-based role-plays with virtual humans to build military cultural competency and helping behaviors in faculty and staff. *Online Learning, 19*, 48–63.

Center for American Progress. (2012). *Easing the transition from combat to the classroom*. https://www.americanprogress.org/wp-content/uploads/issues/2012/04/pdf/student_veterans.pdf.

Center for First Generation College Success. (2016). *First generation college students*. https://firstgen.naspa.org/files/dmfile/FactSheet-01.pdf.

Chiara, N. A., Chiara, M., & Ashlock, M. Z. (2020). What is impacting the military-affiliated learner experience in Higher Education. In V. McDermott, L. H. Hernández, & A. May (Eds.), *Supporting the military-affiliated learner: Communication approaches to military pedagogy & education* (pp. 47–65). Lexington Books.

Church, T. E. (2009). Returning veterans on campus with war related injuries and the long road back home. *Journal of Postsecondary Education and Disability, 22*(1), 43–52.

Cobos, A. (2020). Two-way communication as a constructivist pedagogy tool for enhancing meaning making in digital learning classrooms with military affiliated learners. In V. McDermott, L. H. Hernández, & A. May (Eds.), *Supporting the military-affiliated learner: Communication approaches to military pedagogy & education* (pp. 257–278). Lexington Books.

Committee on Veterans Affairs. (2019, November 15). *Tester, colleagues introduce first-ever bipartisan bill to protect student veterans from predatory for-profit institutions*. https://www.veterans.senate.gov/newsroom/minority-news/tester-colleagues-introduce-first-ever-bipartisan-bill-to-protect-student-veterans-from-predatory-for-profit-institutions.

Congressional Budget Office [CBO]. (2019, May). *The Post-9/11 GI Bill: Beneficiaries, choices, and cost*. https://www.cbo.gov/system/files/2019-05/55179-Post911GIBill.pdf.

Department of Defense [DoD]. (2012). *Supporting our military families: Best practices for streamlining occupational licensing across state lines*. https://download.militaryonesource.mil/12038/MOS/Reports/Occupational-Licensing-and-Military-Spouses-Report.pdf.

Department of Defense [DoD]. (2019). *Military spouse licensure: State best practices and strategies for achieving reciprocity*. https://download.militaryonesource.mil/12038/MOS/Reports/military-spouse-licensure-report-2019.pdf.

DiRamio, D., Ackerman, R., & Mitchell, R. L. (2008). From combat to campus: Voices of student-veterans. *Journal of Student Affairs Research and Practice, 45*, 73–102.

Doenges, T. J. (2011). *Calling and meaningful work among student military veterans: Impact on well-being and experiences on campus*. [Doctoral Dissertation, Colorado State University]. Researchgate.

Draper, D. A. (2019, July 14). Veterans healthcare: Opportunities remain to improve appointment scheduling within VA and through community care. *U.S. Government Accountability Office*. https://www.gao.gov/assets/710/700431.pdf.

Drummet, A. R., Coleman, M., & Cable, S. (2003). Military families under stress: Implications for family life education. *Family Relations, 52*(3), 279–287.

Elliott, M., Gonzalez, C., & Larsen, B. (2011). U.S. military veterans transition to College: Combat, PTSD, and alienation on campus. *Journal of Student Affairs Research and Practice, 48*, 279–296. doi:10.2202/1949-6605.6293.

Harry, W. (2017). *Colmery Veterans Educational Assistance Act of 2017*. 3218 U.S.C. § 115c.

Hernández, L. H. (2019). "We get it done because we have to:" Military spouses with advanced degrees, career and educational experiences, and grit amongst uncertainty. In L. H. Hernández & J. N. Belding (Eds.), *Military spouses with graduate degrees: Interdisciplinary approaches to thriving amidst uncertainty* (pp. 239–258). Lanham, MD: Lexington Books.

Hernández, L. H., & Belding, J. N. (2020). Best practices in supporting military-affiliated learners: Suggestions from communication and psychological sciences. In V. McDermott, L. H. Hernández, & A. May (Eds.), *Supporting the military-affiliated learner: Communication approaches to military pedagogy & education* (pp. 231–256). Lexington Books.

Hernández, L. H., De Los Santos Upton, S., & Aguirre, A. A. (2020). Nepantla, liminality, and teaching and learning in the space between: A borderlands approach to military-affiliated learner pedagogy. In V. McDermott, L. H. Hernández, & A. May (Eds.), *Supporting the military-affiliated learner: Communication approaches to military pedagogy & education* (pp. 133–157). Lexington Books.

Higher Education Opportunity Act. (2010). 20 U.S.C. § 1091c. https://www.govinfo.gov/content/pkg/USCODE-2015-title20/pdf/USCODE-2015-title20-chap28-subchapIV-partF-sec1091c.pdf.

Huebner, A. J., Mancini, J. A., Wilcox, R. M., Grass, S. R., & Grass, G. A. (2007). Parental deployment and youth in military families: Exploring uncertainty and ambiguous loss. *Family Relations, 56*, 112–122.

Howe, W. T. (2020). Supporting military veterans in the graduate classroom: A case study of political beliefs and discussion. In V. McDermott, L. H. Hernández, & A. May (Eds.), *Supporting the military-affiliated learner: Communication approaches to military pedagogy & education* (pp. 67–86). Lexington Books.

Iraq and Afghanistan Vets of America. [IAVA]. (n.d.). *Storm the hill*. https://iava.org/storm-the-hill/.

Kezar, A. (2019). *Creating a diverse student success infrastructure: Keys to catalyzing cultural change for today's student*. American Council on Education. https://www.acenet.edu/Documents/Creating-a-Diverse-Student-Success-Infrastructure.pdf.

King, B. M., Carr, D. C., & Taylor, M. G. (2020). Loneliness following widowhood: The role of the military and social support. *The Journal of Gerontology: Series B*, gbz164, doi:10.1093/geronb/gbz164.

Kirchner, M. J., Coryell, L., & Yelich Biniecki, S. M. (2014). Promising practices for engaging student Veterans. *Quality Approaches in Higher Education, 5*(1), 12–18. http://asq.org/edu/2014/09/promising-practices-for-engaging-student-veterans.pdf.

Lederman, D. (2020, April 22). *How teaching changed in the (forced) shift to remote learning*. Inside HigherEd. https://www.insidehighered.com/digital-learning/article/2020/04/22/how-professors-changed-their-teaching-springs-shift-remote.

Lee, V., & Looney, A. (2019). *Understanding the 90/10 Rule: How reliant are public, private, and for-profit institutions on federal aid?* The Brooking Institution. https://www.brookings.edu/wp-content/uploads/2019/01/ES_20190116_Looney-90-10.pdf.

Ludewig, A. (2020). Collaborative apprenticeship approaches to writing instruction in the military. In V. McDermott, L. H. Hernández, & A. May (Eds.), *Supporting the military-affiliated learner: Communication approaches to military pedagogy & education* (pp. 311–336). Lexington Books.

May, A., & McDermott, V. (2020). Understanding dual identity formation: Identifying commonalities and resolving "differentness" for military affiliated learners. In V. McDermott, L. H. Hernández, & A. May (Eds.), *Supporting the military-affiliated learner: Communication approaches to military pedagogy & education* (pp. 11–45). Lexington Books.

McGregor, B. (2020). "So we hit it with a hammer, and it worked": How active-duty marines communicate, collaborate, and forge their own learning communities in social media. In V. McDermott, L. H. Hernández, & A. May (Eds.), *Supporting the military-affiliated learner: Communication approaches to military pedagogy & education* (pp. 279–309). Lexington Books.

Midwestern Higher Education Compact. (2016). *Valuing military learning: A guide to prior military learning assessment and more*. https://www.mhec.org/sites/default/files/resources/20160803MCMC_Guide_to_Military_PLA.pdf.

Military Family Research Institute. (2015a). *Appreciating military service realities: Higher learning institution policies regarding deployment and GI Bill benefits*. https://www.mfri.purdue.edu/wp-content/uploads/resources/landscape/MFRI_LandscapeInHigherEducation-Appreciating.pdf.

Military Family Research Institute. (2015b). *Rallying the troops: Enlisting the support of faculty, staff, and community veterans*. https://www.mfri.purdue.edu/wp-content/uploads/resources/landscape/MFRI_LandscapeInHigherEducation-Rallying.pdf.

Military Spouse Career Education Act. (2020). 7112 U.S.C. § 116c. https://www.congress.gov/bill/116th-congress/house-bill/7112.

Moore, R. G. (2019). *Military veterans' academic success and satisfaction through social media peer mentorship involvement*. (Publication No. 22588045). [Doctoral Dissertation, Capella University]. ProQuest Dissertations & Theses Global.

Moullet, Z., & Belding, J. N. (2020). Supporting the military-affiliated learner's medical school aspirations. In V. McDermott, L. H. Hernández, & A. May (Eds.), *Supporting the military-affiliated learner: Communication approaches to military pedagogy & education* (pp. 159–180). Lexington Books.

National Center for Education Statistics. (2020, May). *Postsecondary institution revenues*. https://nces.ed.gov/programs/coe/indicator_cud.asp.

National Communication Association [NCA]. (n.d.). *NCA interest groups*. https://www.natcom.org/about-nca/membership-and-interest-groups/nca-interest-groups.

Nichols-Casebolt, A. (2012). The Green Zone: A program to support military students on campus. *About Campus*. https://www.naspa.org/images/uploads/kcs/Nichols_The_Green_Zone_A_program_to_Support_Military_Students_on_Campus_2012.pdf.

Pearrow, M., & Cosgrove, L. (2009). The aftermath of combat-related PTSD: Toward an understanding of transgenerational trauma. *Communication Disorders Quarterly, 30*(2), 77–82. doi:10.1177/1525740108328227.

Pérez, F. G. (2020). Speaking chicana/o secrets to military students: An autoethnography of teaching senior military noncommissioned officers. In V. McDermott, L. H. Hernández, & A. May (Eds.), *Supporting the military-affiliated learner: Communication approaches to military pedagogy & education* (pp. 109–131). Lexington Books.

Postsecondary National Policy Institute. (2018). *Veterans in higher education*. https://pnpi.org/first-generation-students.

Protect Veterans' Education and Taxpayer Spending Act of 2019. (2019). 2857 U.S.C. § 116c.

Radford, A. W., Bentz, A., Dekker, R., & Paslov, J. (2016). *After the Post-9/11 GI Bill: A profile of military service members and veterans enrolled in undergraduate and graduate education*. https://nces.ed.gov/pubs2016/2016435.pdf.

Reyes, A. T., Kearney, C. A., Isla, K., & Bryant, R. (2017). Student veterans' construction and enactment of resilience: Constructivist theory study. *Journal of Psychiatry Mental Health Nursing, 25*, 37–48. doi:10.1111/jpm.12437.

Rossetto, K. R., & Owlett, J. S. (2020). Challenges for military partners pursuing college and resilience-based strategies to support their academic success. In V. McDermott, L. H. Hernandez, & A. May (Eds.), *Supporting the military-affiliated learner: Communication approaches to military pedagogy & education* (pp. 87–106). Lexington Books.

Runge, C. E., MacKenzie A., & McGuire, A. C. L. (2014). Spouses of military members' experiences and insights: Qualitative responses to an open-ended question in a survey of health and wellbeing. *PLoS One, 10*(2), e0118433. doi:10.1371/journal.pone.0118433.

Ryu, M. (2013). *Credit for prior learning from the student, campus, and industries perspectives*. https://www.acenet.edu/Documents/Credit-for-Prior-Learning-Issue-Brief.pdf.

Schreger, C., & Kimble, M. (2017). Assessing civilian perceptions of combat veterans: An IAT study. *Psychological Trauma: Theory, Research, Practice and Policy, 9*(Suppl 1), 12–18. doi:10.1037/tra0000191.

Sciullo, N. J. (2020). Nuclear war!: Theorizing the negative impacts of policy debate's tropogical mistreatment of war for military-affiliated learners. In V. McDermott, L. H. Hernández, & A. May (Eds.), *Supporting the military-affiliated learner: Communication approaches to military pedagogy & education* (pp. 181–200). Lexington Books.

Smith, B. (2013). *Mentoring at-risk students through the hidden curriculum of higher education*. Lexington.

Southwell, K. H., Whiteman, S. D., MacDermid Wadsworth, K. M., & Barry, A. E. (2018). The use of university services and student retention: Differential links for student service members or veterans and civilian students. *Journal of College Student Retention: Research, Theory, & Practice, 19*(4), 394–412. doi:10.1177/1521025116636133.

Steele, J. L. (2012). *Military veteran's experiences in for-profit higher education*. https://www.rand.org/content/dam/rand/pubs/testimonies/2012/RAND_CT376.pdf.

Steele, J. L., Salcedo, N., & Coley, J. (2011). *Service members in school: Military veterans' experiences using the Post-9/11 GI Bill and pursuing postsecondary education*. Rand National Defense Research Institute. https://www.rand.org/content/dam/rand/pubs/monographs/2011/RAND_MG1083.pdf.

Student Veterans of America [SVA]. (n.d.). Student Veterans of America, a national coalition of over 500 chapters, supports H.R. *3447 – Veterans' Education Protection Act*. https://studentveterans.org/media-news/press-releases/229-sva-releases-findings-from-the-million-records-project.html.

Sportsman, M. A., & Thomas, L. (2015). Coming home to school: Challenges and strategies for effective teaching with military Veterans. *InSight: A Journal of Scholarly Teaching, 10,* 43–55. https://eric.ed.gov/?id=EJ1074051.

St. Pierre, C. (2020). Adapt and overcome: Military-affiliated learners and transfer of learning. In V. McDermott, L. H. Hernández, & A. May (Eds.), *Supporting the military-affiliated learner: Communication approaches to military pedagogy & education* (pp. 337–371). Lexington Books.

Syverud, K. (2019, June 21). Commentary: Universities need to better serve military students and veterans. *U.S. News*. https://www.usnews.com/news/education-news/articles/2019-06-21/commentary-higher-education-must-better-serve-military-students-and-veterans.

Tam-Seto, L. (2019). Need for cultural competency in the mentorship of female veterans during civilian transitions. *Journal of Military, Veteran and Family Health, 5*(1), 115–118. doi:10.3138/jmvfh.2017-0049.

Thompson, G. (2019). *Relationship between self-development, mentorship, and senior military officer moral judgment.* (Publication No. 1725). [Doctoral Dissertation, University of Southern Mississippi]. The Aquila Digital Community.

Uniformed Services University. (n.d.). *Military culture: Enhancing clinical competence course description*. https://deploymentpsych.org/content/military-culture-enhancing-clinical-competence.

U.S. Department of the Army. (2016). *Army innovation strategy: 2017–2021*. https://api.army.mil/e2/c/downloads/493916.pdf.

Velez, M., McDermott, V., & Marlin, K. W. (2020). The hard sell: Supporting MALs through resource centers. In V. McDermott, L. H. Hernández, & A. May (Eds.), *Supporting the military-affiliated learner: Communication approaches to military pedagogy & education* (pp. 201–228). Lexington Books.

Veterans Affairs. (2020, February 11). *Education and training*. https://www.benefits.va.gov/GIBILL/resources/benefits_resources/rate_tables.asp#ch33.

Villagran, M., Canzona, M. R., & Ledford, C. J. (2013). The milspouse battle rhythm: Communicating resilience throughout the deployment cycle. *Health Communication, 28*(8), 778–788.

Wagnon, E. (2020, June 23). *MOAA endorses bill to expand education and career opportunities for military spouses*. Military Officers Association of America. https://www.moaa.org/content/publications-and-media/news-articles/2020-news-articles/advocacy/moaa-endorses-bill-to-expand-education-and-career-opportunities-for-military-spouses/.

Whiteman, S. D., Barry, A. E., Mroczek, D. K., & MacDermid Wadsworth, S. (2013). The development and implications of peer emotional support for student service members/veterans and civilian college students. *Journal of Counseling Psychology, 60*(2), 265. doi:10.1037/a0031650.

Williams, R., Mariani, J., Routh, A., Keyal, A., & Hill, M. (2020). *Military spouse unemployment: Exploring solutions to a local problem of national importance*. https://www2.deloitte.com/us/en/insights/industry/public-sector/military-spouse-unemployment.html?id=us:2el:3pr:4di6601:5awa:6di:MMDDYY:&pkid=1006981.

Zhao, C., & Kuh, G. (2004). Adding value: Learning communities and student engagement. *Research in Higher Education, 45*, 115–138.

Index

About the Editors and Contributors

Victoria McDermott (MA University of Alaska Fairbanks) is a current PhD student and instructor at the University of Maryland and adjunct faculty at the University of Alaska Fairbanks. She is an enlisted military spouse of four years. Her research explores the communication of power in traditionally masculine environments specifically within military and science fields (i.e., field stations and marine laboratories). Previously, Victoria has researched the potential impact of the clash of military and academic cultures on spouses' educational experiences with a chapter in *Military Spouses with Graduate Degrees: Interdisciplinary Approaches to Thriving amidst Uncertainty* (2019). Other work has included paper presentations, panels, and short courses at the National Communication Association Conference in 2018, 2019, and 2020 and the National Communication Association Conference in 2019 regarding the varied experiences of MALs such as supporting MALs in public speaking centers, the effect of reintegration on marital communication, and the impact of military elopement on family communication. Victoria has witnessed the struggles MALs face in academia as a military spouse with a graduate degree, from her husband's experience as an enlisted service member attempting to juggle the military and college, and from working with a variety of MALs throughout her years of teaching. Thus, she strives to act as an advocate on college campuses for MALs. Victoria looks forward to continuing her career researching military culture, military-affiliated experiences, and developing strategies to improve MALs integration into higher education.

Leandra Hinojosa Hernández (PhD, Texas A&M University) is an assistant professor in the Department of Communication at Utah Valley University. She is the coauthor of *Challenging Reproductive Control and Gendered Violence* *in the Américas: Intersectionality, Power, & Struggles for Rights* (2019) and co-editor of the following books: *This Bridge We Call Communication: Anzaldúan Approaches to Theory, Method, and Praxis* (2019), *Military Spouses with Graduate Degrees: Interdisciplinary Approaches to Thriving amidst Uncertainty* (2019), and *Latina/o/x Communication Studies: Theories, Methods, and Practice* (2019). Leandra has published peer-reviewed scholarly journal articles in outlets such as *Women's Studies in Communication, Communication Research, Women & Language, Frontiers in Communication, Journal of Media and Religion*, and *Health Communication.* As a MAL herself (the spouse of a Navy veteran), she is passionate about advocating for equitable MAL education, mentorship, retention, and success.

Amy R. May (PhD, University of Wisconsin-Milwakuee) has spent roughly 20 years married to an Army officer, navigating the complexities of military life while struggling to maintain an academic career as a trailing spouse. As a student, she struggled to find institutions willing to support her as a military spouse pursuing a graduate degree. As an educator in the classroom, she witnesses MALs struggle to "fit in" and find their way in a system not fully "woke" to their unique needs. She has taught at four major colleges and universities. At each one, MALs were directed to an overwhelmed and understaffed veterans resource center. She advocates for a more holistic approach to support MALs, and as a resource center director, she strives to maintain culturally responsive coaching practices to create a safe space for growth and development. Finally, as a dual citizen of the academy and military, she uses her voice to educate peers, advocate for institutional resources, and design instructional strategies to empower MALs in the classroom. Dr. May is currently an assistant professor at Shippensburg University in the Department of Human Communication Studies. Her research has been featured in several peer-reviewed journals, including *Journal of GLTB Family Studies, Teaching in Higher Education,* and *Patient Education and Counseling.* She is currently serving as co-editor on the forthcoming edited volume *Women of the Wild: Challenging Gender Disparities in Remote Field Stations and Marine Laboratories.*

* * *

Arthur A. Aguirre, PhD, received both his Bachelor's (2007) and Master's (2011) of Arts in Communication from the University of Texas at El Paso (UTEP) and his doctorate in Communication from the University of New Mexico (2015). His research has focused on a variety of topics within communication and leadership including intercultural communication, organizational communication, public relations, leadership studies, fantasy sports, and interpersonal communication. He has presented his work on fantasy sports, pop culture and communication, food and communication, and leadership pedagogy at various conferences, and these projects are being edited for journal submission. He is also currently working on a book on leadership communication and team building. He is currently an assistant professor of instruction at UTEP in the Leadership and Community Engagement Program and also works as a consultant through his own company Crowd and Town Creative. Outside of academia and consulting, Art loves spending time with his family, his dog, Poppy, and is the host of The Forefront with Art Aguirre podcast.

Mary Z. Ashlock is an associate professor in the Department of Communication at the University of Louisville where she has worked for the past fourteen years. She earned her BS degree (1984) in Education, a MS (1986) in College Student Personnel, and a PhD in Communication (1989) from Florida State University. Mary has worked in higher education and business for thirty years. She was a senior consultant in the areas of teamwork, leadership development and project management. Mary's military affiliation began as a child growing up as an Army dependent and then as an Air Force spouse. In her role as a consultant, she facilitated The Citadel [The Military College of South Carolina] Board of Visitors meeting in preparation to admit women to The Citadel. Currently, Mary's son is in his third year at the U.S.A.F. Academy and her daughter is a national Air Force ROTC college scholarship recipient. Mary is active in promoting efforts for the National Communication Association "Communication and Military" division. During Mary's tenure at the university, she has been named as a Faculty Favorite and Faculty Mentor including the Kentucky Communication Association Annual Undergraduate Mentor Award. Mary has published numerous articles, book chapters and co-edited several books. Her background includes work in motivational communication, customer service, and human resources in a variety of service and manufacturing industries where she specializes in teaching and research in organizational and interpersonal communication. Her current research focuses on millennials in higher education and organizations.

Jennifer N. Belding, PhD, earned her doctorate in social psychology from The Ohio State University. While in traditional academic settings, she scientifically studied the interplay of attitudes, persuasion, motivation, and goals. Drawing on her unique perspective that people can be persuaded to achieve their goals, she studied when and how people are receptive to negative feedback, how people are persuaded by themselves or others to achieve goals, and how the strength of one's goals influence their behavior. In nontraditional academic settings, such as being a contractor at a military research command, Dr. Belding has studied issues directly relevant to the health and wellbeing of our service members, including epidemiological investigations related to blast injury and traumatic brain injury, as well as developing and evaluating interventions to improve leadership, team communication, and clinical practice. She currently serves as a Lecturer for the University of California, San Diego, Department of Psychology. In her teaching career, she has taught courses on introductory psychology, social psychology, research methodology, social cognition, stereotyping, prejudice, and discrimination, judgment and decision making, teaching of psychology, and a post-baccalaureate class on the psychological, sociological, and biological foundations of behavior for premedical students. Dr. Belding is married to LT Christopher Martin, an orthopedic surgery resident in the United States Navy, and together they share one son and are expecting a daughter in the near future.

Michael "Mick" Chiara currently serves as the Chief of Protocol for Fort Knox, Kentucky. He is a twenty-year retired veteran of the U.S. Army where he was an Army Musician (trombone) and served in six different Army bands around the United States, Germany, and the Republic of South Korea. During his Army career, he also served in three different staff positions as a training course and doctrine writer, noncommissioned officer advanced courses instructor, and capped off his career in charge of the national recruiting program for Army musicians. He holds a Bachelor's degree in Communication from University of Louisville and a Master's degree in Business Communication and Organizational Leadership from Spalding University. Since retiring from the Army he has also worked as a strategic communications and outreach specialist for Army marketing, and as a recruiting operations specialist and instructor for U.S. Army Cadet Command where he taught pre-command and recruiting operations officer courses for military science instructors and recruiting officers. He volunteered with the local Boy Scouts of America district and served over six years as the Scoutmaster for Troop 128, one of the Boy Scout troops on Fort Knox. He is married to the former Dr. Nancy Neer, who is also a retired Army musician, and they have two adult children.

Nancy A. Chiara is currently the Metadata Librarian for Jefferson County Public Schools in Louisville, Kentucky where she has worked for the past thirteen years. She also teaches graduate education courses in information and technology subjects. She has worked in libraries in some capacity for over forty years including academic, public, military, and school libraries. She connects her role as an educator to her military experience because the ideals of teamwork, comradery, and a shared mission which are embedded in the military environment are also components that foster learning and engagement. Nancy's military career began 1989 after she graduated with an undergraduate degree in music. She enlisted in the U.S. Army as a bandsman and served for twenty-two years; ten years active duty, and twelve years in the reserve components. While on active duty she had the opportunity to combine her library skills with her military occupational specialty (MOS) and served as Reference Librarian at the Armed Forces School of Music, the training institution for Army, Navy and Marine Corps bandsmen. Nancy took advantage of the educational opportunities available on the military installations where she was stationed and also the financial benefits for service members while attending school in a civilian environment. She earned a Bachelor of Music in Performance from Columbus State University, a Master of Science in Library and Information Studies from Florida State University, a second masters in Music Education from Eastern Kentucky University, and an EdD in Educational Technology from Morehead State University.

April Cobos is a lecturer in the English Department at Christopher Newport University in Newport News, Virginia. She received her PhD from Old Dominion University in rhetoric and composition. Her current research interests include women and gender in professional workplaces, military-affiliated learners, teaching with twenty-first century tools and technologies, as well as teaching grant writing, professional and technical writing, and civic engagement. She has taught traditional, online, hybrid, and live-streaming composition, rhetoric, professional writing, and writing for civic engagement at various universities and community colleges with close military affiliations over the past fifteen years.

Sarah De Los Santos Upton (PhD, University of New Mexico) is an assistant professor in the Department of Communication at the University of Texas at El Paso. Her research and teaching explore nepantla identity, borderland pedagogy, and reproductive justice. She is the coauthor of *Challenging Reproductive Control and Gendered Violence in the Américas: Intersectionality, Power, and Struggles for Rights*, winner of the NCA Feminist and Women's Studies Division's 2018 Bonnie Ritter Book Award. She is also the co-editor of *Latino/a Communication Studies: Theories, Methods, and Practice.* Her research is published in the journals *Departures in Critical Qualitative Research*, *Women's Studies in Communication, Development in Practice*, *Action Research*, and *Frontiers in Communication* as well as various edited volumes.

William Howe is an assistant professor in the Department of Communication at the University of Kentucky with a focus on Organizational Communication. William received his Doctor of Philosophy degree from the University of Oklahoma in Organizational Communication and he received both his Bachelor and Master of Science degrees from Texas Christian University in Communication Studies. William focuses primarily on High-Reliability Organizations such as the military, police, emergency weather forecasters, etc. William has published nine articles to date and has presented over twenty competitive papers at conferences, including two top papers. William was awarded Master's Thesis of the Year by the Master's Division of the National Communication for his examination military entry. He has also the Stanley L. Saxton Applied Research Award for his work with military veterans. Before beginning his academic career, William served in the U.S. Army as a combat medic. He deployed in support of Operation Iraqi Freedom.

Ashley Ludewig currently serves as Interim Director of Composition at Auburn University in Auburn, Alabama. She earned her PhD in Rhetoric and Composition from the University of Louisville in 2017. Her dissertation, The Military Meets the University: Mapping Issues of Literacy Sponsorship across Military and Academic Settings uses a literacy studies framework to explore the relationship between student veterans' experiences with military writing and academic writing. The project offers an in-depth analysis of some common military literacy practices and describes paths and barriers to knowledge transfer between military and academic contexts. Dr. Ludewig's research interests include first-year writing pedagogy, writing program administration and assessment, and writing in the disciplines.

Kenneth W. (Broken) Marlin is a current adjunct instructor at Embry-Riddle Aeronautical University. Mr. Marlin is a graduate of Cameron University in Lawton, Oklahoma. His degrees include a Bachelor's in English and a Master's in Teaching. He, currently, is pursuing a Doctorate in Linguistics. Mr. Marlin is married and has about thirty-seven kids and grandkids. He is retired from the U.S. Army. He is a native Floridian, living in the Pensacola area, a fervent swimmer, all year long, and a beach enthusiast, extraordinaire. His work has been published, and he never has an empty reading pile. Mr. Marlin or, as his students call him, Mr. M, has taught for ERAU for fourteen years. He teaches all levels of English Composition, Introduction to Literature, Business Communication, and Technical Writing. His expertise, in addition to those, is speech and languages. He speaks thirteen of those. Additionally, he teaches math in another world.

Bree McGregor is a doctoral candidate* in the Writing & Rhetoric PhD Program at George Mason University, and she is the spouse of an active-duty Marine. She researches topics related to military service members and veterans, digital communication, and theories of social learning. Her publications include "Cultivating Critical Makers: Crafting with Paper-Electronic Circuits in an Online First Year Composition Course" in Re-Programmable Rhetorics (Forthcoming, 2019); "Becoming the Shaman: A Review of Rhetorical Delivery and Digital Technologies: Networks, Affect, Electracy" in Enculturation: A Journal of Rhetoric, Writing, and Culture (May 9, 2017); and "Writing Faculty on the Marine Corps Base: Building Strong Classroom Communities Through Engagement and Advocacy," co-authored with Lourdes Fernandez, in Reflections: A Journal of Public Rhetoric, Civic Writing, and Service Learning (December 2016). Bree has taught undergraduate writing courses on Marine Corps bases and college campuses, and she currently teaches technical and professional writing courses online for George Mason University.

Zea Moullet, MEd, earned her Master of Education degree from Oregon State University, studying College Student Services Administration. She has spent much of her career in student services roles at a variety of universities, and currently serves as the Director of Admissions for California Health Sciences University College of Osteopathic Medicine. Prior to CHSU-COM, Zea spent time working with the Navy's Fleet and Family Support Center. There, she supported Service members and their families through deployments, frequent moves, and navigating the transition into the civilian work force, including pursuing higher education. This breadth of roles has allowed her to develop a respect for the diversity of life experiences that can lead to success, both as a student and a professional. Zea has a deep appreciation for military medicine, as her husband Paul is a Doctor of Osteopathic Medicine (DO) and a Radiologist in the U.S. Navy. Paul served five years in the Navy before college, then attended medical school on a Health Professions Scholarship (HPSP) through the Navy. This afforded him the perspective of attending college and medical school as both a veteran and future Sailor. It also allowed Zea to witness firsthand many challenges common to Military-Affiliated learners.

Jennifer S. Owlett, PhD (Purdue University, 2014) is an assistant professor of Communication at William Paterson University. Her research largely focuses on the ways in which couples and families use supportive communication and social media during times of stress. Previous projects have tended to include a focus on military couples and families. Her secondary line of research explores the connections between interpersonal communication theory and pedagogy. Her work has been shared with multiple academic and the military audiences including the International Association for Relationship Research, the International Public Relations Research Symposium (BledCom), the Military Family Research Institute, and Operation Purple Camp. Previous publications have appeared in several outlets including the *Journal of Family Communication, Communication Teacher,* and the *Basic Course Communication Annual.*

Frank G. Pérez (PhD, University of New Mexico) is an associate professor of Communication and Chicano Studies at the University of Texas at El Paso. His research examines the erasure and vilification of Mexican/Americans in U.S. public memory with a focus on West Texas and the Southwest. He is the author (with Carlos F. Ortega) of Deconstructing Eurocentric Tourism and Heritage Narratives in Mexican American Communities: Juan de Oñate as a West Texas Icon (2020) and the 2019 NCA Córdova-Puchot Scholar of the Year. Pérez is currently a collaborator on a National Institute of Minority Health and Health Disparities-funded project on responsible alcohol prevention titled, WayToServe Español. His research has appeared in Aztlán: A Journal of Chicano Studies, Communication for Social Change Journal, Communication Yearbook, and other scholarly outlets.

Kelly R. Rossetto, PhD (University of Texas at Austin, 2009) is an associate professor in the Department of Communication and Media at Boise State University. Her research centers on how individuals and families navigate stressful events (e.g., military deployment, college transition, bereavement), online and face-to-face, to maintain resilience. Dr. Rossetto has over fifteen years of experience teaching and researching coping and resilience in relationships. She has completed two major research projects where she investigated coping and support among military partners and spouses. In the past year, she has begun researching and developing pedagogical approaches that aid student resilience. Her research has been funded through Boise State University and the NCAA. Examples of Dr. Rossetto's published work can be found in *Journal of Family Communication, Communication Monographs, Personal Relationships, and Journal of Social and Personal Relationships.*

Nick J. Sciullo, JD, PhD is an assistant professor of Communications at Texas A&M University, Kingsville. He is an awarding-winning debate coach and educator who was recognized as the Forensic Educator of the Year by the Argumentation and Forensics Division of the Southern States Communication Association in 2018 and the Coach of the Year for the South East Division of the Cross Examination Debate Association (CEDA) in 2018. That year his teams won the Florida Intercollegiate Forensics Association State Champions in Lincoln-Douglas and Parliamentary Debate. He also coached two CEDA All-Americans and seven CEDA Debate Scholars in 2018. In 2016, he and his debate team were awarded the Bradley Garrison Award by the Cross Examination Debate Association as the best new debate team in the country.

Sciullo's communication scholarship appears in *Review of Communication, Communication Education, Communication Teacher, Argumentation & Advocacy, Departures in Critical Qualitative Research, Canadian Journal of Communication, Communication Currents, The Forensic of Pi Kappa Delta, Kentucky Journal of Communication*, and *Journal of the Speech & Theatre Association of Missouri*. In 2015, he won the Best Dissertation Award from the Critical and Cultural Studies Division of the National Communication Association. And, in 2014, he won the Robert Bostrom Young Scholar Award from the Southern States Communication Association. He works on race, critical communication pedagogy, law and the humanities, and debate and argumentation.

Margaret C. Stewart (PhD, Indiana University of Pennsylvania, 2013) is an associate professor of Communication Studies in the School of Communication at University of North Florida in Jacksonville. She currently instructs courses in Principles of Communication Studies (Communication Theory), Mediated Communication, Public Speaking for Professionals, and Strategic Social Media. Dr. Stewart is also a certified social media strategist (SMS) and trainer for the National Institute of Social Media (NISM) and a consultant for Socially Inspired, LLC. Her research expertise is in the areas of social media and emerging communication technologies. She primarily explores their interpersonal and organizational effects among military-affiliated and sports-athlete populations, as well as in crisis communication and management, audience engagement, and mediated deception. She is the pioneering academic researcher on social listening and how it is used in strategic social media by businesses and institutions and its role in interpersonal relationships. She has published various book chapters and academic journals publications on her active research agenda, including venues such as *Communication Reports, Computer Mediated Communication in Business: Theory to Practice, International Journal of Listening, The Journal of Digital and Social Media Marketing,* and *Computers in Human Behavior*. She also holds a Master's Degree in Professional Communication from La Salle University in Philadelphia, PA.

Catherine St. Pierre is a military spouse currently living in Okinawa, Japan. She is an adjunct faculty member teaching Writing with University of Maryland Global Campus, Asia. She earned both her Master's and PhD in Rhetoric, Composition, and Literacy Studies through the Department of English at The Ohio State University. Her dissertation was titled *Uniforms and universities: A qualitative study of post-9/11 Marine student veterans' literacy practices*. She has research interests in veterans' studies, composition, and literacy studies.

Meghan Velez earned her PhD in Rhetoric and Composition from Florida State University in 2020. She is currently an Instructor of Communication at Embry-Riddle Aeronautical University Worldwide, where she teaches courses in first-year composition, rhetorical theory, and the history of communication technologies. She also serves as Assistant Director of the Virtual Environment for Communication: Teaching, Outreach, and Research (VECTOR). In this capacity, she leads an online communication and multiliteracy center and co-facilitates faculty development in teaching and research related to communication. Meghan's research interests include writing program administration, communication and writing assessment, WAC/WID faculty development, and digital media policy and law. Her previous work has appeared in Women's Studies in Communication and The Routledge Handbook of Comparative World Rhetorics.

Carol M. Walker, passionate about providing consumable education environments so all learners may succeed, Walker has dedicated her career to the research and implementation of instructional design technologies and media communication. She has extensively studied both the positive and negative impacts of technology on learners. This commitment has provided professional opportunities as an educator in the seventh grade classroom, community college, and at the undergraduate and graduate college level. As an assistant professor and graduate coordinator of Instructional Technology MEd program, Walker has had the privilege of supporting many students as they face the challenges in today's educational environment. It has been a special privilege to serve veterans as they struggle to adapt to civilian education. Along with the theoretical applications of technologies for learning, her research has also focused on cyberbullying. This has provided the opportunity to publish numerous articles and chapters on the impact of cyberbullying and facilitated consulting opportunities for seminars to educate youth and college students on the importance of "Staying Safe on the Web."

www.ingramcontent.com/pod-product-compliance
Lightning Source LLC
LaVergne TN
LVHW091532200926
848483LV00002B/381